Water Pollution: Prevention and Treatment

Water Pollution: Prevention and Treatment

Edited by Vincent Emerson

SYRAWOOD
PUBLISHING HOUSE

New York

Published by Syrawood Publishing House,
750 Third Avenue, 9th Floor,
New York, NY 10017, USA
www.syrawoodpublishinghouse.com

Water Pollution: Prevention and Treatment
Edited by Vincent Emerson

© 2019 Syrawood Publishing House

International Standard Book Number: 978-1-68286-730-3 (Hardback)

Cataloging-in-Publication Data

Water pollution : prevention and treatment / edited by Vincent Emerson.
   p. cm.
Includes bibliographical references and index.
ISBN 978-1-68286-730-3
1. Water--Pollution. 2. Pollution prevention. 3. Water--Purification.
I. Emerson, Vincent.
TD420 .W38 2019
628.168--dc23

TABLE OF CONTENTS

PREFACE

Water pollution is the contamination of water as a result of human activities. It is primarily caused by the discharge of inadequately treated wastewater into natural water streams leading to environmental degradation and impacts on public health. Other effects include eutrophication, acidity, anoxia, etc. Common contaminants contributing to water pollution include chemicals and pathogens. Water pollution can be classified as surface water, groundwater and marine pollution. The measurement of water pollution is done by analyzing water samples with a variety of physical, chemical and biological tests. Prevention of pollution can be achieved by the adoption of appropriate sewage and industrial treatments, control of urban runoff, agricultural wastewater management, erosion and sediment control, etc. This book provides comprehensive insights into the field of water pollution. It discusses some existing theories and innovative concepts revolving around water pollution, its treatment and mitigation strategies. This book is a vital tool for all researching and studying this field.

The information contained in this book is the result of intensive hard work done by researchers in this field. All due efforts have been made to make this book serve as a complete guiding source for students and researchers. The topics in this book have been comprehensively explained to help readers understand the growing trends in the field.

I would like to thank the entire group of writers who made sincere efforts in this book and my family who supported me in my efforts of working on this book. I take this opportunity to thank all those who have been a guiding force throughout my life.

Editor

Heavy metals and hydrocarbon concentrations in water, sediments and tissue of *Cyclope neritea* from two sites in Suez Canal, Egypt and histopathological effects

Hesham M Sharaf[1] and Abdalla M Shehata[2*]

Abstract

Heavy metals and hydrocarbons are of the most common marine pollutants around the world. The present study aimed to assess the concentration of petroleum hydrocarbons and heavy metals in tissues of the snail *cyclope neritea*, water and sediments from two sites of the study area (Temsah lake and Suez canal) represent polluted and unpolluted sites respectively. The results showed that, the levels of the heavy metals (Pb, Cd, Co, Mg and Zn) in the polluted area have reached harmful limits recorded globally. Lead in water, sediment and tissue of the snail reached to 0.95 ppm, 4.54 ppm and 7.93 ppm respectively. Cadmium reached 0.31 ppm, 1.15 ppm and 3.08 ppm in the corresponding samples. Cobalt was not detected in water, but it reached 1.42 ppm and 10.36 ppm in the sediment and snails tissue respectively. Magnesium in water, sediment and tissue of the snail reached 3.73 ppm, 9.44 ppm and12.6 ppm respectively. Zinc reached 0.11 ppm, 3.89 ppm and 12.60ppm in the corresponding samples. Meanwhile, hydrocarbons in the polluted area (site1) reached 110.10 µg/L, 980.15 µg/g and 228.00 µg/g in water sediment and digestive gland tissues of the snails respectively. Whereas, hydrocarbons in the unpolluted area (site2) were estimated as 14.20 µg/L, 55.60 µg/g and 22.66 µg/g in water, sediment and tissue of the snails respectively. The combination of histopathological image with monitoring of the metal level in the digestive gland of the present snail provides an important tool for early detection of impending environmental problems and potential public health issues. Petroleum hydrocarbons are toxic to the marine fauna when present above certain limit in the marine water. The major detoxification organ in molluscs is the digestive gland, which has been used as a bioindicator organ for toxicity assessment. The effect of high crude oil on the digestive gland tubules of exposed snails when examined microscopically reveals a series of histological changes which indicates that the cellular compensatory mechanism is activated by hydrocarbons. These changes include vacuolation and presence of pyknotic nuclei.

Keywords: Heavy metals, Hydrocarbons, Timsah Lake, Suez canal, *Cyclope neritea*

Introduction

Adverse anthropogenic effects on the coastal environment include eutrophication, heavy metals, organic, oil spills and microbial pollution. Consequently, levels of contaminants in the marine environment are increasing continuously. In order to establish adequate coastal management programs, it is important to characterize the environment of concern chemically. The extent of contamination can be assessed by measuring pollutant concentration in water, sediments and exposed animal tissues samples. Trace metals can be divided to essential and non essential elements. Essential elements occur naturally in all organisms, essential elements in high doses can be poisonous causing hazardous effects on organisms. Non essential elements do not have any positive effects on organisms and they are harmful already in low doses. They can inhibit an essential element to bind

* Correspondence: ashehataeg@yahoo.com
[2]Chemistry Department, Faculty of Science, Suez Canal University, Al-arish, Egypt
Full list of author information is available at the end of the article

to enzyme and disturb the normal enzymatic function in the body [1-9].

As compared with the open sea, lagoons are more subject to pollution, particularly by heavy metals from industrial, agricultural and urban origin. Near shore sediments are found in a wide variety of environments (bays, lagoons). The water sediments interface is more important to biological fauna as compared with surface sediments, since meiofauna live about the reduced zone in sediment. Therefore, the composition of sediment has a significant influence on the living conditions of marine organisms. The trace metal results were obtained by sediment analysis, unlike sea water analysis, where the detection limit and contamination risks are significantly reduced [9].

It is well known that molluscs accumulate organic and metal pollutants at concentrations several order of magnitude above those observed in the field environment. Fewer studies have been done on gastropod molluscs, some of which are considered as useful biomonitors of certain metals [10]. Most metals are generally concentrated many times within an organism's soft tissue.

Advancement in technology as well as increase in population have led to environmental concerns relating from indiscriminate dumping of refuse and discharge of industrial effluents, petroleum waste water and crude oil spills replete with most common heavy metals in our environment [7,11].

Histopathological patterns represent a rapid, sensitive, reliable and comparatively inexpensive tool for assessment of stress response to xenobiotics [12-14]. These cascades of stress related responses including histopathology are now increasingly being used as biomarkers of environmental stress since they provide a definite biological endpoint of historical exposure [15-17].

Heavy metal pollution of terrestrial and aquatic ecosystems has long been recognized as a serious environmental concern. This is largely due to their non biodegradability and tendency to accumulate in plants and animals tissues. As a result, metal bioaccumulation is a major route through which increased levels of the pollutants are transferred across food chains, creating public health problems [18-20]. Therefore, it is important to determine the bioaccumulation capacity for heavy metals by certain organisms in order to assess potential risk to human health.

Several authors [9,13,21] have reported the importance of molluscs as good indicators for monitoring heavy metal pollution even through abnormally high environmental concentrations, since heavy metals affect numerous biological processes involved in the development and maintenance of molluscan populations such as feeding, growth, reproduction and general physiological activities [22].

Contamination of the sea with petroleum hydrocarbons, especially in shipping channels and ports, where crude and refined petroleum products are transported in significant quantities of oil entering the water column and sediment. The hepatopancreas of molluscs is the major site of petroleum hydrocarbons (PHC) detoxification. The main route for elimination of hydrocarbon metabolism is through faecal matter [23].

This study aimed to investigate the concentration of Pb, Cd, Co, Mg, Zn and hydrocarbons in water, sediment and soft tissues of *Cyclope neritea* snail from two different sites at Suez Canal, Egypt, map(1). The first site is

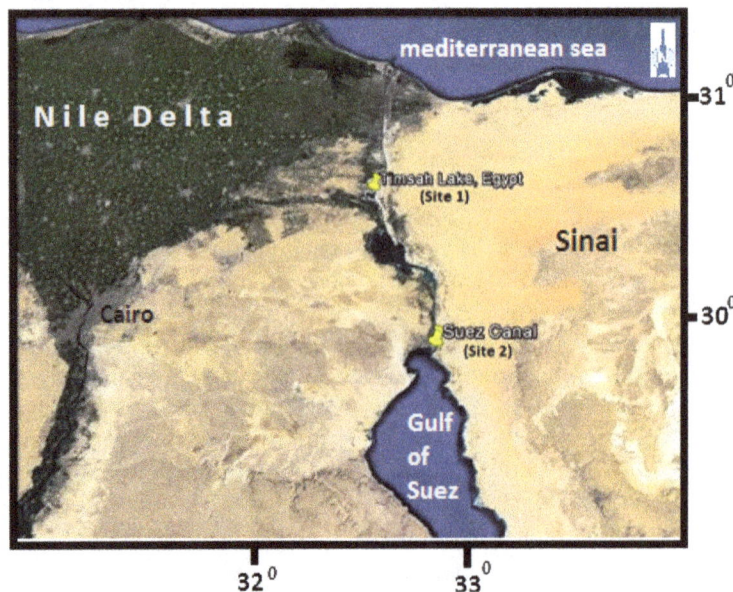

Figure 1 A map of Suez Canal showing the two investigated sites. The first site; Timsah Lack and the second site; Suez Canal.

Table 1 Concentration of heavy metals (ppm) in water, sediment and digestive gland tissue of *Cyclope neritea* snails (Mean ± SD) at the two investigated sites

Metal	Timsah Lake (site1)			Suez Canal (site2)		
	Mean ± SD			Mean ± SD		
	Water	Sediment	Digestive gland tissue	Water	Sediment	Digestive gland tissue
Pb	0.95 ± 0.04	4.54 ± 0.05	7.93 ± 0.03	0.92 ± 1.04	2.11 ± 0.82	3.93 ± 0.50
Cd	0.31 ± 0.04	1.15 ± 0.17	3.08 ± 0.20	0.50 ± 0.02	1.00 ± 0.13	1.72 ± 0.07
Co	0.00 ± 0.00	1.42 ± 0.17	10.63 ± 0.21	0.00 ± 0.00	0.67 ± 0.96	2.66 ± 0.25
Mg	3.73 ± 0.20	9.44 ± 1.18	12.6 ± 0.96	2.63 ± 0.01	3.92 ± 0.37	5.86 ± 0.15
Zn	0.11 ± 0.04	3.89 ± 0.20	12.6 ± 0.96	0.13 ± 0.01	1.33 ± 1.00	4.44 ± 0.80

located in Timsah Lake which receives effluents from many industrial sources site (1) where as the second site is Suez Canal where there is limited industrial or anthropogenic activity site (2).

Moreover, an attempt has been made to elucidate the response of the digestive gland of *Cyclope neritea snails* to heavy metals and PHCs and to identify histopathological biomarkers.

Materials and methods

Heavy metal content (Pb, Cd, Co, Mg and Zn) and hydrocarbons were analyzed in the water, sediment and soft tissues of *Cyclope neritea* snails from two different sites at Suez canal, the first was polluted, site1 (Timsah lake) while the second unpolluted site2 (Suez Canal) from the main canal (Figure 1).

Water and sediment

From both sites water samples were collected directly in a precleaned polyethylene bottles then sealed until analysis, sediment samples were collected using a precleaned PVC corer and immediately kept in polyethylene

bags which were sealed and kept in an ice box for further analysis in the laboratory. Sediment samples were washed with bidistilled water and dried at 60°C then ground in a glass mortar and reduced into fine particles.

Samples collections

Snails were collected from both sites by hand picking. The soft tissues were removed from snail's shells with a sharp knife and dried at 60°C. The dried tissue was ground into fine powder and stored in a desiccator for further analysis.

Chemicals and method of analysis

Estimation of trace metals in water and sediment was carried out by suitable volume of water or digestion of one gram of sediment with conc. Nitric acid (HNO3) and conc. Perchloric acid (HClO4) (4:l) and analyzed in Atomic absorption specterophotometers perkin-Elmer, Analyzed TM 300 (USA) at suitable current and wave length for all studied heavy metals [24,25]. The values were expressed in ppm and the standard deviation was calculated. The hydrocarbons in all samples were

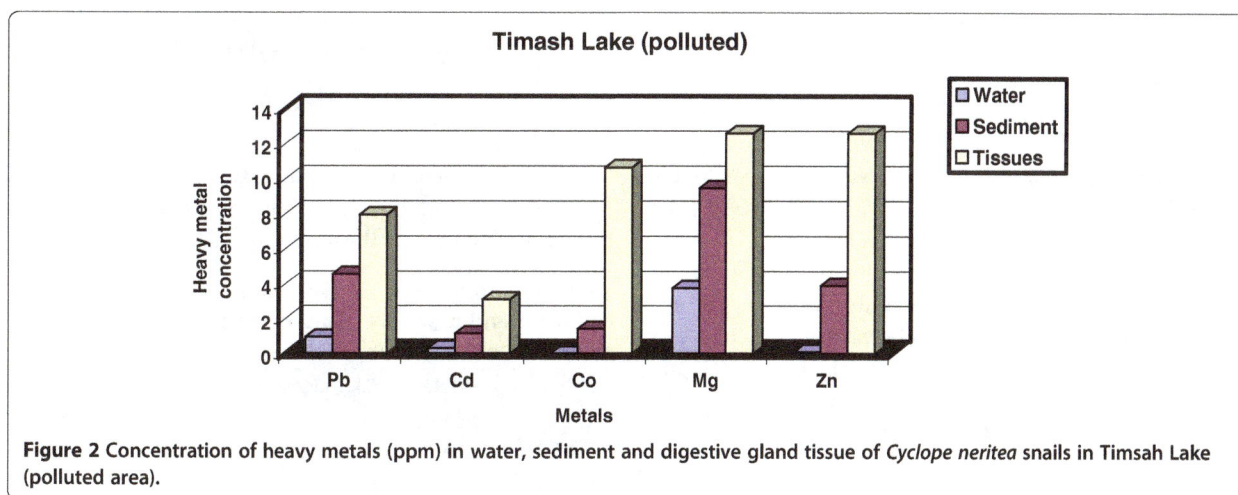

Figure 2 Concentration of heavy metals (ppm) in water, sediment and digestive gland tissue of *Cyclope neritea* snails in Timsah Lake (polluted area).

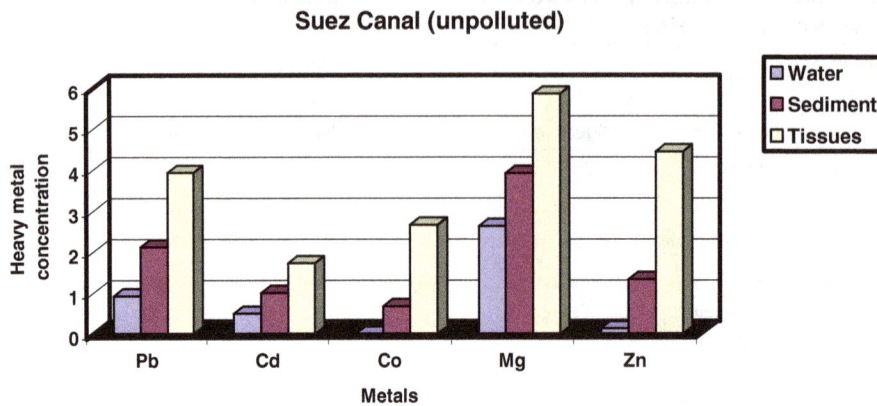

Figure 3 Concentration of heavy metals (ppm) in water, sediment and digestive gland tissue of *Cyclope neritea* snails in Suez Canal (unpolluted area).

extracted by Soxhlet instrument and measured by a Spectrofluorophotometer.

Statistical analysis

SPSS13 for Windows software was used for the statistical analysis of the heavy metal contents in water, sediment and digestive gland tissue of snails at the two investigated sites by using Mann–Whitney test.

For histolopathological examination the digestive gland of *Cyclope neritea* was taken out of their shells and dropped immediately into Alcoholic Bowin's, fixative then processed in the usual manner for histological investigation and stained with Haematoxylin and Eosin.

Results

Mean ± SD of the heavy metals at the two investigated sites were tabulated in Table 1. Statistical Mann–Whitney test was applied for all parameters at the two sites.

Heavy metals in Site 1 (Timsah Lake, polluted area) are shown in Table 1, Figure 2. The results revealed that the values of heavy metals in water are 0.95 ± 0.04, 0.31 ± 0.04, 0.00 ± 0.00, 3.73 ± 0.20 and 0.11 ± 0.04 ppm for Pb, Cd, Co, Mg and Zn respectively. Meanwhile, heavy metals values in sediment are 4.54 ± 0.05, 1.15 ± 0.17, 1.42 ± 0.17, 9.44 ± 1.18 and 3.89 ± 0.20 ppm for Pb, Cd, Co, Mg and Zn respectively. Besides, the values of heavy metals in digestive gland tissues are 7.93 ± 0.03, 3.08 ± 0.20, 10.63 ± 0.21, 12.6 ± 0.96 and 12.6 ± 0.96 ppm for Pb, Cd, Co, Mg and Zn respectively.

On other hand, heavy metals in Site2 (Suez Canal, unpolluted area) are shown in Table 1, Figure 3. The results revealed that, the values of heavy metals in water are 0.92 ± 1.04, 0.50 ± 0.02, 0.00 ± 0.00, 2.63 ± 0.01 and 0.13 ± 0.01 ppm for Pb, Cd, Co, Mg and Zn respectively. Meanwhile, heavy metals values in sediment are 2.11 ± 0.82, 1.00 ± 0.13, 0.67 ± 0.96, 3.92 ± 0.37 and 1.33 ± 1.00 ppm for Pb, Cd, Co, Mg and Zn respectively. And

the values of heavy metals in digestive gland tissues are 3.93 ± 0.50, 1.72 ± 0.07, 2.66 ± 0.25, 5.86 ± 0.15 and 4.44 ± 0.80 ppm for Pb, Cd, Co, Mg and Zn respectively.

The Mann–Whitney test was made between the heavy metals concentrations of the digestive gland tissue of the snails and both water and sediment in the two sites, asymptotic significance (2-tailed) = 1.00 and 0.18 respectively was not significant at $P < 0.05$ level. Also, the Mann–Whitney test was made between the heavy metals concentrations of the digestive gland tissue of the snails in the two sites, asymptotic significance (2-tailed) = 0.047 was significant at $P < 0.05$ level.

On the other hand, the concentrations of petroleum hydrocarbons were tabulated in Table 2, Figure 4. The results revealed that concentration of petroleum hydrocarbons are 110.1, 980.15 and 228.00 ug/g for water, sediment and digestive gland tissues respectively in site (1). Meanwhile, they are 14.2, 55.6 and 22.66 ug/g for water, sediment and digestive gland tissues respectively in site (2). The Mann–Whitney test was made between the petroleum hydrocarbons concentrations of the digestive gland tissues and both of water and sediment from the two sits, asymptotic significance (2-tailed) = 0.0499 was significant at $P < 0.05$ level.

Histpathological changes

The digestive gland of normal *C. neritea* is a large, tubulo-acinar gland which occupies the greater part of the cavity of the shell spire. The gland is covered by

Table 2 Concentration of hydrocarbons in water, sediment and digestive gland tissue of *Cyclope neritea* snails in the study area

Site	Water (µg/L)	Sediment (µg/g)	Digestive gland tissue (µg/g)
Timsah Lake	110.10	980.15	228.00
Suez Canal	14.20	55.60	22.66

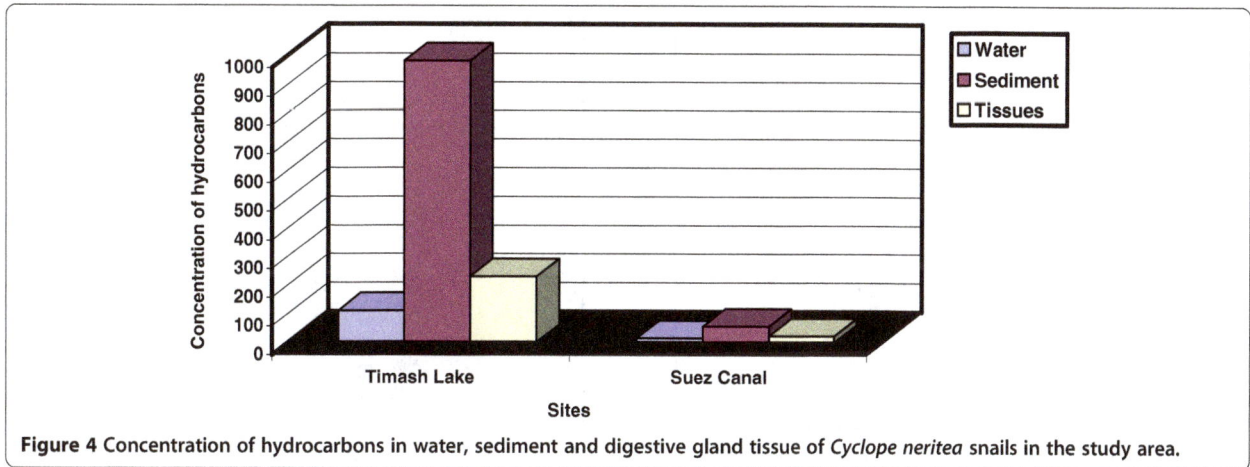

Figure 4 Concentration of hydrocarbons in water, sediment and digestive gland tissue of *Cyclope neritea* snails in the study area.

squamous epithelium resting on a thin layer of fibrous connective tissue Figure 5. The digestive gland tubules are lined with simple epithelium. This epithelium consists of two main cell types, digestive cells and secretory cells.

The digestive cells Figure 5 are by far the most numerous elements in the wall of the digestive gland tubules. They are long columnar or cube shaped with domed distal apices and flat bases by which they rest on a very thin basement membrane. They vary greatly in length within the same tubule. The nucleus is basal, usually oval but may be spheroidal or even irregular. Inside the major part of the cell body the cytoplasm is lightly stained and shows various degrees of vaculation and different contents.

The secretory cells Figure 5 are present in much smaller number than the digestive cells. They are shorter pyramidal or cone-shaped, but may sometimes be columnar. They are markedly shorter than the digestive cells and therefore appear wedged in between groups of the digestive cells. Their cytoplasm is basophilic and the body of the excretory cells is usually crowded with numerous spherules (ex.s) of regular form but different sizes. Each excretory cell contains oval or spherical excretory bodies (dark granules) which are usually chromophobic, usually retain their natural colour and appear located inside colourless vacuoles.

These bodies may represent the final stage in elaboration of the excretory material within the excretory cells. The digestive gland obtained from the polluted area shows an increase in the secretion of the digestive cells, where the cytoplasm becomes more acidophilic and their nuclei start unusual division Figure 6. It can be seen that the dark granules increased greatly in number and most of the digestive cells become completely degenerated and lysing while most of the tubules are damaged Figures 7, 8 and 9. At much higher concentrations of pollutants the digestive gland tubules become

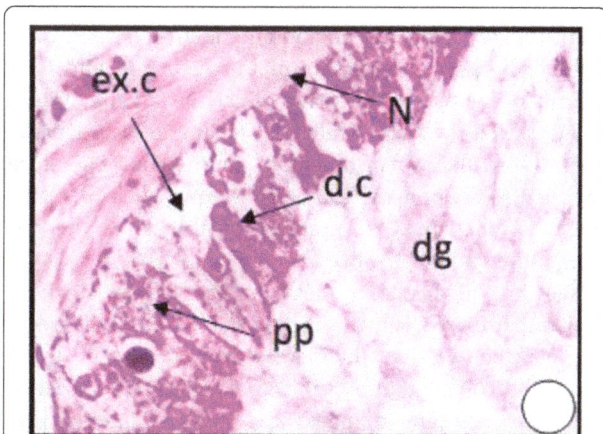

Figure 5 A photomicrograph of T.S. of digestive gland of *Cyclope neritea* from the unpolluted site (Suez Canal) showing digestive tubules without hydrocarbon precipitations in the digestive gland cells. N (Nucleus); dg (digestive gland); d.c (digestive cell); ex.c (excretory cell); PP (hydrocarbon precipitations). X400.

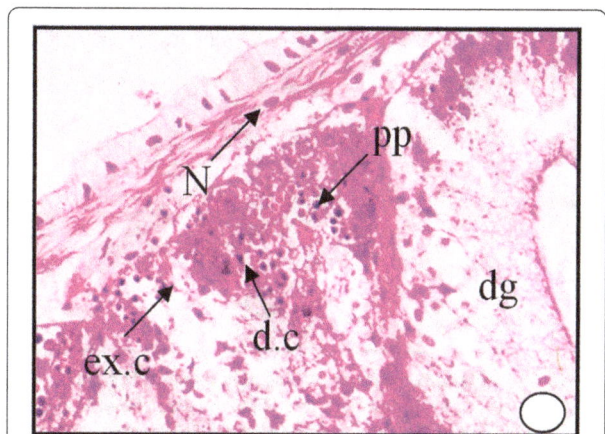

Figure 6 Digestive tubules T.S. of digestive gland in snail *Cyclope neritea* from polluted area (Timsah Lake) showing hydrocarbon precipitations in the digestive gland cells. X200.

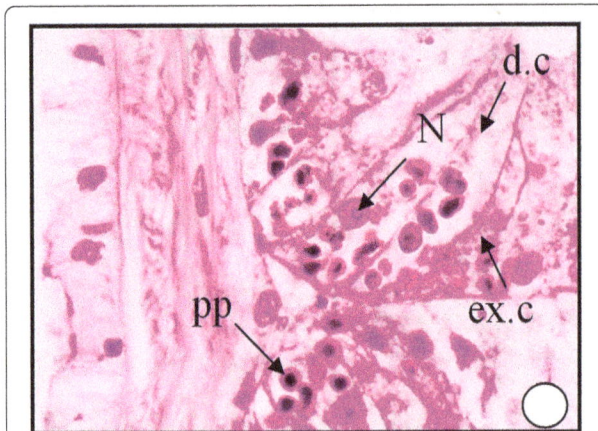

Figure 7 Higher magnification of Digestive tubules in T.S. of digestive gland of the snail *Cyclope neritea* from polluted area (Timsah Lake) showing hydrocarbon precipitations in the digestive gland cells. N (Nucleus); dg (digestive gland); d.c (digestive cell); ex.c (excretory cell); PP (hydrocarbon precipitations); V (vacuolation). X400.

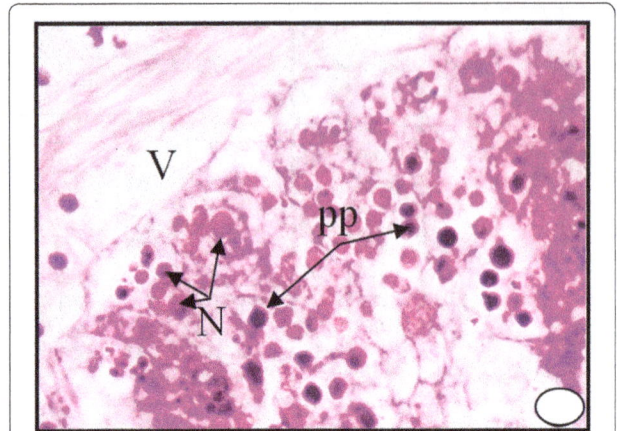

Figure 9 A photomicrograph of digestive gland of the snail *Cyclope neritea* from polluted area (Timsah Lake) showing that most of the digestive cells became polygonal and bi and multinucleated and finally degenerated and lysed due to hydrocarbon precipitations. N (Nucleus); PP (hydrocarbon precipitations); V (vacuolation). X400.

mostly damaged, necrotic and the cells take cluster shape and become bi and multinucleated and finally lysed Figures 8 and 9.

Discussion

This study shed light on some of the environmental factors that might have an impact directly or indirectly on the concentration ability of heavy metals into the bodies of the investigated host. The concentration of water with a wide range of heavy metals has become a matter of great concern over the last few decades [26] and a lot of studies have been published on the heavy metals at all

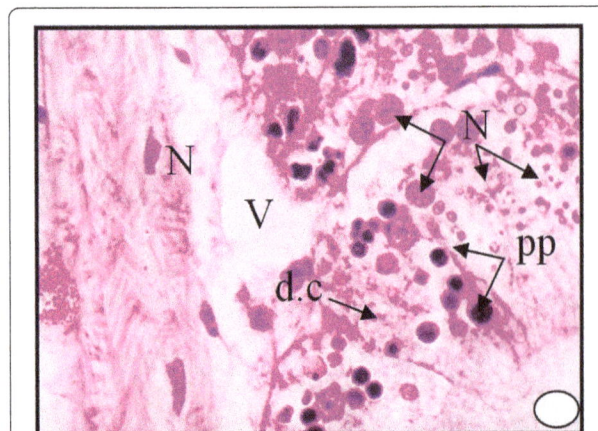

Figure 8 A photomicrograph of digestive gland of the snail *Cyclope neritea* from polluted area (Timsah Lake) showing the accumulations of hydrocarbons particles which lead to damaging and vacuolation of most digestive cells of the digestive gland cells. N (Nucleus); d.c (digestive cell); PP (hydrocarbon precipitations); V (vacuolation). X400.

levels of aquatic ecosystem [27-30]. Many authors associated the heavy pollution in water with industrial and municipal discharges. These heavy metals may be taken up by living organisms, deposited in the sediments or remain for some period in the water itself [31,32].

In the present study concentrations of heavy metals in water, sediment and tissues of molluscs from two sites of Suez Canal were determined. The highest concentrations for all studied heavy metals were recorded at site (1) when compared with site (2). The differences in heavy metals concentration between site 1 and site 2 might be attributed to the highly discharge of mixture of industrial, municipal and agricultural drains into site (1), this opinion agrees with [33,34]. Several authors reported that the variation in heavy metals concentration in water might be attributed to the contaminated sediment; these sediments reflect the quality of water current and form the major repository of heavy metals in aquatic system. They added that the rate of accumulation depends mainly on the environmental parameters. Therefore, sediments can be used to detect the presence of contamination that does not remain soluble after the discharge into water [35,36].

The data of the heavy metals in the present work revealed that there is a highly significant ($P < 0.01$) difference between the two sites, the increase in values of site 1 may be attributed to a number of factors such as industrial effluents, agricultural drainage and waste municipal [31]. Water of the two sites showed higher concentrations of Lead (0.95, 0.92 ppm); Cadmium (0.31, 0.50 ppm), Cobalt (0.00, 0.00 ppm); Manganese (3.73, 2.63 ppm); and Zinc (0.11, 0.13 ppm). Meanwhile, sediment of the two sites showed higher concentrations of Lead (4.54, 2.11 ppm);

Cadmium (1.15, 1.00 ppm), Cobalt (1.42, 0.67 ppm); Manganese (9.44, 3.92 ppm); and Zinc (3.89, 1.33 ppm), Also, heavy metals concentration in the digestive gland tissues of *C. neritea* snail of the two sites showed higher concentrations of Lead (7.93, 3.93 ppm); Cadmium (3.08, 1.72 ppm), Cobalt (10.63, 2.66 ppm); Manganese (12.6, 5.86 ppm); and Zinc (12.6, 4.44 ppm), these finding are in agreement with [37] where they determined the levels of heavy metals Lead (Pb), Cobalt (Co), Manganese (Mg), Zinc (Zn) and Cadmium (Cd) in coastal water, sediment and soft tissues of the gastropod limpet *Patella caerulea* and the bivalvae *Barbatus barbatus* from seven different stations in the western coast of the Gulf of Suez. The highest accumulated metals were Fe, Zn and Mn in both *Patella caerulea* than *Barbatus barbatus*. In the present study, metals such as Pb, Mg and Zn, exhibited modreate concentrations. Cadmium and Cobalt metals were below the detectable limit at both sites.

In concordance with the present study, several relevant studies have been made earlier. The heavy metal accumulation in the gastropod *Cerithium scabridium* from Kuwait coast has been analyzed [38] where the concentration of the cadmium in the gills ranged between 7.06 ppm and 0.09 ppm, which is comparatively lower than that in the present study. The concentrations of the analyzed heavy metals exhibited variations in water, sediments and tissues of the studied animal from both sites.

The assessment and description of toxicants on organisms is attracting the interest of the researchers working with marine organisms [2,9,39]. Accumulated hydrocarbons were found to interact with cellular hydrocarbons leading to different histopathological lesions with the highest necrosis index of cells [18]. Cellular and subcellular histopathological changes have proved to be reliable biomarkers in toxicological assays [20]. In the present study, the accumulations of hydrocarbons lead to damaging and vacuolation of most digestive cells, where they became polygonal and bi and multinucleated and finally degenerated and lysed. These results agree with those of [2,17,40-42].

Further studies are still needed to investigate the effect of heavy metal contamination on the biochemical parameters and fine structure to evaluate their potentiality as sentinel organisms to heavy metal contamination in aquatic habitats.

Competing interests
The authors declare that they have no competing interests.

Authors' contributions
HS and AS are the main investigators, performed all experimental work and paper writing, carried out data analysis the authors read and approved the final manuscript.

Authors' information
Sharaf H.M, Zoology Department, Faculty of Science, Zagazig University, Egypt.

Shehata A.M.A, Chemistry Department, Faculty of Science, Suez Canal University, Al-arish, Egypt.

Acknowledgements
The authors acknowledge Prof. Dr. Abdallah Ibrahium, Faculty of Science, Ain Shams, Cairo, Egypt, for his guidance, help, read and approved the final manuscript.

Author details
[1]Department of Zoology, Faculty of Science, Zagazig University, Zagazig, Egypt. [2]Chemistry Department, Faculty of Science, Suez Canal University, Al-arish, Egypt.

References
1. Samiulallah Y. Biological effects of marine oil pollution. Oil Petrochem Pollut. 1985;2:235–64.
2. Alyami M Y. Ecological Study and Petroleum Hydrocarbons Estimation of Bivalve (Tridacna maxima) from the coastal area of Jeddah Red Sea, Msc. Faculty of Marine Science, King Abdulaziz University 2002.
3. Azza khaled. Heavy metals concentrations in certain tissues of five commercially important fishes from El-max bay, Alexandria, Egypt, Environment Division, National institute of Oceanography and fisheries, kayet bay Alexandria, Egypt 2003.
4. Hassn MA, Heba M, Al Edresi AM. Background level of heavy metal in Dissolved particulate passes of water and sediment of Al Hodeidah Rea Sea coast of Yemen, Faculty of marine science and Environment Alhodeidah university city republic of Yemen JKAU. Mar Sci. 2004;15:53–71.
5. Bu-olayan AH, Thomas BV. Validating species diversity of benthic organisms to trace metal pollution in Kuwait bay, of the Arabian Gulf. Appl Ecol Environ Res. 2005;3(2):93–100.
6. Al Sayad HA, Dairi M. Metal accumulation in the edible snail *Turbo coronatus* (Gmelin)) from different 10 cautions in Bahrain, Arab Gulf. J Sci Res. 2006;24:48–57.
7. Gbaruko BC, Friday OU. Bioaccumulation of heavy metals in some fauna and flora. Int J Environ Sci Tech. 2007;4(2):197–202.
8. Chukwujindu MA, Iwegbue Francis O, Arimoro Godwin E, Nwa jh Osa Eguavoen. Heavy metal content in the African giant snail *Archachatina marginata* (Swainson,1821) (Gastropoda: Pulmonata: Achatinidae) in Southern Nigeria. Folia Malacologica. 2008;16(1):31–4.
9. Kesavan K, Murugan V, Venkatesan V, Vijay Kumar BS. Heavy metal accumulation in mollusks and sediment from uppanar estuary, southeast coast of India. An Int J Mar Sci. 2013;29(2):15–21.
10. Bryan GW, Langston WJ, Hummerstone LG, Burt GR, Ho YB. An assessment of the gastropod, *Litorina lgtorea* ,as an indicator of heavy metal contamination in United Kingdom estuaries. J Mar Biol Assoc UK. 1983;63:327–45.
11. Wills J. A survey of offshore oil field drilling wastes and disposal techniques to reduce the ecological impact of sea dumping. Sakhalin Environ. Watch. 2000;13:23–9.
12. Balkas IT, Tugrel S, Salhogln I. Trace metals levels in fish and crustacean from Northern Mediterranean coastal waters. Mar Environ Res. 1982;6:281–9.
13. Bryan GW, Langston WJ. Bioavailability, accumulation and effects of heavy metals in sediments with special reference to United Kingdom estuaries: a review. Eviron Pollut. 1992;76(2):89–131.
14. Otitoloju AA, Don-Pedro KN. Integrated laboratory and filed assessments of heavy metals accumulation in edible periwinkle, *Tympanotonus fuscatus* var radula (L.). Ecotoxicol Environ Saf. 2004;57(3):354–62.
15. Matthiessen P, Thain JE, Law RJ, Fileman TW. Attempts to assess the environmental hazard posed by complex mixtures of organic chemicals in UK estuaries. Mar Pollut Bull. 1993;26:90–5.
16. Stentiford GD, Feist SW. A histopsthological survey of shore crab (*Carcinus maenas*) and brown shrimp (*Crangon crangon*) from six estuaries in the United Kingdom. J Inert Pathol. 2005;88:136–46.
17. Ravera O, Beone GM, Trincherine PR, Riccardi N. Seasonal variations in metal content of two Union *Pictorum mancus* (Mollusca, Unionidae) populations from two lakes of different tropic state. J Limnol. 2007;66(1):28–39.
18. Berthou FG, Balouet G, Bodennee, Marchand M. The occurrence of hydrocarbons and histopathological abnormalities in oysters for seven years

following the wreck of Amoco Cadiz in Brittany (France). Mar Environ Res. 1987;23:103–33.

19. Tuzen M. Determination of heavy metals in fish samples of the middle Black Sea (Turky) by graphite furnance atomic absorption spectromertry. Food Chem. 2003;80:19–23.

20. Otitoloju AA, Ajikobi DO, Egonmwan RI. Histopathology and bioaccumulation of heavy metals (Cu&Pb) in the giant land snail, *Archachatina marginata* (Swainson). Open Environ Pollut Toxicol J. 2009;1:79–88.

21. Kiffney PM, Clement WH. Bioaccumulation of heavy metals by benthic Invertebrates at the Arkansas River. Colorado. Environ Toxicol Chem. 1993;12(8):1507–17.

22. Coughtre PJ, Martin MH. The distribution of Pb, Zn and Cu within the pulmonate molluscs *Helix aspersa*. Oeclogia. 1976;23:315–22.

23. Lee RF, Sauerheber R, Benson A. Petroleum hydrocarbons uptake and discharge by the marine mussel *Mytilus edulis*. Science. 1972;177:344–6.

24. Chester R, Hughes MJ. A chemical technique for separation of ferromanganese minerals, carbonate minerals and absorbed trace elements from pelagic sediments. Chem Geol. 1967;2:249–62.

25. Clescerl L, Greenberg A, Eaton A. Standard methods for the examination of waters and wastewaters. 20th edition. Washington DC: American Public Health Association; 1999. p. 1325. ISBN 0875532357.

26. Yilmaz F, Ozdemir N, Demirak A, Tuna AL. Heavy metal levels in two fish species *Leuciscus cepalus* and *Lepomis gibbosus*. Food Chem. 2007;100 (2):830–5.

27. Ravera O. Monitoring of the aquatic environment by species accumulator of pollutants: a review. J Limnol. 2001;60(Suppl1):63–78.

28. Rainbow PS. Trace metal concentrations in aquatic invertebrates, why so what? Enivron Pollut. 2002;120:497–507.

29. Dugong G, Pera LL, Bruzzese A, Pellicano T, Turco V. Concentration of Cd (II), Cu (II), Pb (II), Se (IV) and Zn (II) in cultured sea bass (Dicentrarchus labrax) tissues from Tyrrhenian Sea and Sicilian Sea by derivative stripping potentiometry. Food Control. 2006;17:146–52.

30. Jayakumar P, Paul VI. Patterns of cadmium accumulation in selected tissues of the catfish *Claiias batrachus* (Linn.) exposed to sublethal concentration of cadmium chloride. Vet Arhiv. 2006;76:167–77.

31. Emara AM, Belal AH. Marine fouling in Suez Canal. Egypt J Aquat Res. 2004;30(A):189–206.

32. Santos IR, Silva-Filho MR, Campos LS. Heavy metals contamination in coastal sediments and soils near the Brazilian Antarctic Station, King George Island. Mar Poll Bull. 2005;50:85–194.

33. Haggag AM, Marie MAS, Zaghloul KH. Seasonal effects of the industrial effluents on the Nile catfish, Clarias gariepinus. J Egypt Ger Soc Zool. 1999;28(A):365–91.

34. Zaghloul KH. Effect of different water sources on some biological and biochemical aspects of the Nile tilapia; *Oreochromis niloticus* and the Nile catfish, *Clarias gariepinus*. Egypt. J Zool. 2000;34:353–77.

35. Chapman PM, Wang FY. Asseccing sediment contamination in estuaries. Environ Toxicol Chem. 2001;20:3–22.

36. Shahat MA, Amer OSO, AbdAllah AT, Abdelsater N, Moustafa MA. The distribution of certain heavy metals between intestinal parasites and their fish hosts in the River Nile at Assuit Province, Egypt. Egypt J Hosp Med. 2011;43:241–57.

37. Mohamed AH, Ahmed ME. Marine molluscs as biomonitors for heavy metal levels in the Gulf of Suez, Red Sea. J Mar Syst. 2006;60:220–34.

38. Bu-olayan AH, Thomas BV. Heavy metal accumulation in the gastropod, *Cerithium scabridum* L., from the Kuwait coast. Environ Mon Assess. 2001;68:187–95.

39. Miriam Paul S, Menon NR. Histopathological changes in the hepatopancreas of the penaeid shrimp *Metapenaeus dobsoni* exposed to petroleum hydrocarbons. J Mar Biol Ass India. 2005;47(2):160–8.

40. Sorenen EMB, Ramirez-Mitchell R, Harlan CW, Bell JS. Cytological changes in the fish liver following chronic, environmental arsenic exposure. Bull Environ Contam Toxicol. 1980;25:93–9.

41. Auffret M. Histopathological changes related to chemical contamination in *Mytilus edulis* from field and experimental conditions. Mar Ecol Prog Ser. 1988;64:101–7.

42. Stentiford GD, Longshaw M, Lyons BP, Jones G, Green M, Feist SW. Histopathological biomarkers in estuarine fish species for the assessment of biological effects of contaminants. Mar Environ Res. 2003;55:137–59.

Performance of a single chamber microbial fuel cell at different organic loads and pH values using purified terephthalic acid wastewater

Seyed Kamran Foad Marashi and Hamid-Reza Kariminia[*]

Abstract

Background: Purified terephthalic acid (PTA) wastewater from a petrochemical complex was utilized as a fuel in the anode of a microbial fuel cell (MFC). Effects of two important parameters including different dilutions of the PTA wastewater and pH on the performance of the MFC were investigated.

Methods: The MFC used was a membrane-less single chamber consisted of a stainless steel mesh as anode electrode and a carbon cloth as cathode electrode. Both power density and current density were calculated based on the projected surface area of the cathode electrode. Power density curve method was used to specify maximum power density and internal resistance of the MFC.

Results: Using 10-times, 4-times and 2-times diluted wastewater as well as the raw wastewater resulted in the maximum power density of 10.5, 43.3, 55.5 and 65.6 mW m^{-2}, respectively. The difference between the power densities at two successive concentrations of the wastewater was considerable in the ohmic resistance zone. It was also observed that voltage vs. initial wastewater concentration follows a Monod-type equation at a specific external resistance in the ohmic zone.
MFC performance at three different pH values (5.5, 7.0 and 8.5) was evaluated. The power generated at pH 8.5 was higher for 40% and 66% than that for pH 7.0 and pH 5.4, respectively.

Conclusions: The best performance of the examined MFC for industrial applications is achievable using the raw wastewater and under alkaline or neutralized condition.

Keywords: Bioelectricity generation, Concentration effect, Microbial fuel cell, Petrochemical wastewater, pH effect, Purified terephthalic acid wastewater

Introduction

Microbial fuel cell (MFC) is a device that converts biochemically released energy from bacterial catalysis of organic and inorganic materials into electrical energy. In MFCs, electricity generation and wastewater treatment can occur, simultaneously. Therefore, MFCs are considered as one of the potential solutions to overcome the crises of energy shortage and environmental pollution. On the other hand, many challenges are still remained for commercialization of MFC technology. Designing a cost-effective system with high power generation is one of the most important challenges. For example, in order to achieve a high performance, expensive materials such as platinum as cathodic catalyst, carbon cloth as cathode or anode electrode, and also a suitable membrane such as nafion are inevitably necessary. Accordingly, more studies are required to find efficient materials both in terms of cost and power production [1].

In recent years, this technology has been studied by many researchers. Generally, different parameters including both operational and designing factors affect the MFC performance. Nature of the substrate, substrate concentration [2], temperature [3], microorganisms' species, alkalinity of anode and cathode chambers [4], external resistance [5] and residence time [6] are among the operational parameters. Anode and cathode material

* Correspondence: kariminia@sharif.ir
Department of Chemical and Petroleum Engineering, Sharif University of Technology, P.O. Box 11155–9465, Azadi Ave., Tehran, Iran

[7], type of membrane [8] and MFC architecture are among the designing parameters.

A wide range of different wastewaters have been examined in MFCs. For instance, domestic wastewater [3,9], landfill leachate [6], coking wastewater [2], confectionery wastewater [8] and cassava mill wastewater [10] can be mentioned. However, wastewaters from petrochemical industries have not been receiving much attention due to complications in their biodegradatiuon. Purified terephthalic acid (PTA) wastewater with a high strength organic content is generated during the production process. PTA is a raw material for manufacturing of many petrochemical products such as polyester textile fibers, polyethylene terephthalate bottles and polyester films. As much as 3–10 m^3 of such a wastewater is usually generated per one ton of PTA as product [11].

In our previous work, we studied the feasibility of utilizing PTA wastewater in a MFC for the first time [12]. In the present work, we investigate the effect of two main characteristics of the wastewater i.e. organic load and pH value that significantly influence the power generation. In the present research, we investigated the following issues:

1) Effect of wastewater concentration on power generation
2) Correlation between voltage and wastewater concentration
3) Effect of different pH values on power generation

Materials and methods
Wastewater and microorganisms
PTA wastewater was obtained from the PTA production plant of Shahid Tondgoyan Petrochemical Company, Mahshahr, Iran. It was kept at 4°C until use. This wastewater had the pH of 4.45 and pollution load of 8000 mg COD L^{-1}. It consisted of following components with given concentrations (mg L^{-1}): acetic acid (AA); 9850, benzoic acid (BA); 318, phthalic acid (PA); 400, terephthalic acid (TA); 389, p-toluic acid (p-Tol); 273, nitrate; 2234.7, and phosphate; 48.

Microorganisms, used in this research were obtained from the sludge of an up-flow anaerobic contact filter existing in the treatment plant of the abovementioned petrochemical company.

MFC assembly and operation
A typical membrane-less single chamber MFC as described previously [13] was used in this study (Figure 1). A stainless steel mesh (30 × 60 cm^2) and a carbon cloth (30% w/w wet proofed (type B-1B, E-TEK) using platinum as catalyst (0.5 mg Pt cm^{-2}) and four diffusion layers with projected area of 0.785 cm^2) were used as anode and cathode electrodes, respectively. Content of

the MFC was mixed gently by a magnetic stirrer to reach a uniform concentration.

Analytical methods and calculations
To measure the voltage, a digital multimeter (m58217, Mastech) was used at specific time intervals. Current was calculated based on Ohm's law as $I = V/R$, where I (mA), V (mV) and R (Ω) stand for current, voltage and external resistance, respectively. To calculate the current density, $i = I/A$ was used where, A (m^2) is the projected surface area of the cathode. Power density (P) was calculated using the following equation: $P = IV/A$. Reactions occurred in the cathode were considered as the basis to calculate current density and power density. To derive the polarization curve, the external resistance was changed from 400 KΩ to 300 Ω and the voltage was measured. Power density curve method was used to obtain maximum power density and internal resistance of the MFC [1]. Chemical oxygen demand (COD), nitrate, phosphate and pH were measured according to standard methods. The concentration of BA, PA, TA, p-Tol and AA was measured by a high performance liquid chromatography (HPLC, Agilent Technologies 1200 series, US) under isocratic conditions at the ambient temperature [12].

Results and discussion
Power production at different concentrations of wastewater
Effect of substrates' concentration required for the microbial activity in the anode chamber was investigated. The substrate as fuel was supplied in the following order, consecutively: 10-times diluted wastewater (C$_1$), 4-times diluted wastewater (C$_2$), 2-times diluted wastewater (C$_3$) and the raw wastewater (C$_4$). Figure 2 shows power density curves of the MFC at different concentrations of the wastewater. Maximum power density was 10.5, 43.3, 55.5 and 65.6 mW m^{-2} for C$_1$, C$_2$, C$_3$ and C$_4$, respectively. Internal resistance of the MFC was 5.6 kΩ for all concentrations except for C$_3$ that was equal to 3.2 kΩ.

Maximum power densities of C$_3$, C$_2$ and C$_1$ wastewaters were 84%, 66% and 16% of the maximum power density of the raw wastewater. Therefore, the maximum power density increased when more concentrated wastewater was used.

Current generated in a MFC is limited by two factors: (i) oxidization rate of substrate by bacteria and (ii) rate of electrons transfer to the electrode surface [1]. Substrate oxidation rate depends on the concentration of substrate which is assumed usually as a first order reaction. Therefore, it is expected to observe a higher current and power at higher concentrations. However, other factors such as mass transfer and biofilm layer thickness can suppress power production. The influence of such factors can be studied when the substrate

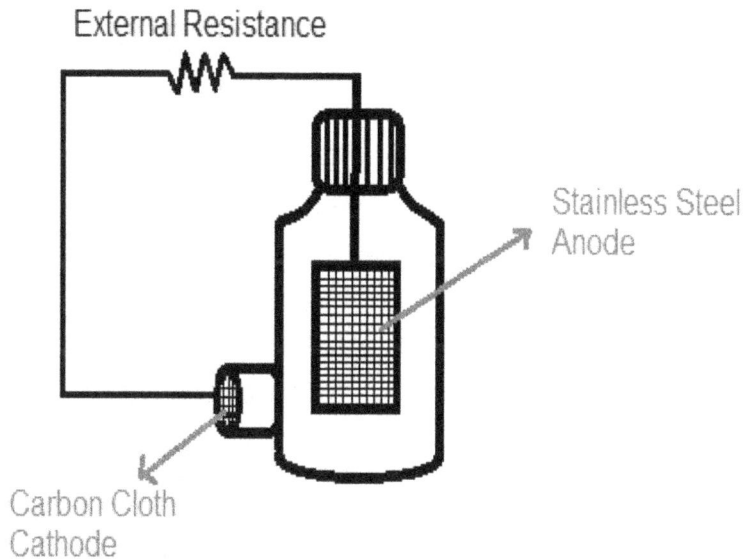

Figure 1 Schematic of the membrane-less single chamber MFC. The reactor was operated in a batch mode at room temperature (22–26°C). The working volume of the MFC was 250 mL. No mediator was added to the anode chamber.

concentration is high enough where oxidation rate is not limited. This observation has been further explained in the next section.

Power differences at different external resistances between two successive concentrations

Increase in the power density due to employing more concentrated wastewater in the anode, was correlated to the external resistance. Figure 3 indicates the difference between the power densities at two successive concentrations of wastewater at definite external resistances.

The maximum power density difference was 32.7, 11.9 and 18.3 mW m^{-2} between C_2 - C_1, C_3 - C_2, and C_4 - C_3, respectively.

According to the polarization curve, mass transfer resistance zone and activation loss zone are observable at low external resistances and large external resistances, respectively. All three curves follow the same trend. The power density difference was negligible at higher external resistances while the cell operated in the activation loss zone. This is the same at very low external resistances when the cell operates in the mass transfer zone.

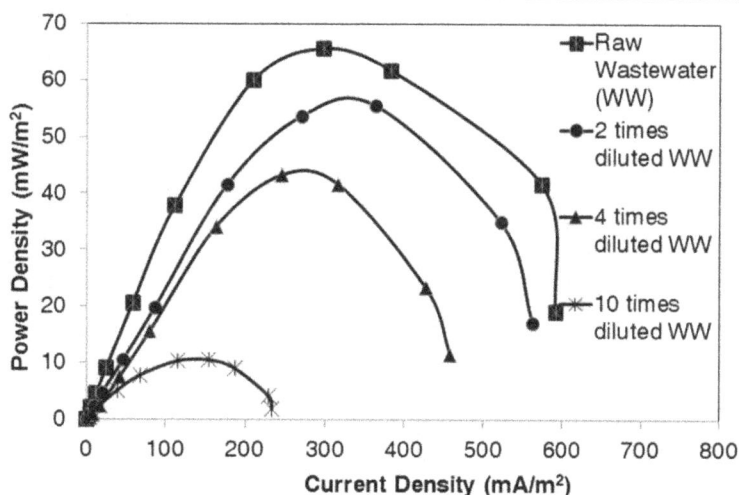

Figure 2 Power density curves at different concentration of PTA wastewater. Power density curves were obtained when the voltage reached to a stable value.

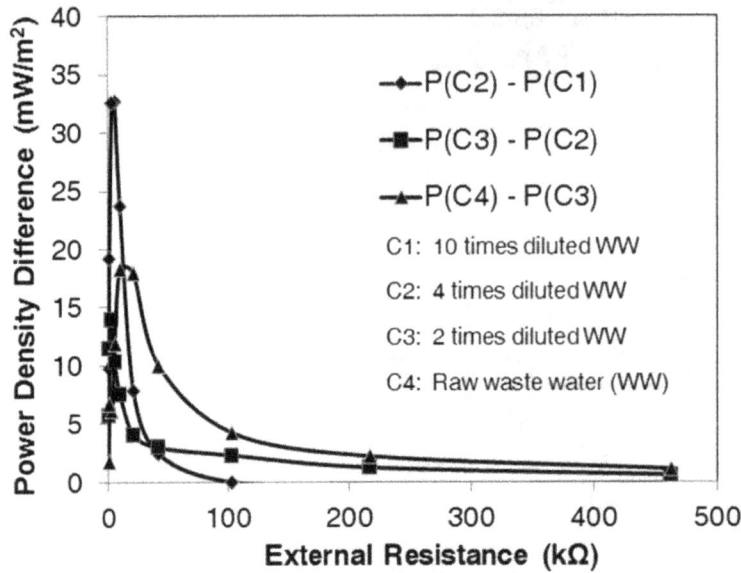

Figure 3 Power density difference between two successive concentrations of wastewater at different external resistances. Maximum power density was 10.5, 43.3, 55.5 and 65.6 mW m^{-2} for C_1, C_2, C_3 and C_4, respectively. The wastewater had the concentration of 8000 mg COD L^{-1}.

This shows that the substrate concentration has a minor effect in these zones. Accordingly, if a MFC is working with an external resistance which leads to operating, either in the activation loss zone or the mass transfer zone, increasing the substrate concentration would not have a significant effect on its performance. In contrary, a considerable power density difference was observed in the ohmic resistance zone. The power density difference reached a maximum value at an external resistance equal or close to the internal resistance of the MFC. Therefore, according to these observations, the maximum power production is reachable when the external resistance is near the internal resistance. Besides, the major positive

effect of concentration increase is visible when the MFC works in the ohmic zone.

Generated voltage versus wastewater concentration in ohmic zone

In this section, the behavior of MFC in the ohmic zone is explored. The generated voltage at different concentration of wastewater for a certain external resistance (in the Ohmic zone) is exhibited in Figure 4. It was observed that the correlation between voltage and concentration is so that at higher concentrations, a little increase occurs in the voltage generation. For example, there is no significant difference between the produced,

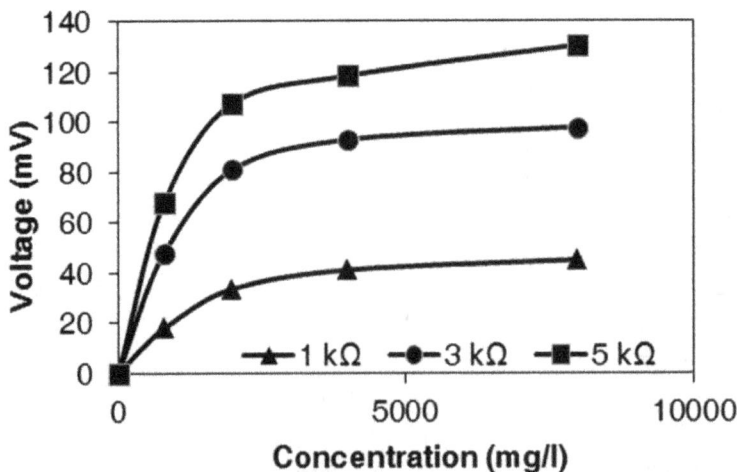

Figure 4 Generated voltage at different concentration of wastewater. Monod-type behavior of voltage against concentration.

Table 1 Calculated constants of Eq. 1 for different external resistances

R_{ext} (kΩ)	V_{max} (mV)	K_s (mg L^{-1})
1	45.0	1148.4
3	98.0	48.0
5	130.5	614.7

voltage when C3 or C4 was applied. One can say that the concentration increase effect is limited even in the ohmic zone which might be as a result of concentration inhibition that halts bacteria metabolism.

The voltages vs. concentration curves suggest a Monod-type equation as follows:

$$V = V_{max}\frac{s}{K_s + s} \tag{1}$$

where, V (mV) and V_{max} (mV) stand for voltage and maximum voltage, respectively; S (mg L^{-1}) represents COD of substrates and K_s (mg L^{-1}) is the half-saturation constant. V_{max} and K_s were calculated for each curve as presented in Table 1.

Maximum achievable voltage versus external resistance follows a linear equation:

$$V_{max} = 21.375\, R_{ext} + 27 \tag{2}$$

Not to be neglected that the above equation is valid only when the MFC operates at the ohmic zone.

Power production at different pH values

pH has a significant effect on the activity of bacteria in terms of removal efficiency and energy production. In order to study the influence of pH, the MFC was fed with 10- times diluted wastewater at three different pH values including 8.5, 7.0 and 5.4, periodically. These pH values were selected based on the optimal range of the pH reported for methane-producing bacteria. It has been observed that these bacteria are active in the pH range of 6.3-7.8 [14]. Presence of methane producers is very possible in our system. The power density curves for different pH values are shown in Figure 5. It was observed that the maximum power density was 12.5, 7.5 and 4.3 mW m^{-2} for the pH values of 8.5, 7.0 and 5.4, respectively.

In general, the higher the pH value, the higher the power density. The produced power at pH 8.5 was higher for 40% and 66% than that for pH 7.0 and pH 5.4, respectively. This observation is consistent with other previous studies [15,16].

Apparently, acidogenic bacteria are active in pH 5.5. Under this condition, hydrogen production would be the dominant mechanism which overcomes the pollutants degradation and a decreased removal rate is expected compared to the neutral or alkaline conditions [14]. Due to the low removal rate, fewer electrons are released and the power production is lowered, consequently. At pH 7.0, methane gas production is the dominant metabolic pathway. This would lead to a less number of released electrons that can contribute in electricity generation and a lower power density is observed, eventually. The increase in power density production at

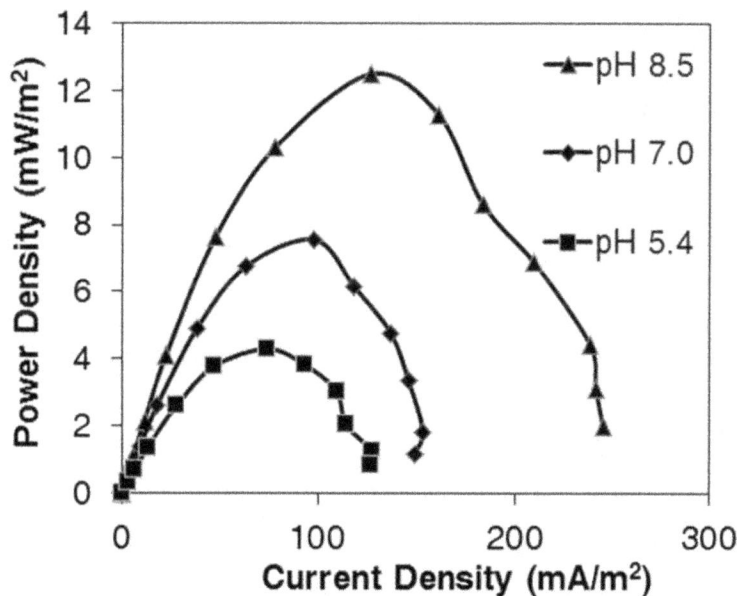

Figure 5 Power density curves at different pH values for 10-times diluted wastewater.

pH 8.5 might be due to the lower activity of methano-genic and acidogenic bacteria. As a result, the electrons released in the oxidation process of the substrates would contribute significantly in electricity generation. However, further studies are required to clarify the occurrence of these phenomena, more precisely.

It can be concluded from the trend of power production at different pH values that alkaline condition provides a favorable situation for the growth of electrogenic bacteria. Previous studies have shown that the electrochemical interaction of bacteria significantly increases under alkaline conditions [15,16], which ultimately leads to a higher power production.

Conclusion

The main purpose of this research was to provide more information and insight into the MFC operation that can pave the way towards practical utilization of MFC technology for the application of real wastewater. Bioelectricity generation using purified terephthalic acid wastewater from a petrochemical plant was successfully conducted in a single chamber microbial fuel cell with a stainless steel mesh as anode electrode.

The influence of wastewater concentration on the MFC performance showed that using the raw wastewater with the concentration of 8000 mg COD L^{-1} results in the highest power density (65.6 mW m^{-2}). Our observations suggest that the best performance is achievable when the MFC operates at the ohmic zone and has an external resistance which is equal to the internal resistance of the cell.

The voltage against different initial concentrations of the wastewater in the ohmic zone followed a Monod-type equation. This observation implies that the concentration increase has a positive effect of electricity generation but it cannot exceed a maximum value.

Performance of the MFC at different pH values was investigated and the highest power density was observed under alkaline condition (pH 8.5) due to inactivation of acidogenic and methanogenic bacteria in favor of more activity for electrogenic bacteria.

Competing interests
The authors declare that they have no competing interests.

Authors' contributions
SKFM carried out the experiments, analyzed the data and drafted the manuscript. HRK designed the study, participated in data analysis and reviewed the article. Both authors read and approved the final manuscript.

Acknowledgements
Authors wish to thank the research office of Sharif University of Technology for the financial support.

References
1. Logan BE. Microbial Fuel Cells. NewJersy: John Wiley & Sons; 2008.
2. Huang L, Yang X, Quan X, Chen J, Yang F. A microbial fuel cell–electro-oxidation system for coking wastewater treatment and bioelectricity generation. J Chem Technol Biot. 2010;85:621–7.
3. Ahn Y, Logan BE. Effectiveness of domestic wastewater treatment using microbial fuel cells at ambient and mesophilic temperatures. Bioresour Technol. 2010;101:469–75.
4. Zhuang L, Zhou S, Li Y, Yuan Y. Enhanced performance of air-cathode two-chamber microbial fuel cells with high-pH anode and low-pH cathode. Bioresour Technol. 2010;101:3514–9.
5. Lyon DY, Buret F, Vogel TM, Monier J-M. Is resistance futile? Changing external resistance does not improve microbial fuel cell performance. Bioelectrochemistry. 2010;78:2–7.
6. Greenman J, Gálvez A, Giusti L, Ieropoulos I. Electricity from landfill leachate using microbial fuel cells: Comparison with a biological aerated filter. Enzyme Microb Technol. 2009;44:112–9.
7. Popov AL, Kim JR, Dinsdale RM, Esteves SR, Guwy AJ. The effect of physico-chemically immobilized methylene blue and neutral red on the anode of microbial fuel cell. Biotechnol Bioprocess Eng. 2012;17:361–70.
8. Sun J, Hu Y, Bi Z, Cao Y. Improved performance of air-cathode single-chamber microbial fuel cell for wastewater treatment using microfiltration membranes and multiple sludge inoculation. J Power Sources. 2009;187:471–9.
9. Rodrigo MA, Ca῀nizares P, Lobato J, Paz CSa R, Linares JJ. Production of electricity from the treatment of urban waste water using a microbial fuel cell. J Power Sources. 2007;169:198–204.
10. Kaewkannetra P, Chiwes W, Chiu TY. Treatment of cassava mill wastewater and production of electricity through microbial fuel cell technology. Fuel. 2011;90:2746–50.
11. Joung JY, Lee HW, Choi H, Lee MW, Park JM. Influences of organic loading disturbances on the performance of anaerobic filter process to treat purified terephthalic acid wastewater. Bioresour Technol. 2009;100:2457–61.
12. Marashi SKF, Kariminia H-R, Savizi ISP. Bimodal electricity generation and aromatic compounds removal from purified terephthalic acid plant wastewater in a microbial fuel cell. Biotechnol Lett. 2013;35:197–203.
13. Marashi, SKF, Kariminia, H-R, Electricity generation from petrochemical wastewater using a membrane-less single chamber microbial fuel cell. Renewable Energy and Distributed Generation (ICREDG), 2012 Second Iranian Conference on (pp. 23-27). IEEE. doi:10.1109/ICREDG.2012.6190462
14. Zhu G-F, Wua P, Wei Q-S, Lin J-y, Gao Y-L, Liu H-N. Biohydrogen production from purified terephthalic acid (PTA) processing wastewater by anaerobic fermentation using mixed microbial communities. Int J Hydrogen Energy. 2010;35:8350–6.
15. Behera M, Jana PS, More TT, Ghangrekar MM. Rice mill wastewater treatment in microbial fuel cells fabricated using proton exchange membrane and earthen pot at different pH. Bioelectrochemistry. 2010;79:228–33.
16. Yuan Y, Zhao B, Zhou S, Zhong S, Zhuang L. Electrocatalytic activity of anodic biofilm responses to pH changes in microbial fuel cells. Bioresour Technol. 2011;102:6887–91.

Biosorption of Hg(II) and Cu(II) by biomass of dried *Sargassum fusiforme* in aquatic solution

Shengmou Huang and Gan Lin[*]

Abstract

The biosorption of heavy metals Hg(II) and Cu(II) from aquatic solution by biomass of dried *Sargassum fusiforme* was studied in the paper. The *Sargassum fusiforme* was able to absorb appreciable amount of mercury and copper from the aquatic solutions within 60 min of contact time with the metal solution and exhibited high removal of mercury and copper at low equilibrium concentrations. The specific adsorption of both Hg(II) and Cu(II) increased at low concentration of biomass and decreased when biomass concentration exceeded 2.0 g/L. The binding of mercury followed Freundlich model while copper supported Langmuir isotherm for adsorption with their r^2 values of 0.971 and 0.923, respectively. The maximum adsorption per unit masses of *Sargassum fusiforme* (mg/L) at equilibrium (q_{max}) for Hg(II) and Cu(II) were calculated to be 30.86 and 7.69 mg/g, respectively. The biosorption by *Sargassum fusiforme* was best described using a pseudo-second-order kinetic model for copper and mercury ions in solution in the study. The adsorption was pH dependent as the maximum mercury biosorption and copper adsorption was happened at solution pH of 8–10.

Keywords: Biosorption, *Sargassum fusiforme*, Mercury, Copper

Background

The heavy metal is among the most common pollutant found in industrial effluents. The major sources of pollution in the aquatic environment are industries such as paint, pulp and paper, oil refining, electrical, rubber, processing, fertilizer, pharmaceutical and battery manufacturing [1,2]. The major effects of mercury and copper poisoning manifest as neurological and renal disturbances as it can easily pass the blood–brain barrier. Environmental contamination with toxic heavy metals is a significant worldwide problem with their successive accumulation in the food chain and continued persistence in the ecosystem which will hurt human beings. Efforts have been made to remove toxic heavy metal from the wastewater and environment by using conserved technologies such as ion exchange or chemical precipitation, which are sometimes inefficient and expensive, particularly for removal of low concentrated heavy metal ions [3], and also leads to produce toxic sludge that adverse the economical feasibility of the treatment methods.

Indeed, many early studies have shown that nonliving biomass may be even more effective than living cells in sequestering metallic elements. Over the past two decades, much effort has been directed at identifying readily available biomass which, in its nonliving state, is capable of effectively removing heavy metals. It has been demonstrated that biosorption is a potential alternative to traditional treatment processes of metal ions removal. Biosorption is a property of certain types of inactive, dead biomass to bind and concentrate heavy metals from even very dilute aqueous solutions [4]. Biomass exhibits this property, acting just as a chemical substance, as an ion exchange of biological origin. Research on biosorption is revealing that it is sometimes a complex phenomenon where the metallic species could be deposited in the solid biosorbent through different sorption processes of ion exchange, complexation, chelation, microprecipitation, etc.

In these years, the role of seagrass in removal of toxic metal ions from the polluted water have taken more importance as they have a high area-to-volume ratio and therefore provide a large contact area for metal binding [5]. Small bacteria, algae, fungi and yeast have been well recognized for heavy metal removal [6]. Up to date, the role of algae has received increased attention to years

* Correspondence: lingan1978@126.com
School of chemical engineering and food science, Hubei University of Arts and Science, Xiangyang 441053, China

because of it are potential for application of environ-mental protection as well as recovery of heavy metals. The affinity with various algae species for binding of heavy metals shows different results [7]. Metal ions with greater electro-negativity and smaller ionic radii are preferably absorbed by algae biomass [8]. Metal accumu-lation capacity of algae biomass is sometimes higher than chemical sorbents therefore algae biomass may serve as an economically feasible to the existing physico-chemical methods of metal removal of wastewaters. The major challenge to biosorption studies is to select the most promising biomass from a large pool of available and inexpensive biomaterials. Contributions to *Sargassum fusiforme* in the metal adsorption is of great concern [9].

In the present study, dried biomass of *Sargassum fusiforme* were used and characterized for its Cu(II) and Hg(II) removal potential from synthetic metal solution. The effect of pH, biomass concentration, initial metal con-centration and contact time was studied for metal sorption procedure as well as heavy metal equilibrium sorption kin-etics was made in lab.

Methods
Biosorbent material
The dry biomass of *Sargassum fusiforme* were purchased from Fuzhou Lantian limited Company, Fujian (China). It was powdered and sieved into less than 1 mm in diameter and dried at 80°C for 12 h with drying oven(01A, Suzhou sinovel oven manufacturing co., LTD). The characteristics of the *Sargassum fusiforme* were determined by F-Sorb type 3400 specific surfaces and pore diameter gauge (Beijing gold spectrum technology co., LTD). The surface BET on the algae was of means with 35-40 m^2/g, total pore volume is of 50-70% and the pore structure is meso-porous material. The FTIR spectrum of the fresh and metal-loaded ones were made by the FTIR920(Tianjin topology instrument co., LTD). The SEM images of the fresh and metal-loaded ones were attained by SU3500 (Hitachi High-Technologies Corporation).

Mercury and copper sorption experiments
Heavy metal solutions were prepared for diluting 200 mg/L of stock solutions, which were made by dissolving copper nitrate and mercury nitrate of analytical grade (Shanghai bo yiu biological technology co., LTD) in double distilled water. Metal sorption studies were carried out to evaluate the capacity of dry biomass of *Sargassum fusiforme* to ad-sorb metal ions from solutions. In batch ones, 100 mL of synthetic metal solutions having different concentrations (10,20,30,40, and 50 mg/L) of copper or mercury were placed in 250 ml Erlenmeyer flasks with a range of biomass concentrations(0.5,1.0,1.5.2.0,2.5, and 3.0 g/L) as biosor-bent. Erlenmeyer flasks were kept under shaking at 120 rpm at 25°C. Samples were taken after 20-120 min and

filtered and analyzed for final metal concentration (C_f) using ICP-AES (HK-2000, Huake Beijing tiancheng tech-nology co., LTD) after an acid digestion [10].

To see the effect of pH on Hg(II) and Cu(II) removal, a range of pH (2–10) was adjusted to 0.1 M NaOH or 0.1 M HCl in 100 ml metal solutions containing fixed concentra-tion of metal at 10 mg/L and biomass of 3.0 g/L followed by contact time of 60 min at rotation of 120 rpm. The metal adsorption(q) with *Sargassum fusiforme* and bior-emoval efficiency(R) were calculated by the following formulae.

$$q = \left(C_i - C_f\right) * \frac{V}{M} \tag{1}$$

$$R(\%) = \frac{C_i - C_f}{C_i} * 100 \tag{2}$$

Where q = metal adsorption (mg/g); M = dry mass of *Sargassum fusiforme*(g); V = volume of initial metal solu-tion used (L); R = bioremoval efficiency (%); C_i = initial con-centration of metal in aquatic solution (mg/L); C_f = final concentration of metal in aquatic solution (mg/L) [11].

Adsorption isotherm
During biosorption, the equilibrium is established be-tween absorbed metal ion on the *Sargassum fusiforme*(q) and unabsorbed metal ions in the solution(C_f). This equilibrium represented by Langmuir and Freundlich adsorption isotherms, are widely used to analyze data for wastewater treatment application [12]. Langmuir equa-tion, which is valid for monolayer sorption onto a sur-face, with identical sites was given by Eq. 3.

$$q = q_{max} * b * \frac{C_f}{1 + bC_f} \tag{3}$$

Where q_{max}(mg/g) is the maximum amount of the metal ion per unit weights of algae to form a complete mono-layer on the surface bound at high C_f(mg/L), and b is a constant related to the affinity of the binding sites(mg/L), q_{max} represents a practical limiting adsorption capacity when the surface is fully covered with metal ions and as-sists in the comparison of adsorption performance [13]. The q_{max} and b can be determined from the liner plot of C_f/q versus C_f. The empirical Freundlich equation based on sorption on a heterogeneous surface is given below by Eq. 4.

$$q = k * C_f^{1/n} \tag{4}$$

The k and n parameters are the constants of the Freundlich isotherm. The k and n are indicators of adsorption capacity and adsorption intensity, respectively. The Eq.4 can be linearized in logarithmic forms and Freundlich constants can be determined by the plot. Freundlich isotherm is also more widely used as it provides no information on the monolayer adsorption capacity [14].

Biosorption kinetics

The experimental biosorption kinetic data was modeled using the pseudo-first-order (Eq.5), and pseudo-second-order models (Eq.6). The linear pseudo-first-order model can be represented by the following equation:

$$\log(q_e - q_t) \log q_e - \frac{K_1}{2.303} * t \qquad (5)$$

The qe (mg/g) and qt (mg/g) are the amounts of adsorbed metal on the sorbent at the equilibrium time and at any time t, respectively, and $K_1(\text{min}^{-1})$ is the rate constant of the pseudo-first-order adsorption process. The linear pseudo-second-order model can be represented by the following equation:

$$t/q_t = \frac{1}{K_2} * q_e^2 + \frac{t}{q_e} \qquad (6)$$

Where K_2 ($g*mg^{-1}*min^{-1}$) is the equilibrium rate constant of pseudo-second-order [11].

Results and discussion

Effect of pH solution

The pH is one of the important parameters in heavy metal sorption by Sargassum fusiforme. Therefore metal sorption studies were carried out at different pH values. Results revealed the maximum biosorption of Hg(II) were at pH 8(70%) and 10(72%) and Cu(II)

demonstrated the same result at pH8(90%) and 10(92%) from aquatic solution containing initial 10 mg/L of metal concentration (Figure 1).

It was observed that Hg(II) and Cu(II) adsorption was less than 10% at pH 2. Enhance adsorption of Hg(II) and Cu(II) ions at higher pH was observed which coincides with earlier findings where in most cases the removal efficiency increased steadily with rise in pH [15]. The adsorption of metal ions was lower at low pH because of high concentration of protons in the solution which competed with metal ions in forming a bond between the active sites on the surface of the algae biomass [16]. Selective sorption of specific metals due to distinct pH optima for their sorption may be due to the chemical composition of cell surfaces. A distinct relationship between pH of aquatic metal solutions and involvement of functional group in binding of Hg(II) and Cu(II) onto Sargassum fusiforme maxima was observed with the involvement of functional groups such as carboxyl, phosphate and hydroxyl [17].

Metal concentration

Biosorption studies carried out for both Hg(II)and Cu(II) in 100 ml solutions containing metals varying from 10 to 50 mg/L with Hg(II)and Cu(II) by 3.0 g/L of biomass exhibited effective role of initial metal concentration on metal removal. A consistent decrease in metal removal was observed by increasing external metal concentration (Figure 2) where 70-74% removal efficiency was reported on 10 to 20 mg/L Hg(II) solutions followed by decline in solutions to 50 mg/L Hg(II)(Figure 2). Similarly decreases in Cu(II) removal efficiency was observed by increasing external metal concentration on 50 mg/L Cu(II) (Figure 2). Rapid metal adsorption profile of Sargassum fusiforme was obtained for both Hg(II) and Cu(II), which is important when the algae are used for biosorption. It

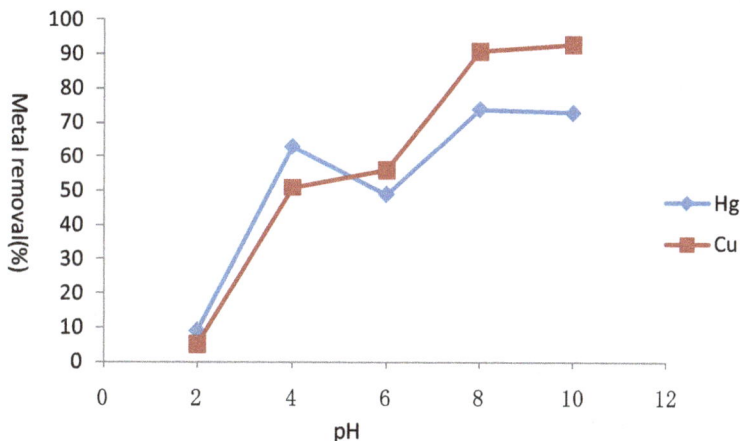

Figure 1 pH dependent sorption of Hg(II) and Cu(II) by Sargassum fusiforme(Hg(II):10 mg/L, Cu(II):10 mg/L), standard deviation is less than 5% in triplicates.

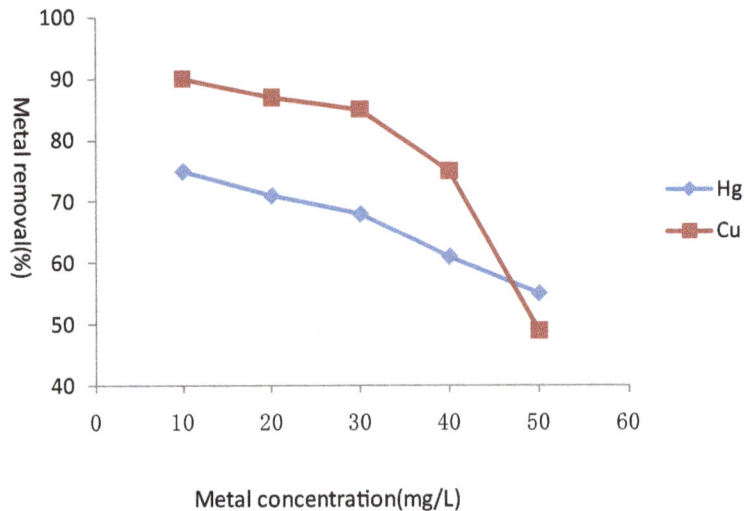

Figure 2 Initial concentration dependent sorption of Hg(II) and Cu(II) by *Sargassum fusiforme*, standard deviation is less than 5% in triplicates.

exhibited rapid biosorption at first 60 min by removing 72% Hg(II) and 90% of Cu(II) from metal solutions. Decreases in metal removal of increasing initial metal concentration was supported by the findings that observed that the removal of metal generally decreases from increasing concentration of metals in the solution [18]. Algae surface has different functional groups of varying the affinity with ionic kinds, therefore decline in metal removal is largely attributed to saturation of adsorption sites [7].

Contact time determination

Rapid metal adsorption profile of *Sargassum fusiforme* was obtained for both Hg(II) and Cu(II), which is important when the material is to be used for biosorption. It exhibited rapid biosorption in first 60 min by removing 71% Hg(II) and 88% of Cu(II) from metal solutions thereafter increase in metal removal was magical. Equilibrium was established between absorbed metal ions after 40 min with maximum removal of 55% and 75% of Hg(II) and Cu(II) (Figure 3).

The *Sargassum fusiforme* biomass showed rapid biosorption at first 60 min. It has been reported that the sorption of heavy metal ions by algae followed mechanism where the metal ion is physically or chemically taken up onto the surface of the algae [19]. In this case, since the algae were dried and biological functions were no longer active, the sorption could only take place on the

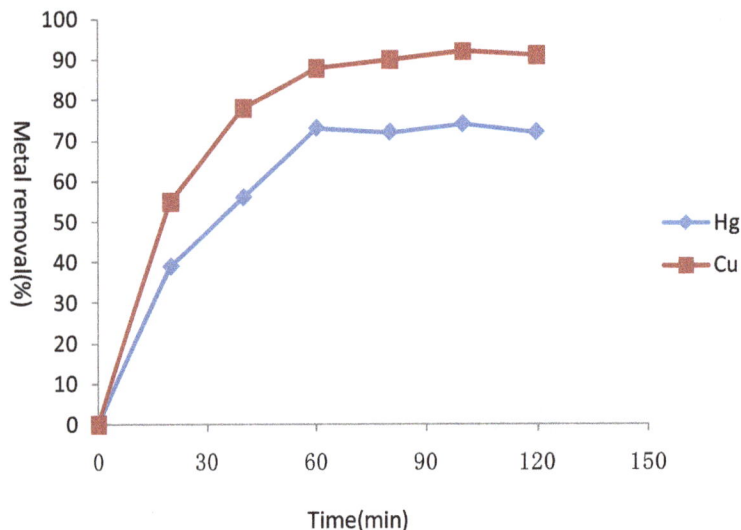

Figure 3 Time dependent removal of Hg(II) and Cu(II) by *Sargassum fusiforme*(Hg(II):10 mg/L, Cu(II):10 mg/L), standard deviation is less than 5% in triplicates.

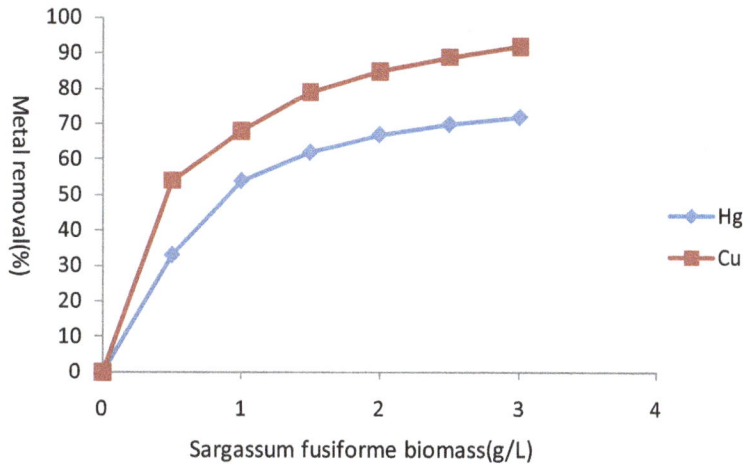

Figure 4 Effect of *Sargassum fusiforme* biomass concentration on Hg(II) and Cu(II) removal(Hg(II):10 mg/L, Cu(II):10 mg/L), standard deviation is less than 5% in triplicates.

cell surface. The increase in Hg(II) and Cu(II) adsorbed by increasing biomass was also expected as a result of increase in available area-to-volume ratio and therefore providing a large contact area for heavy metal binding [20].

Biomass determination

Increase in metal removal efficiency from 10 mg/L manual metal solutions to both Hg(II) and Cu(II) was observed on increasing biomass concentration (Figure 4).

Hg(II) adsorption was increased to 71% by increasing biomass concentration on 0 to 3.0 g/L in 10 mg/L of mercury containing synthetic solutions whereas, same trend was observed in case of Cu(II) with the increase of 90%. The *Sargassum fusiforme* biomass displayed its equilibrium for Hg(II) removal of 3.0 g/L of biomass concentration whereas, a continuous increase in Cu(II) removal of increasing biomass concentration was observed of 3.0 g/L. Metal adsorption studies from solution mass balance revealed a decline in sorption of Hg(II) and Cu(II) respectively by increasing biomass from 0 to 3.0 g/L after 60 min of contact time. The other probable explanations for such a relationship between biomass concentration and adsorption may be limited availability of metal, increased electrostatic interactions between binding sites and reduced mixing at higher biomass concentration [21,22].

The FTIR and SEM analysis

With the determination of functional matrices and group in the heavy metal biosorption of the algae, it is useful for the study of FTIR spectrums. The figures are shown in Figure 5. With the determination of biosorption kinetics and isotherms in the heavy metal biosorption of the algae, the use of SEM images is shown in Figure 6 for the study.

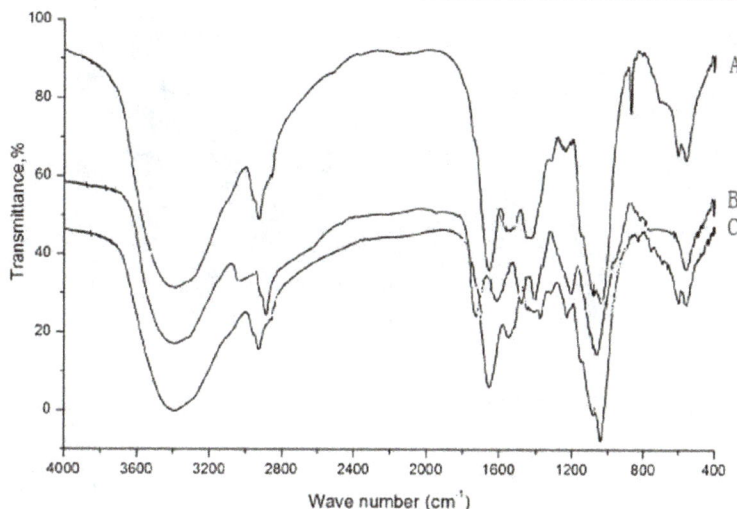

Figure 5 The FTIR spectrum of the adsorbent for fresh and metal loaded ones. **A**: fresh, **B**: Cu(II) loaded, **C**: Hg(II) loaded.

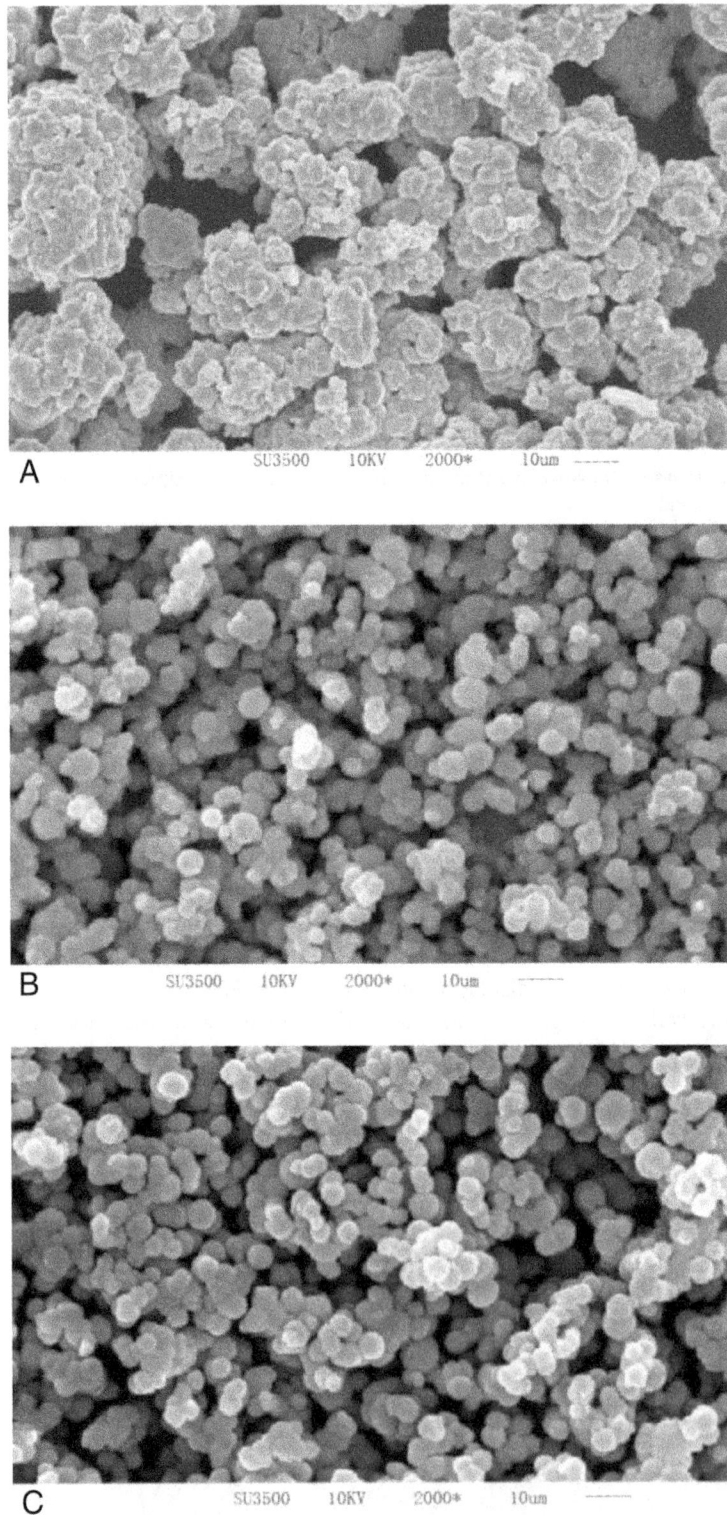

Figure 6 The SEM images of the adsorbent for fresh and metal loaded ones. **A:** fresh, **B:** Cu(II) loaded, **C:** Hg(II) loaded.

Modeling study

A linear regression of the experimental results from Hg(II) and Cu(II) differed in terms of Hg(II) adsorption fitted better to Freundlich isotherm and Cu(II) to Langmuir isotherms with r^2 values of 0.971 and 0.823 respectively (Table 1). The maximum adsorption per unit masses of

Table 1 Langmuir and Freundlich parameters for the sorption of the test metals by *Sargassum fusiforme*

Metal	Langmuir isotherm			Freundlich isotherm		
	b(L/mg)	r^2	qmax(g/mg)	K(mg/g)	n	r^2
Hg(II)	0.988	0.668	30.86	2.012	3.89	0.971
Cu(II)	1.269	0.923	7.69	0.899	5.41	0.765

Sargassum fusiforme (mg/L), at equilibrium (q_{max}) for Hg(II) and Cu(II) were calculated to be 30.86 and 7.69 mg/g respectively (Table 1).

The sorption isotherm is the relationship between equilibrium concentration on the solution and equilibrium concentration of solute in the sorbent at constant temperature where either Freundlich or Langmuir model can describe the biosorption equilibrium of copper and mercury [23]. An extremely high r^2 value of Freundlich isotherm for Hg(II) sorption indicated that ion exchange interaction takes place between metal ion and the biosorbent, while Cu(II) follows Langmuir isotherm and thus supported physicochemical interactions with each other.

For biomasses of *Sargassum fusiforme,* the kinetics of copper and mercury biosorption were analyzed using pseudo-first-order and pseudo-second-order models. All the constants and regression coefficients are shown in Table 2. In the present study, biosorption by *Sargassum fusiforme* was best described using a pseudo-second-order kinetic models for copper and mercury ions in solution. This adsorption kinetic is typical of the adsorption of divalent metals onto biosorbents.

Conclusions

The goal of the study was to explore and find out the potential use of algae biomass as a low cost sorbent for the removal of heavy metals from aquatic solutions. The heavy metals Hg(II) and Cu(II) from aquatic solution was able to be dealt with dried *Sargassum fusiforme*. The seagrass adsorbed appreciable amount of mercury and copper from the aquatic solutions within 60 min at low equilibrium concentrations. The specific adsorption of both Hg(II) and Cu(II) increased at low concentration while decreased when biomass concentration exceeded 2.0 g/L. The binding of mercury followed Freundlich model but copper supported Langmuir isotherm for adsorption with their r^2

Table 2 First and second order adsorption rate constants for Hg(II) and Cu(II)

Metal	Pseudo-first-order		Pseudo-second-order	
	$K_1(min^{-1})$	r^2	$K_2(g*mg^{-1}*min^{-1})$	r^2
Hg(II)	$5.3*10^{-3}$	0.745	$9.15*10^{-3}$	0.957
Cu(II)	$3.1*10^{-3}$	0.615	$8.84*10^{-3}$	0.963

values of 0.971 and 0.923, respectively. The maximum adsorption per unit masses of *Sargassum fusiforme* (mg/L), at equilibrium (q_{max}) for Hg(II) and Cu(II) were calculated to be 30.86 and 7.69 mg/g respectively. The adsorption was pH dependent as the maximum mercury biosorption utilized at pH 8 and 10 and Cu(II) adsorption was at pH 8 and 10. The biosorption by *Sargassum fusiforme* was best described using a pseudo-second-order kinetic model for copper and mercury ions in solution in the study. The present paper emphasizes the *Sargassum fusiforme* is an ideal candidate and can be designed as a practical and economical process for wastewater treatment polluted by heavy metals.

Competing interests
The authors declare that they have no competing interests.

Authors' contributions
HSM and LG conceived the work, performed the experiments, designed the experiments, analyzed the data, and wrote the manuscript. Both authors read and approved the final manuscript.

Acknowledgments
We greatly thank Chen Guantong from FAU for providing researching materials in this investigation. We are also grateful to Zhang Danfeng for the helpful comments and linguistic revision of the manuscript. This work was supported by grants from the ministry of education from Hubei province, People's Republic of China(project No. Q20122506).

References
1. Rezaee A, Ramavandi B, Ganati F. Biosorption of mercury by biomass of filamentous algae *Spirogyra* species. J Biol Sci. 2006;6(4):695–700.
2. Khodaverdiloo H, Samadi A. Batch equilibrium study on sorption, desorption, and immobilisation of cadmium in some semi-arid zone soils as affected by soil properties. Soil Res. 2011;49:444–54.
3. Lu WB, Shi JJ, Wang CH, Chang JS. Biosorption of lead, copper and cadmium by indigenous isolate *Enterobacter* sp. Processing high heavy metal resistance. J Hazard Mater. 2006;134:80–6.
4. Rezaee A, Ramavandi B, Ganati F. Equilibrium and spectroscopic studies on biosorption of mercury by algae biomass. Pak J Biol Sci. 2006;9(4):777–82.
5. Pandiyan S, Mahendradas D. Application of bacteria to remove Ni(II) Ions from aqueous solution. Eur J Sci Res. 2011;52:345–58.
6. Ekmekyapar F, Aslan A, Bayhan YK, Cakici A. Biosorption of copper (II) by nonliving lichen biomass of *Cladonia rangiformis* hoffm. J Hazard Mater. 2006;B137:293–8.
7. Zouboulis AI, Loukidou MX, Matis KA. Biosorption of toxic metals from aqueous solution by bacterial strains isolated from metal-polluted soils. Process Biochem. 2004;39:909–16.
8. Anjana K, Kaushik A, Kiran B, Nisha R. Biosorption of Cr(VI) by immobilized biomass of two indigenous strains of cyanobacteria isolated from metal contaminated soil. J Hazard Mater. 2007;148:383–6.
9. Naik UC, Srivastava S, Thakur IS. Isolation and characterization of *Bacillus cereus* IST105 from electroplating effluent for detoxification of hexavalent chromium. Environ Sci Pollut Res Int. 2011;19:3005–14.
10. Vijayaraghavan K, Jegan J, Palanivenu K, Velan M. Biosorption of cobalt(II) and nickel(II) by seaweeds: batch and column studies. Sep Purif Technol. 2005;44:53–9.
11. Ok YS, Yang JE, Zhang YS, Kim SJ, Chung DY. Heavy metal adsorption by a formulated zeolite-Portland cement mixture. J Hazard Mater. 2007;147:91–6.
12. Kaewsarn P, Yu Q. Cadmium (II) removal from aqueous solutions by pretreated biomass of marine alga *Padina* sp. Environ Pollut. 2001;112:209–13.
13. Chatterjee SK, Bhattacharjee I, Chandra G. Biosorption of heavy metals from industrial waste water by *Geobacillus thermode* nitrificans. J Hazard Mater. 2010;175:117–25.

14. Lodeiro P, Barriada JL, Herrero R, Sastre deVicente ME. The marine macroalga *Cystoseira baccata* as biosorbent for cadmium (II) and lead (II) removal: kinetic and equilibrium studies. Environ Pollut. 2006;142:264–73.

15. Vilar VJP, Botelho CMS, Boaventura RAR. Lead uptake by algae *Gelidium* and composite material particles in a packed bed column. Chem Eng J. 2008;144:420–30.

16. Costodes VCT, Fauduet H, Porte C, Delacroix A. Removal of Cd(II) and Pb(II) ions from aqueous solutions by adsorption onto sawdust of *Pinus sylvestris*. J Hazard Mater. 2003;105:121–42.

17. Pabst MW, Miller CD, Dimkpa CO, Anderson AJ, McLean JE. Defining the surface adsorption and internalization of copper and cadmium in a soil bacterium *Pseudomonas putida*. Chemosphere. 2010;81:904–10.

18. Sari A, Tuzen M. Biosorption of Pb(II) and Cd(II) from aqueous solution using green alga (*Ulva lactuca*) biomass. J Hazard Mater. 2008;152:302–8.

19. Oliveira SM, Pessenda LC, Gouveia SE, Favaro DI. Heavy metal concentrations in soils from a remote oceanic island, Fernando de Noronha. Braz An Acad Bras Cienc. 2011;83:1193–206.

20. Pardo R, Herguedas M, Barrado E, Vega M. Biosorption of cadmium, copper, lead and zinc by inactive biomass of *Pseudomonas putida*. Anal Bioanal Chem. 2003;376:26–32.

21. Wang XS, Qin Y. Removal of Ni(II), Zn(II) and Cr(VI) from aqueous solution by *Alternanthera philoxeroides* biomass. J Hazard Mater. 2006;138:582–8.

22. Xiong J, He Z, Liu D, Mahmood Q, Yang X. The role of bacteria in the heavy metals removal and growth of *Sedum alfredii* Hance in an aqueous medium. Chemosphere. 2008;70:489–94.

23. Mata YN, Blazquez ML, Ballester A, Gonzalez F, Munoz JA. Characterization of the biosorption of cadmium, lead and copper with the brown alga *Fucus vesiculosus*. J Hazard Mater. 2008;158:316–23.

4-chlorophenol removal from water using graphite and graphene oxides as photocatalysts

Karina Bustos-Ramírez[1,2,3], Carlos Eduardo Barrera-Díaz[1], Miguel De Icaza-Herrera[3], Ana Laura Martínez-Hernández[2,3], Reyna Natividad-Rangel[1] and Carlos Velasco-Santos[2,3*]

Abstract

Graphite and graphene oxides have been studied amply in the last decade, due to their diverse properties and possible applications. Recently, their functionality as photocatalytic materials in water splitting was reported. Research in these materials is increasing due to their band gap values around 1.8-4 eV, and therefore, these are comparable with other photocatalysts currently used in heterogeneous photocatalytic processes. Thus, this research reports the photocatalytic effectiveness of graphite oxide (GO) and graphene oxide (GEO) in the degradation of 4-chlorophenol (4-CP) in water. Under the conditions defined for this research, 92 and 97% of 4-CP were degraded with GO and GEO respectively, also 97% of total organic carbon was removed. In addition, by-products of 4-CP that produce a yellow solution obtained only using photolysis are eliminated by photocatalyst process with GO and GEO. The degradation of 4-CP was monitored by UV-Vis spectroscopy, High Performance Liquid Chromatography (HPLC) and Chemical Oxygen Demand (COD). Thus, photocatalytic activity to remove 4-CP from water employing GO and GEO without doping is successfully showed, and therefore, a new gate in research for these materials is opened.

Keywords: 4-chlorophenol, Photocatalyst, Graphite oxide, Graphene oxide, Water pollution

Background

With increasing global air and water pollutions, photocatalysis has attracted considerable attention because this method provides a promising pathway to attenuate environmental pollution problems, mainly due to the capacity of photocatalyst to degrade organic contaminants [1].

Chlorophenols represent an important group of organic water pollutants due to their toxicity and low biodegradability. They are considered priority pollutants and are employed in numerous industrial processes; in particular, 4-CP is involved in the synthesis of many pesticides, pharmaceuticals and dyes [2-5].

Most polluting organic compounds, including 4-CP, are difficult to degrade, so, it is important to develop more effective methods to promote their degradation [6]. Different biological, physical and chemical methods have been applied for chlorophenol degradation [7,8].

Other established alternative methods, such as Advanced Oxidation Processes (AOP) [9], have been reported to be effective for the degradation of soluble organic contaminants from water, providing almost total degradation. Several technologies are included under AOP such as photo-Fenton, ozonation, photocatalysis, etc. [10,11].

Thus, the remediation of wastewater with chlorophenols using photocatalysis has been widely studied, but not in the case of 4-CP; however, the complexity of degradation processes with heterogeneous catalysts and the high sensitivity of the reaction to experimental conditions represent two main disadvantages, that make it difficult to compare different related systems. Several studies have evaluated the effects of different substituent groups in the aromatic ring in phenolic compounds, as well as the influence of metal ions and/or oxidant compounds, on the adsorption equilibrium and kinetic parameters of degradation during UV irradiation of a suspension in a photocatalytic process with a semiconductor.

The photocatalytic activity is governed by the catalyst, which is a semiconductor (e.g., TiO_2 (the most used), ZnO and Fe_2O_3) [12]. These compounds have two primary characteristics: the first, their band edge energy; semiconductors

* Correspondence: cylaura@gmail.com
[2]División de Estudios de Posgrado e Investigación, Instituto Tecnológico de Querétaro, Av. Tecnológico s/n Esq. M. Escobedo, Col. Centro Histórico, 76000 Querétaro, México
[3]Centro de Física Aplicada y Tecnología Avanzada, Universidad Nacional Autónoma de México, Boulevard Juriquilla 3001, 76230 Querétaro, México
Full list of author information is available at the end of the article

with more negative band edge energy than the reduction potential of water (or protons), and that remain stable when are in contact with water can be considered as appropriate catalyst materials [13]; the second characteristic is the particle size, of primary importance in heterogeneous photocatalysis because it is directly related to the efficiency of the catalyst through the definition of its specific surface area [12].

An increasing amount of works has been addressed in searching alternative materials to TiO_2. Anyway, the information gathered indicates that exist a valid alternative to the use of modified-TiO_2 for photocatalytic process, although the intrinsic electronic and physico-chemical properties of some compounds reported in the literature suggest more investigation. In the degradation of 4-CP, the TiO_2 has been applied in the following cases: 1) Photocatalytic degradation method using 1% silver loaded TiO_2 (Ag-TiO_2), the degradation/transformation of 4-CP is extremely high compared to neat TiO_2, this latter requires longer exposure time to obtain the same degree of degradation than the compound Ag-TiO_2 [14]; 2) Cobalt-modified TiO_2 (Co-TiO_2) is able to bring about the photocatalytic degradation of pollutants such as 4-chlorophenol, the best photoactivity was observed for an amount of cobalt between 0.2% and 0.5% Co/TiO_2 w/w [15]; 3) the mesoporous anatase is able to degrade 100% 4-CP, while Degussa P-25 degraded 57% in time of 180 min, the enhanced photocatalytic activity of the mesoporous titania samples when compared to Degussa P-25 was related to smaller crystallite size, presence of pure anatase phase, higher average pore diameter, and surface area [16]; these are some published works, but in the case of research applying graphite oxide and graphene oxide without doping to degrade photocatalytically 4-CP, it is non-existent until this time.

In the development of new photocatalysts, the graphite oxide (GO), a polymer-like semiconductor made of only carbon, oxygen and hydrogen, has a large exposed area and can be extensively dispersed in water on the molecular scale. These structural features, both due to their chemical and physical aspects, suggest the favourable role of graphite oxide as a photocatalyst. Therefore, graphene oxide (GEO) derivatives of GO can also be good photocatalysts. The difference between them is the number of stacked layers. Both materials (GO and GEO) have photocatalytic activity, as it was reported recently and showed by water splitting and hydrogen production [12,17,18].

Also, nanostructured semiconductor materials are anticipated as new photocatalysts to open up new opportunities and employ the renewable energy sources, such as: Iodine-doped TiO_2 nanoparticles, titanium oxynitride porous thin films, TiO_2/SnO_2 nanofibers, square Bi_2WO_6 nanoplates, and Si nanowires. All these materials have been explored and exhibited interesting photoactivities. These materials

with optical gap of 2.7 eV have been recognized as a very promising metal free photocatalyst, and its photocatalytic activity is confirmed to be even higher than the commercial nitrogen-doped TiO_2. But like the TiO_2, the rapid recombination of electrons and holes is one of the main reasons for the low photocatalytic efficiency of these kinds of photocatalysts [19-21]. But in the case of GO and GEO, the anti-recombination is a given characteristic in current works of research related to the topic of photocatalysts. This effect can be attributed to their nanometric thickness, surface area and the presence of oxygenated groups. Also, both carbon structures can be doped with different materials or atoms to improve their performance as photocatalysts [22]. Additionally, reduced graphene oxide (RGO) modified with different compounds has been used successfully as a photocatalytic material in the remediation treatment of heavy metals and organic compounds [23-26]. Also, a recent work has studied about the photocatalytic effectivity of graphene oxide in the removal of 4-CP by computational method [27]. However, the photocatalytic activities of unmodified GO and GEO have not been studied experimentally, for the removal of organic compounds in water. Thus, it is important to verify the photocatalytic activity of these unmodified carbon materials for the removal of organic compounds in water. GEO has attracted great interest because of its easy availability in bulk quantities, readiness for functionalization by chemical reaction, good dispersion in water and high biocompatibility [26]. Thus, in this research is showed for the first time the capacity of graphene and graphite oxides to remove 4-CP from aqueous solutions under UV-light. Both undoped oxides (GEO and GO) present sufficient activity to remove the organic pollutant from water.

Materials and methods
Materials
Graphite (Electron Microscopy Sciences, No. 70230), distilled water (J.T. Baker), 4-chlorophenol (≥99% Sigma Aldrich), Sulfuric acid (H_2SO_4, Baker, 98%), potassium permanganate ($KMnO_4$, Merck), hydrogen peroxide (H_2O_2, Baker, 30%) and distilled water (H_2O) were used as received. Graphite oxide (GO) and graphene oxide (GEO) were synthesized from graphite by a methodology reported in a previous work [28], and described briefly below.

Synthesis of graphite and graphene oxides
First, H_2SO_4 (46 mL) was added into the reaction flask maintained at 0°C (±2°C) (ice bath), then graphite (2 g) and $KMnO_4$ (6 g) were added slowly. After an increase in temperature to 35°C (±2°C), the mixture was stirred by a magnetic stirring bar and mixed for 2 h. Later, the

excess water was incorporated to the mixture and H_2O_2 (10 ml) was added until there was no gas production.

Then, filtration was performed with distilled water in a glass filter, and the obtained brown GO was dried in an oven (Barnstead Thermoline, Model 3478) at 65°C for 12 h. After that, a solution containing 100 mg of dried GO in 10 mL of H_2O was prepared. This solution was sonicated for 3 h at room temperature with the assistance of an ultrasonic bath (Branson 1510R-MTH) at 55 degassing units, in order to obtain graphene oxide sheets (GEO).

Adsorption test
Adsorption test consisted on placing an aqueous solution of 4-CP with a concentration of 30 ppm and neutral pH into a tubular glass reactor, adding graphite oxide or graphene oxide (0.8 g/L in both cases) with magnetic stirring (1000 rpm) and then, the reactor was covered with aluminium foil to prevent contact with external light. The reaction time was 100 minutes and aliquots were taken at 20, 40, 60, 80 and 100 minutes.

Photolysis test
For these experiments, 30 mL of 4-CP solution (30 ppm) at neutral pH were placed in the same reactor used for the adsorption test and induced with UV irradiation from a lamp (Pencil UV lamp, 254 nm, 5.5 W). Light source was located in the centre of the vessel along with a magnetic stirrer (1000 rpm). The reactor temperature was maintained at 24°C +/- 2°C and the reactor was covered with aluminium foil to prevent contact with external light. The reaction time was 100 minutes and aliquots were taken after 20, 40, 60, 80 and 100 minutes.

Photocatalyst test
The photocatalytic efficiencies of the GO and GEO were determined in the same tubular glass reactor used for the photolysis test with the same continuous stirring (1000 rpm). The total reaction volume was 30 ml. The tests were performed using 0.8 g/L of graphite and graphene oxides, with 30 ppm of 4-CP at neutral pH. The reactor temperature was maintained at 24°C +/- 2°C and aluminium foil was placed around the reactor. A UV lamp (Pencil UV lamp, 254 nm, 5.5 W) was placed in the centre of the reactor to provide UV light radiation. The reaction time was 100 minutes and aliquots were taken after 20, 40, 60, 80 and 100 minutes.

Chemical Oxygen Demand (COD)
Method 8000/reactor digestion method in the high range (0–1500 mg/L), programmed in the HACH DR5000 spectrophotometer was used to determine the COD in the initial and final solutions of 4-CP.

Characterization
Fourier transform infrared spectroscopy (FT-IR) of GO and GEO samples was performed using a Bruker-Vector 33 with a scanning range of 4000-500 cm^{-1} with resolution of 1 cm^{-1}.

Raman spectroscopy of the carbon samples was carried on a Micro-Raman (Dilor, Lab Ram), with measurements at 488 nm incident laser light with a spectral resolution of 1 cm^{-1}.

UV-vis spectroscopy was carried out on a HACH DR5000 spectrophotometer at wavelengths of 200–1100 nm to determine the absorption bands characteristic of the 4-CP solution and monitor the progress of 4-CP removal from the solution by adsorption, photolysis and the photocatalytic processes.

HPLC (High Performance Liquid Chromatography) was performed with a mobile phase of H_2O_2 with 5 mmol H_2SO_4 and methanol (80:20), a flow rate of 1 mL/min and an Ascentis Express C_{18} 3 cm × 4.6 mm (2.7 μm) SUPELCO column for the determination of intermediate compounds.

Mineralization of 4-CP was followed by measuring the chemical oxygen demand (COD) in a typical reaction. An aliquot was taken at the end of the reaction and the COD was measured by a colorimetric method on the HACH DR5000 spectrophotometer.

Results and discussion
FT-IR spectroscopy
The effect of chemical modification due to reaction of graphite with acids is observed in the FT-IR spectra (Figure 1). Pure graphite is inactive as is observed in spectrum (a), whereas GO and GEO spectra (b and c respectively) show several distinctive signals such as: 1732 cm^{-1}, $v(C = O)$ at carboxyl groups; 1620 cm^{-1}, $v(C = C)$; and 1065 cm^{-1}, $v(C-O)$, giving evidence for the presence of oxygen-containing groups, caused by the chemical reaction and reported in previous work [28].

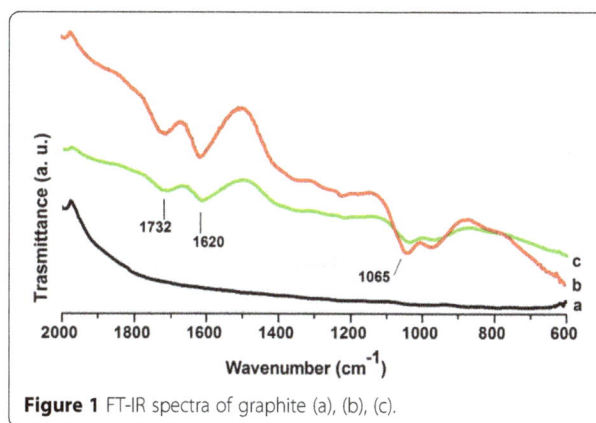

Figure 1 FT-IR spectra of graphite (a), (b), (c).

Raman spectroscopy

Raman spectroscopy shows the characteristic bands assigned to the carbon materials, indicating the modification in the hybridization and defects produced at the surface of materials synthetized by chemical reaction. Figure 2, shows distinctive signals: a strong band close to 1580 cm^{-1} (G band), a line around 1350 cm^{-1} (D band) and 2D band in the region of 2700 cm^{-1}, characteristic for GO and GEO, these bands have been reported by different authors and also in a previous work of this research group [28].

UV-vis spectroscopy

Figure 3 shows the UV-vis spectrum of 4-CP; two characteristic bands of this compound as aqueous solution are observed at 280 and 225 nm [3,29].

The UV-vis spectra in Figure 4 correspond to GO and GEO and show the maximum absorption peaks at 227 and 305 nm that each material exhibits, the bands are attributed to the π-π* electronic transitions of the aromatic C-C bonds and the n-π* transition of the C = O bond, respectively [25]. These spectra are also used to obtain the band gap values for both oxides. Tauc's method is applied, and the calculated values are around 2–4.7 eV for graphite oxide and 1.8–4 eV for graphene oxide; these values are comparable with other semiconductors used as photocatalysts; in addition, these results are very similar to those reported in previous work [13].

Adsorption test

To confirm the degradation of 4-CP due to photocatalysis, samples of graphite and graphene oxides are tested as adsorbents of this pollutant. Figure 5 shows 4-CP removal using GO (A-GO) as adsorbent. In this process, it is observed that the absorption intensity of the band at 280 nm produced by the solution after adsorption with GO is more intense in comparison with the band of only 4-CP (sample 0), indicating that under these conditions GO adsorbs molecules of 4-CP; however, the solution

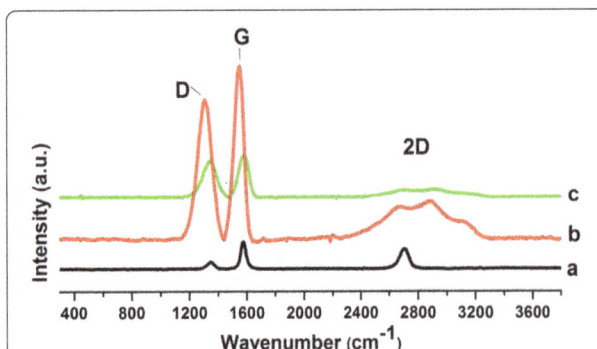

Figure 3 UV-Vis spectrum of 4-chlorophenol (4-CP).

shows a light yellow colour, indicating that the adsorption process may produce some by-products, since it cannot degrade the benzene ring of the 4-CP. The interaction between the two compounds in water may lead to derivatives and by-products of 4-CP, corroborated by the yellow colour of the solution and the increment in absorption of the corresponding band at 225 nm for the benzene ring. Thus, this effect indicates that 4-CP is being adsorbing via link in the GO by C–Cl linkages, with 29% of the 4-CP ultimately removed (calculated by calibration curves used in each experiment). In the case of GEO in the adsorption test (A-GEO), Figure 6 shows the spectra of the carbon samples, where the band at 280 nm decreases; although, in the same way, it can be seen that the band is not eliminated, this could indicate that the functional groups present in the graphene oxide adsorb the complete molecule in a different way than GO. GEO is more efficient in adsorption and it can be considered that no intermediate groups are generated. In the case of GO, the oxygen-containing functional groups between layers do not interact with the 4-CP. However, in 2 dimensions GEO

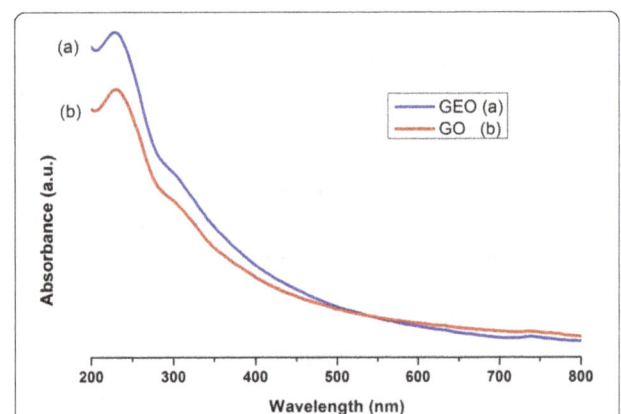

Figure 2 Raman spectra of: graphite (a), graphite oxide (b), graphene oxide (c).

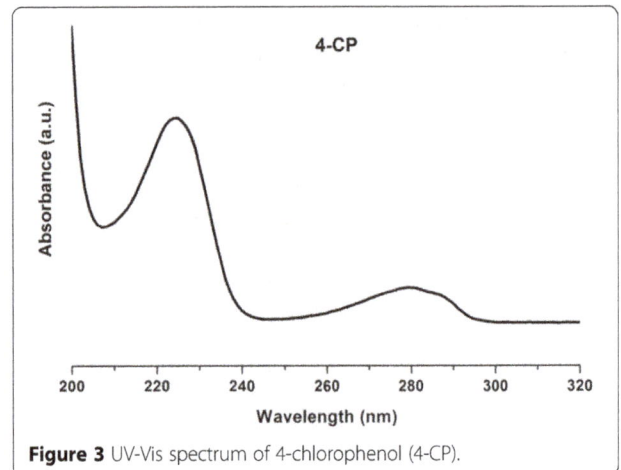

Figure 4 UV-Vis spectra of graphene oxide-GEO (a) and Graphite Oxide-GO (b).

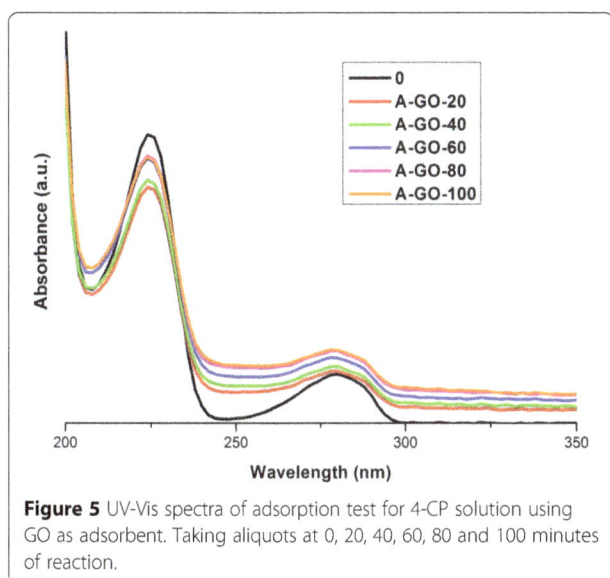

Figure 5 UV-Vis spectra of adsorption test for 4-CP solution using GO as adsorbent. Taking aliquots at 0, 20, 40, 60, 80 and 100 minutes of reaction.

produces better adsorption. The band at 225 nm in the GEO absorption spectra decreases more than in the GO spectra. GEO has a higher capacity as an adsorbent material, with 36% of the removed pollutant (4-CP). These adsorption results are useful when comparing the performance of GO and GEO as photocatalytic materials. Those tests are presented in the next sections.

Photolysis test

Photo dissociation of the C-Cl bond in 4-CP has been reported using UV laser excitation [30]. The photochemical reaction is the first stage in the C-Cl bond cleavage of 4-CP, with subsequent decomposition of the intermediate ion or radical [31]. It is important to consider the band at 280 nm, which defines the generation of intermediate compounds, in the progression of 4-chlorophenol degradation.

Figure 7 shows UV-vis spectra obtained from the photolysis test, which corroborate the interaction between the UV light and the water. It is observed, that the characteristic bands in the UV-vis spectrum of 4-CP are affected, and degradation is evident. This indicates that the wavelength emitted by the used lamp is capable of activate OH groups in the solution, causing oxidation due to air trapped in the reactor where the reaction takes place and breaking the C-Cl bonds present in 4-CP. Also, as presumed, the generation of intermediate compounds is apparent, because in the spectrum obtained we can see that the band at 280 nm has a significant shift and increased absorption. It is important to remember, that the by-products in many cases are more toxic, so that the photolysis process alone is not a sufficient method to completely degrade 4-CP.

Photocatalytic activity

Figures 8 and 9 show the UV-vis spectroscopy obtained during the photocatalytic tests using GO (P-GO) and GEO (P-GEO) as photocatalysts, respectively. The degradation of the absorption bands of 4-CP are clearly observed. Comparing the absorption spectra of the two photocatalysts, it can be observed that the degradation of 4-CP is very similar, but after 20 minutes of photocatalysis reaction, the band at 225 nm diminish more considerably with GEO (Figure 9) than with GO (Figure 8). This effect, observed in both spectra, implies that in the GEO the energy caused by UV light at 254 nm is able to quickly activate the movement of electrons, generating OH radicals in the carbon surface, allowing better contaminant degradation. However, in GO, this effect is less significant. It is suggested, that the size (carbon dimension) and disposition of functional groups play an important role in 4-CP degradation. Also, the spectra of Figures 8 and 9 show that the band at 280 nm disappears from all the samples. This

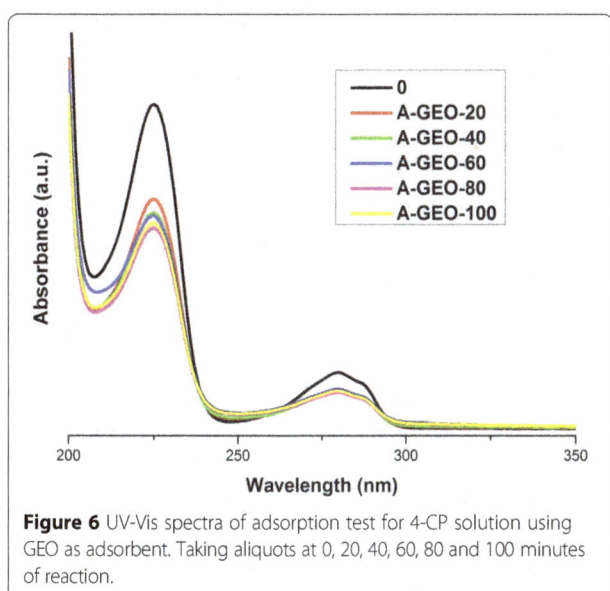

Figure 6 UV-Vis spectra of adsorption test for 4-CP solution using GEO as adsorbent. Taking aliquots at 0, 20, 40, 60, 80 and 100 minutes of reaction.

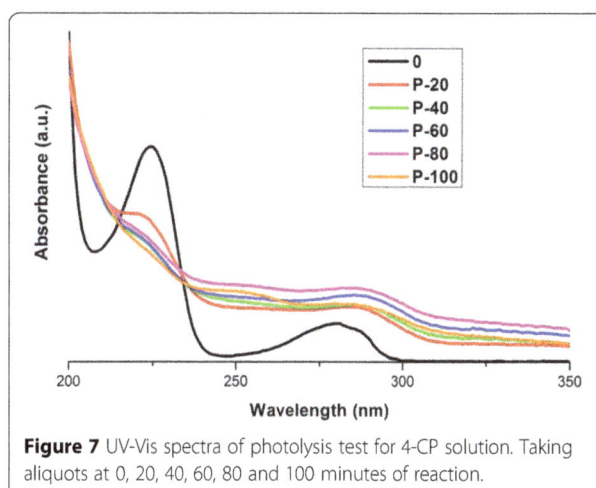

Figure 7 UV-Vis spectra of photolysis test for 4-CP solution. Taking aliquots at 0, 20, 40, 60, 80 and 100 minutes of reaction.

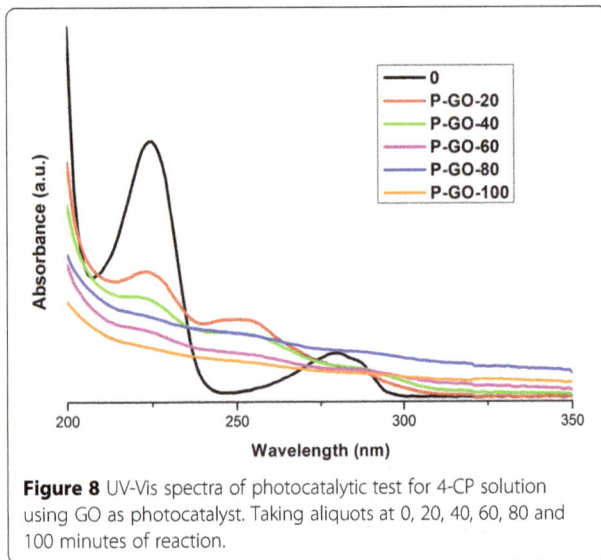

Figure 8 UV-Vis spectra of photocatalytic test for 4-CP solution using GO as photocatalyst. Taking aliquots at 0, 20, 40, 60, 80 and 100 minutes of reaction.

effect is attributed to a minimal generation of intermediate compounds, a good photocatalytic response and the degradation of the pollutant by both materials (GEO, GO). In addition, these results showed that degradation obtained via photocatalysis using GEO is faster than that achieved with GO.

As it was mentioned before, a common characteristic of the presence of by-products generated by chlorophenols, as result of degradation process, is the yellow colouration of the solution, which indicates that intermediates are generated; however, if the final solution is transparent, it indicates that by-products generated are not excessive. Compounds such as hydroquinone and benzoquinone give a yellow colour to the final solution of a treated chlorophenol. Figure 10 shows the final solutions obtained by photolysis (Figure 10a) and photocatalysis

with GO (Figure 10b) and GEO (Figure 10c), respectively, it is observed that the yellow colour is prominent after photolysis, but the solution is only slightly yellow after photocatalysis performed with GO and practically colourless in photocatalysis with GEO, indicating that the best process for removing 4-CP is photocatalysis with GEO.

On the other hand, in Figure 11 is observed over 50% of 4-CP degradation at 20 minutes by the sample of GEO; with photocatalysis using GO, the percentage of 4-CP removed is very similar, indicating that both photocatalysis tests have superior removal compared to the photolysis process. Photolysis at 100 minutes reaches 50% of 4-CP degradation in 100 minutes, while GO and GEO achieve 78% and 81% of removal in the photocatalysis test, respectively. The characteristic bands (225 nm and 280 nm) disappear independently in the spectra of GO and GEO in the photocatalyst tests, it is standard to calculate the removal percentage with the band at 225 nm, which is the distinguishing band for this quantification [29-32]. Also in the Figure 11, it can be observed that the degradation process is faster in photocatalysis with GEO than using the other two processes. This can be attributed to structural features of the GEO that have already been mentioned in this paper, as well as in other researches [13,17,18,20,33]; thus, it is presumed that the oxygen-containing functional groups present in the GEO sheets function as oxidizing radicals in the process and accelerate the degradation of 4-CP, influenced by the nanometric size of GEO and carbon material in 2 dimension. The enhanced photocatalytic activity for GEO could be attributed to the ability to capture and transport electrons, and to promote charge separation. It is known that the higher separation efficiency of electron–hole pairs will enhance photocatalytic activity and results in a large number of holes participating in the photocatalytic process [34]. On the other hand, it is important to note that electron-hole recombination could be lower in GEO compared to GO. However, again, the advantages of corrugated sheets of GEO favour the available surface area; this is the great advantage of this graphitic material as a photocatalyst compared to conventional photocatalysts.

Degradation of 4-CP, using TiO_2 as photocatalysts activated in UV zone supported in different materials such as: activated carbon, silica, zeolite have been reported [35,36]. Also graphene has been used as supported of other materials to produce photocatalytic effect in other researches [37]. However, carbon materials without doping, or specifically graphene oxide materials as photocatalysts have not been reported to degrade 4-CP. Other materials activated in UV zone used as photocatalysts in the degradation of 4-CP are enlisted in the Table 1 in comparison with the carbon materials (GO and GEO) analysed in this research. It is observed that GO and

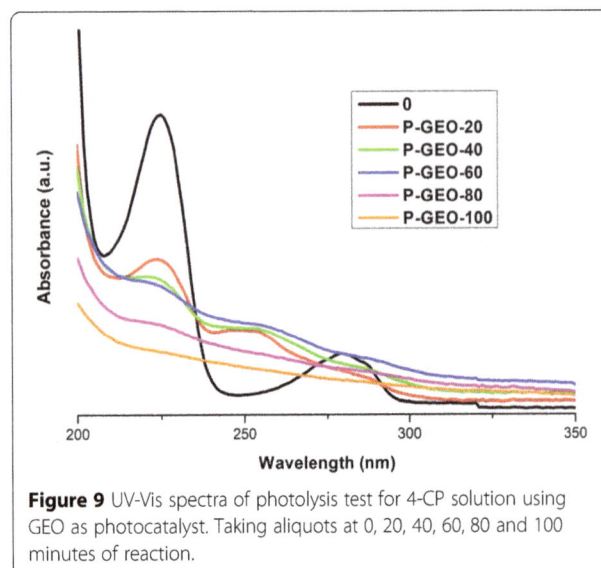

Figure 9 UV-Vis spectra of photolysis test for 4-CP solution using GEO as photocatalyst. Taking aliquots at 0, 20, 40, 60, 80 and 100 minutes of reaction.

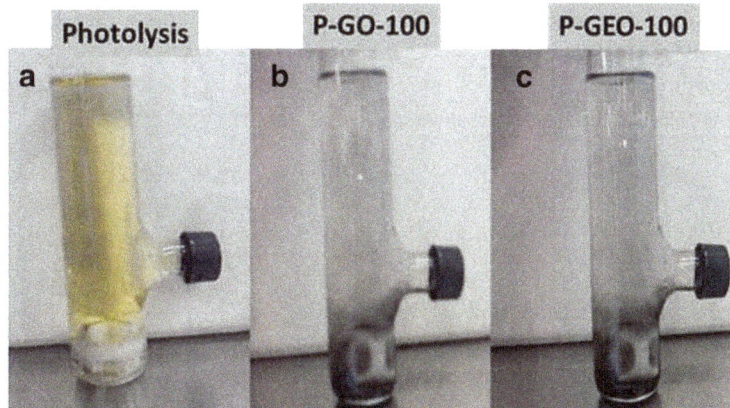

Figure 10 Pictures of the final solutions from the Photolysis. **(a)**, Photocatalytic process with graphite oxide (P-GO) **(b)** and Photocatalytic process with graphene oxide (P-GEO) **(c)**. Differences in colour between yellow and colourless is observed, indicating the generation of by-products (yellow solution).

GEO are efficient materials for the degradation of this organic molecule in comparison with other photocatalysts, either compounds or hybrid materials.

Intermediate degradation of 4-chlorophenol analysed by HPLC

Table 2 shows the concentrations of by-products generated in the reactions of photolysis and photocatalysis as determined by HPLC. In the photocatalytic degradation with both catalysts and with photolysis, aromatic intermediates and carboxylic acids are detected. Three aromatic intermediates are generated during 4-CP degradation, namely benzoquinone, 4-chlorocatechol and hydroquinone. It is known that these three aromatic compounds are commonly the intermediates generated in the TiO_2 photocatalytic degradation of 4-chlorophenol [14,31]. The HPLC analysis was performed on the final sample from each

reaction, after 100 min of photolysis or photocatalysis with GO and GEO. Considering the initial 4-CP concentration of 30 ppm, all three processes gave successful degradation. The concentration of benzoquinone in the three cases is the same, 4.8 ppm. The concentration of 4-chlorocatechol is very low, since it is determined to be less than 1 ppm in the final solution in the case of photocatalysis and above 1 ppm in photolysis, although the difference of 0.3 ppm reflects the more effectiveness of the photocatalytic process employed. In the Table 3 it can be observed that the intermediate hydroquinone is present in the samples treated with photolysis and GO as photocatalyst; however, the sample treated with GEO as a photocatalyst does not contain this intermediate. These results indicate that the degradation of 4-CP is more effective with photocatalysis, and GEO is more effective in the degradation process.

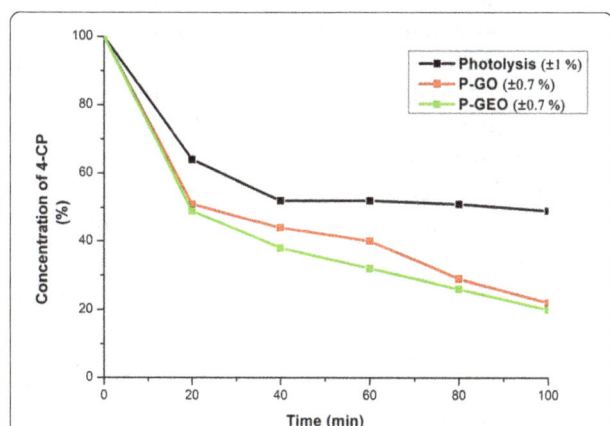

Figure 11 Degradation percentage (%) of concentration of 4-CP solution in the photolysis. Photocatalytic process with graphite oxide (P-GO), and Photocatalytic process with graphene oxide (P-GEO), at 100 min of reaction time. Values of percentage error are included in parentheses.

Table 1 Materials used as photocatalyst to removal 4-CP at different reaction times

Catalysts	Removal of 4-CP (%)	Reaction time (min)	Reference
Ag–TiO2	98	60	[14]
TP2.5 (P-modified TiO2)	78	270	[32]
TiO2-325 mesh	97	180	[35]
MgO TiO2	100	120	[38]
An(P25):Ru(P25)	80	120	[39]
Mesoporuos titania	77-100	180	[16]
ZnO	75	75	[40]
TiO2 synthesized	84	75	[40]
ZnO-TiO2	93	75	[40]
Graphite oxide (GO)	92	100	*This research*
Graphene oxide (GEO)	97	100	*This research*

Table 2 Concentration in ppm of aromatic intermediate compounds generated during the degradation of a 4-CP solution in photolysis and photocatalysis

HPLC	Aromatics compounds (ppm)		
	Benzoquinone	4-chlorocatechol	4-chlorophenol
Photolysis	4.8	1.1	0.7
P-GO	4.8	0.8	0.4
P-GEO	4.8	0.8	0.2

Also, in Table 3, the carboxylic acids generated in the 4-CP degradation process are showed. The highest concentration obtained is oxalic acid, at 5.1 ppm after photolysis; however, in the photocatalytic reactions with GO and GEO the concentration of oxalic acid is 3.8 and 3.3 ppm, respectively. Moreover, in the production of succinic acid the lowest concentration is present in the solution from the photocatalytic process with GEO. The by-products generated in our experimental processes are very similar with those found in other reports, although the amounts or concentrations of these compounds are not reported previously [29,38].

Mineralization of 4-chlorophenol evaluated by COD

The COD test is used to measure the total quantity of oxygen-consuming substances during the complete chemical breakdown of organic substances in water. Figure 12 illustrates the mineralization of 4-CP percentages achieved by photolysis, and photocatalysis with P-GO and P-GEO. It is possible to observe that, the photolysis process eliminates 60% of organic matter, but the photocatalytic processes remove more than 90% (92% and 97% for P-GO and P-GEO, respectively). In addition it is worthy of indicate that the by-products generated in the photocatalytic reaction processes are practically eliminated, in contrast with the photolysis process, because the intermediate compounds are present at the same time in the reactions. Thus, the fact that the photolysis process alone is not enough to degrade 4-CP and the intermediate compounds is corroborated.

Table 3 Concentration in ppm of intermediate compounds (carboxylic acids) generated during the degradation of a 4-CP solution by photolysis and photocatalysis

HPLC	Carboxylic acids (ppm)			
	Oxalic acid	Formic acid	Succinic acid	Hydroquinone
Photolysis	5.1	2.4	3.3	2.7
P-GO	3.8	0.5	2.4	1
P-GEO	3.3	0.5	1.5	0

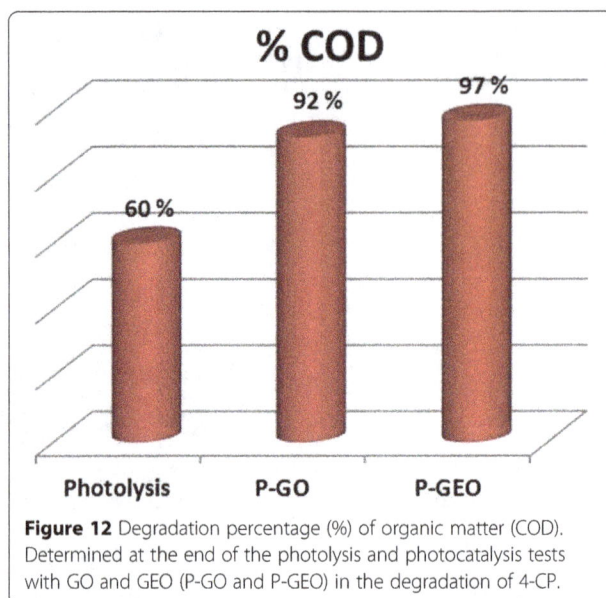

Figure 12 Degradation percentage (%) of organic matter (COD). Determined at the end of the photolysis and photocatalysis tests with GO and GEO (P-GO and P-GEO) in the degradation of 4-CP.

Conclusions

According with the results showed in this research, it can be concluded that GO and GEO are materials with sufficient capacity to degrade 4-CP present in water as contaminant. This fact promises a new trend in the application of carbon nanomaterials without doping as photocatalysts. UV determinations showed that the degradations of 4-CP in 92% and 97% using GO and GEO at 100 min respectively, are superior to photolysis (50%) during the same time, it was found that the reaction mechanism is favourable for 4-CP degradation with carbon oxides. The main by-products generated during the photolysis, and photocatalytic method with GO and GEO, correspond to aromatic compounds and carboxylic acids. In the photocatalytic process the by-products generated were lower in comparison with photolysis process, and better results were obtained with photocatalysis performed with GEO, in this case hydroquinone was not observed at the end of the reaction. Also, mineralization measured by COD indicates that up to 97% of organic matter is removed through photocatalysis using GEO. In addition, research findings show that nanometric size plays an important role in photocatalytic processes, reflected in the fact that GEO show better results in eliminating 4-CP than GO. Materials such as GO and GEO that have recently been used in photocatalytic applications such as water splitting [13,17] are showed in this research as an efficient alternative to eliminate 4-CP in water, using a small quantity of photocatalyst and a short reaction time. Thus, both materials are suggested as effective for their use in advanced oxidation processes. Thus, it can be concluded that, photocatalysis with GO and GEO is simple and effective, because, avoids the use of any oxidizing agent. Furthermore, their lower rates of recombination could be an important factor in this photocatalytic process. Therefore, the highly efficient degradation of 4-CP using GO and GEO as photocatalytic materials could be extended to other organic materials with aromatic rings, and thereby a new line of research in photocatalysis with carbon nanomaterials focused in the treatment of contaminated water, depending on the structural features of these materials, their electronic properties, dimensions and functional groups in the materials' surface.

Competing interests
The authors declare that they have no competing interests.

Authors' contributions
The authors contributed equally to this work. All authors read and approved the final manuscript.

Acknowledgments
The authors are grateful to Mr. Eduardo Martín del Campo for his assistance with the HPLC analyses. Ph.D. student Bustos-Ramírez Karina is thankful to CONACYT for grant no. 218707 and to Centro en Conjunto de Investigación en Quimica Sustentable UAEM-UNAM, especially to Dr. Gabriela Roa for the use of facilities in her laboratory. Also, Velasco-Santos Carlos appreciates the support of TecNM. Article in Memoriam Dr. Adolfo M. Espindola-Gonzalez.

Author details
[1]Centro Conjunto de Investigación en Química Sustentable, UAEM-UNAM. Km.12 de la carretera Toluca-Atlacomulco, San Cayetano 50200, Estado de México, México. [2]División de Estudios de Posgrado e Investigación, Instituto Tecnológico de Querétaro, Av. Tecnológico s/n Esq. M. Escobedo, Col. Centro Histórico, 76000 Querétaro, México. [3]Centro de Física Aplicada y Tecnología Avanzada, Universidad Nacional Autónoma de México, Boulevard Juriquilla 3001, 76230 Querétaro, México.

References
1. Hou Y, Li X, Zhao Q, Chen G. ZnFe2O4 multi-porous microbricks/graphene hybrid photocatalyst: facile synthesis, improved activity and photocatalytic mechanism. Appl Catal B-Environ. 2013;142–143:80–8.
2. Shinde SS, Bhosale CH, Rajpure KY. Photocatalytic oxidation of salicylic acid and 4-chlorophenol in aqueous solutions mediated by modified AlFe2O3 catalyst under sunlight. J Mol Catal A Chem. 2011;347:65–72.
3. Taherian S, Entezari HM, Ghows N. Sono-catalytic degradation and fast mineralization of p-chlorophenol: La0.7Sr0.3MnO3 as a nano-magnetic green catalyst. Ultrason Sonochem. 2013;20:1419–27.
4. Czaplicka M. Sources and transformations of chlorophenols in the natural environment. Sci Total Environ. 2004;322:21–39.
5. Seung-Gun C, Yoon-Seok C, Jae-Woo C, Kyung-Youl B, Seok-Won H, Seong-Taek Y, et al. Photocatalytic degradation of chlorophenols using star block copolymers: removal efficiency by-products and toxicity of catalyst. Chem Eng J. 2013;215–216:921–8.
6. Benitez FJ, Beltran-Heredia J, Acero LJ, Rubio Javier F. Contribution of free radicals to chlorophenol decomposition by several advanced oxidation processes. Chemosphere. 2000;41:1271–7.
7. Olaniran OA, Igbinosa OE. Chlorophenols and other related derivatives of environmental concern: Properties, distribution and microbial degradation processes. Chemosphere. 2001;83:1297–306.
8. Tamer E, Hamid Z, Aly MA, Ossama TE, Bo M, Benoit G. Sequential UV-biological degradation of chlorophenols. Chemosphere. 2006;63:277–84.
9. Chen D, Ray KA. Photocatalytic kinetics of phenol and its derivatives over UV irradiated TiO2. Appl Catal B-Environ. 1999;23:143–57.
10. Pera-Titus M, Garcia-Molina V, Baños AM, Gimenez J, Esplugas S. Degradation of chlorophenols by means of advanced oxidation processes: a general review. Appl Catal B-Environ. 2004;47:219–56.
11. Muruganandham M, Suri RPS, Jafari Sh, Sillanpää M, Gang-Juan L, Wu JJ, et al. Recent Developments in Homogeneous Advanced Oxidation Processes for Water and Wastewater Treatment. Int J Photoenergy. 2014; doi:10.1155/2014/821674.
12. Ahmed S, Rasul MG, Martens WN, Brown R, Hashib MA. Heterogeneous photocatalytic degradation of phenols in wastewater: A review on current status and developments. Desalination. 2010;261:3–18.
13. Te-Fu Y, Jhih-Ming S, Ching C, Ting-Hsiang C, Hsisheng T. Graphite oxide as a photocatalyst for hydrogen production from water. Adv Funct Mat. 2010;20:2255–62.
14. Temel NK, Sökmen M. New catalyst systems for the degradation of chlorophenols. Desalination. 2011;281:209–14.
15. Amadelli R, Samiolo L, Maldotti A, Molinari A, Valigi A, Gazzoli D. Preparation, Characterisation, and Photocatalytic Behaviour of Co-TiO2 with Visible Light Response. Int J Photoenergy. 2008; doi:10.1155/2008/853753.
16. Avilés-García O, Espino-Valencia J, Romero R, Rico- Cerda J L, Natividad R. Oxidation of 4-Chlorophenol by Mesoporous Titania: Effect of Surface Morphological Characteristics. Int J Photoenergy. 2014; doi:10.1155/2014/210751.
17. Te-Fu Y, Jaroslav C, Chih-Yun C, Ching C, Hsishen T. Roles of graphene oxide in photocatalytic wáter splitting. Materials Today. 2013;16:78–84.
18. Te-Fu Y, Fei-Fan C, Chien-Te H, Hsisheng T. Graphite oxide with different oxygenated levels for hydrogen and oxygen production from water under illumination: the band positions of graphite oxide. J Phys Chem C. 2011;115:22587–97.
19. Sheath P, Majumder M. A Comparative Review of Graphene Oxide and Titanium Dioxide as Photocatalysts in Photocatalytic Systems [online]. In: Chemeca 2011: Engineering a Better World: Sydney Hilton Hotel, NSW, Australia, 18-21 September 2011. Barton, A.C.T.: Engineers Australia; 2011. p. 2346–57.
20. Jiang X, Nisar J, Pathak B, Zhao J, Ahuja R. Graphene oxide as a chemically tunable 2-D material for visible-light photocatalyst applications. J Catal. 2013;299:204–9.
21. Lling-Lling T, Wee-Jun O, Siang-Piao C, Mohamed RA. Reduced graphene oxide-TiO2 nanocomposite as a promising visible-light-active photocatalyst for the conversion of carbon dioxide. Nanoscale Res Lett. 2013;8:1–9.
22. Tingshun J, Zhangfeng T, Meiru J, Qian Z, Xiaoqi F, Hengbo Y. Preparation and photocatalytic property of TiO2-graphite oxide intercalated composite. Catal Commun. 2012;28:47–51.
23. Guardia L, Villar-Rodil S, Paredes JI, Rozada R, Martínez-Alonso A, Tascón JMD. UV light exposure of aqueous graphene oxide suspensions to promote their direct reduction, formation of graphene–metal nanoparticle hybrids and dye degradation. Carbon. 2012;50:1014–24.
24. Guoxin H, Bo T. Photocatalytic mechanism of graphene/titanate nanotube photocatalyst under visible-light irradiation. Mater Chem Phys. 2013;138:608–14.
25. Jianfeng S, Na L, Mingxin Y. Supramolecular photocatalyst of RGO-cyclodextrin-TiO2. J Alloy Compd. 2013;580:239–44.
26. Xiangang H, Li M, Jianping W, Qixing Z. Covalently synthesized graphene oxide-aptamer nanosheets for efficient visible-light photocatalysis of nucleic acids and proteins of viruses. Carbon. 2012;50:2772–81.
27. Cortes AD, Sanhueza L, Wrighton K. Removal of 4-chlorophenol using graphene, graphene oxide, and a-doped graphene (A 5 N, B): a computational study. Int J Quantum Chem. 2013;113:1931–9.
28. Bustos-Ramirez K, Martinez-Hernandez AL, Martinez-Barrera G, de Icaza-Herrera M, Castaño MV, Velasco-Santos C. Covalently bonded chitosan on graphene oxide via redox reaction. Materials. 2013;6:911–26.
29. Satuf ML. Photocatalytic degradation of 4-chlorophenol: a kinetic study. Appl Catal B-Environ. 2008;82:37–49.
30. Czaplicka M. Photo-degradation of chlorophenols in aqueous solution. J Hazard Mater-B. 2006;134:45–59.
31. Sharma S, Mukhopadhyay M, Murthy ZVP. Rate parameter estimation for 4-chlorophenol degradation by UV and organic oxidants. J Ind Eng Chem. 2012;18:249–54.
32. Elghniji K, Hentati O, Mlaik N, Mahfoudh A, Ksibi M. Photocatalytic degradation of 4-chlorophenol under P-modified TiO2/UV system: Kinetics, intermediates, phytotoxicity and acute toxicity. J Environ Sci. 2013;24:479–87.
33. Penghui S, Ruijing S, Shaobo Z, Mincong Z, Dengxin L, Shihong X. Supported cobalt oxide on graphene oxide: Highly efficient catalysts for the removal of Orange II from water. J Hazard Mater. 2012;230:331–9.
34. Li K, Xiong J, Chen T, Yan L, Dai Y, Song D, et al. Preparation of graphene/TiO2 composites by nonionic surfactant strategy and their simulated sunlight and visible light photocatalytic activity towards representative aqueous POPs degradation. J Hazard Mater. 2013;250–251:19–28.
35. Pino E, Encinas MV. Photocatalytic degradation of chlorophenols on TiO2-325 mesh and TiO2-P25. An extended kinetic study of photodegradation under competitive conditions. J Photochem Photobiol A. 2012;242:20–7.

36. Naeem K, Ouyang F. Influence of supports on photocatalytic degradation of phenol and 4-chlorophenol in aqueous suspensions of titanium dioxide. J Environ Sci. 2013;25(2):399–404.

37. Muhd JN, Bagheri S. Graphene supported heterogeneous catalysts: an overview. Int J Hydrogen Energ. 2015;40:948–79.

38. Gs P, Kambur A. Removal of 4-chlorophenol from wastewater: preparation, characterization and photocatalytic activity of alkaline earth oxide doped TiO2. Appl Catal B-Environ. 2013;129:409–15.

39. Apopei P, Catrinescu C, Teodosiu C, Royer R. Mixed-phase TiO2 photocatalysts: crystalline phase isolation and reconstruction, characterization and photocatalytic activity in the oxidation of4-chlorophenolfromaqueouseffluents. Appl Catal B-Environ. 2014;160–161:374–82.

40. Gs P, Kambur A. Significant enhancement of photocatalytic activity over bifuntional ZnO-TiO2 catalyst for 4-chlorophenol degradation. Chemosphere. 2014;105:152–9.

5

Influence of dissolved organic matter in natural and simulated water on the photochemical decomposition of butylparaben

Marta Gmurek[*], Magdalena Olak-Kucharczyk and Stanisław Ledakowicz

Abstract

Background: In the last few decades the quality of natural water has often deteriorated as a variety of novel pollutants have contaminated rivers and lakes. Trace amounts of some man-made chemicals can be hazardous to plants, animals as well as human health as carcinogens, mutagens or endocrine disruptors. Light radiation may help in its decomposition, aided by naturally occurring colored organic compounds (humic substances) in the water. The aim of these studies was to check the influence of presence of organic and inorganic matter on the removal of endocrine disrupting compound - butylparaben (BP) from water.

Methods: Photochemical decomposition of BP in aqueous solution using: photolysis by ultraviolet-C (UVC) and visible (VIS) irradiation, advanced oxidation in H_2O_2/UV system and photosensitized oxidation was examined. The degradation processes were carried out in different type of water matrix: natural water from Sulejow Reservoir, simulated natural water with humic acids and buffered solution.

Results: The presence of dissolved organic matter in water did not influence much on UVC photolysis and increases only about 8% of BP depletion rate in H_2O_2/UV system. While during visible light photolysis and photosensitized oxidation the addition of natural water matrix causes the acceleration of reaction rate by 16% and 36%, respectively. Moreover BP degradation proceeds via singlet oxygen generated from humic substances.

Conclusions: Butylparaben undergoes both direct and indirect photodegradation in aqueous solution under UVC and visible radiation. The efficiency of the H_2O_2/UV process, photolysis as well as photosensitized oxidation processes is strongly dependent on composition of the water.

Keywords: Photochemical degradation, Natural and synthetic water, Butylparaben, Humic acid, Hydroxyl radicals, Singlet oxygen

Background

In the last few decades the quality of natural water has often worsened owing to contamination with man-made trace organic chemicals. These are sometimes carcinogens, mutagens or endocrine disruptors. Such substances are often not removed in traditional wastewater treatment, are not easily biodegradable, and may accumulate in organisms [1-4]. Polluted water is a threat to human and environmental health. Contaminations have also influence on the condition and population sizes of animals and plants in natural waters. Therefore, the discovery of efficient degradation methods for these pollutants is of wide interest [5-8].

Photochemical degradation of these substances has been attempted and may be influenced by the array of dissolved organic matter (DOM), e.g., humic and fulvic acids, and by nitrate as well as other inorganic ions [9]. Hydrogen peroxide is also present in natural water, and generated mostly by solar radiation and microbial processes with concentrations in the range 10^{-6} - 10^{-5} M [10]. Reactive oxygen species (ROS) are also generated in the environment through photochemical processes and may be crucial to photochemical degradation. They include singlet oxygen ($^1O_2$), hydroxyl radical ($^•OH$), superoxide

* Correspondence: marta.gmurek@p.lodz.pl
Faculty of Process and Environmental Engineering, Department of Bioprocess Engineering, Lodz University of Technology, Wolczanska 213, 90-924 Lodz, Poland

radicals ($HO_2^{\bullet}/O_2^{\bullet}$), and peroxyl radical ($^{\bullet}OOR$) as well as non-ROS transients, e.g., carbon radicals ($CH_3^{\bullet}$) and triplet exited states of DOM ($^3DOM^*$) [11-15]. The production of these diverse array of reactive species is driven primarily by abiotic photochemical reactions involving naturally occurring organic and sometimes inorganic coloured substances (chromophores) [16].

The absorption of solar radiation between UVB and visible wavelengths by DOM in natural waters initiates a series of complex photochemical reactions that improve water -purification. The chromophores present within DOM become excited from their singlet ground state to their first excited singlet stated ($^1DOM^{\bullet}$). Upon absorption of light, undergo intersystem crossing (ISC) to form their triplet excited state ($^3DOM^{\bullet}$), and then interact with molecular oxygen to form either singlet oxygen ($^1O_2$) or superoxide ($O_2^{\bullet}$). Hydroxyl radicals ($^{\bullet}OH$), another common reactive oxygen species, can be formed from the interaction of water with $^3DOM^{\bullet}$, the photolysis of hydrogen peroxide (H_2O_2), or the photolysis of nitrate (NO_3^-) [16].

As a result of the reactions described in Figure 1., concentrations of reactive oxygen species in natural waters have been detected in the range between 10^{-15} - 5×10^{-13} M, 10^{-15} - 10^{-12} M, 10^{-9} - 10^{-8} M, 10^{-18} - 2×10^{-16} M, for $^3DOM^{\bullet}$, $^1O_2$, $O_2^{\bullet}$, and $^{\bullet}OH$, respectively [16]. The ROS can help to purify the aquatic environment of bioactive pollutants derived from human activities e.g. pharmaceuticals and personal care products, or, in some instances, convert them to less toxic substances. The reaction sequences followed by well-established concepts of direct and sensitized photooxidations, also potentially coupled to thermal autooxidation processes [16]. Although the role of reactive oxygen species in water purification is known, the mechanism

of these processes is still not entirely clear. This work examined degradation of an endocrine disrupting compound (butylparaben (BP)) in aqueous solution using photolysis by Ultraviolet C and visible irradiation, advanced oxidation in a H_2O_2/UV system and photosensitized oxidation (POx). Degradation was carried out in different waters matrix: natural water from Sulejow Reservoir, simulated natural water with humic acids (HA), and buffered solution. The aim was to establish the influence of organic and inorganic matter on the removal of butylparaben from the water.

Materials and methods
Chemicals and reagents
Butylparaben (BP) (>99%, Table 1), synthetic humic acid (HA, technical), tert-butanol and sodium azide (99%) were purchased from Fluka. Hydrogen peroxide (30%), disodium phosphate (Na_2HPO_4), monopotassium phosphate (KH_2PO_4) and potassium nitrate (KNO_3) all p.a. were purchased from POCh, Poland. Meso-tetra (4-sulphonatophenyl)porphin ($TPPS_4$) was bought from Sigma-Aldrich. The element composition of Fluka humic acid (HA, cat. no. 53680) has been reported to contain 48.36% of C, 26.91% of O, 4.24% of H, 0.78% of N and 0.78% of S [17]. The atomic ratios described by Rodrigues et. al. [17], are 1.04 (H/C), 0.42 (O/C), and 0.012 (N/C) which are within the values reported in the literature for soil HA, with the exception of the N/C ratio. All chemicals were used as received.

Model solutions and natural water samples
Aqueous solutions of butylparaben were made by diluting it into: natural water from Sulejow Reservoir (RW), simulated natural water with humic acids (SN1^{pH7} and SN1^{pH9}) or with nitrate ions (SN2) and buffered solution (BS). The concentration of butylparaben was 8×10^{-5} M. In experiments using the H_2O_2/UV system, an optimal concentration of hydrogen peroxide of 0.01 M was used. In buffer solution experiments with an xenon arc lamp (XBO), $TPPS_4$ was used as a photosensitizer at an optimal concentration of 2×10^{-5} M [18]. Buffer solutions were prepared from deionized water, purified using a Millipore Milli-Q Plus System (>18.2 MΩ). The pH of

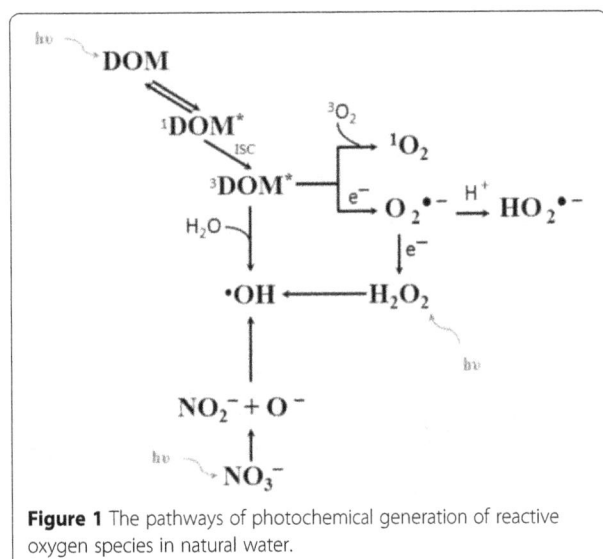

Figure 1 The pathways of photochemical generation of reactive oxygen species in natural water.

Table 1 Physicochemical properties of butylparaben [19,20,23]

	CAS No.	94-26-8
	molecular formula	$C_{11}H_{14}O_3$
	molar mass	194.23
	solubility in water	0.77 µM at 25°C; 7.5 µM at 80°C
	log K_{OW}	3.5
	pKa	8.24

the deionized solution was adjusted by adding phosphate buffer: Na_2HPO_4-KH_2PO_4. For more details, see [18,19]. Synthetic solutions with organic matter ($SN1^{pH7}$ and $SN1^{pH9}$) and nitrate (SN2) were made by diluting humic acids and potassium nitrate, respectively.

Natural water was collected in summer from the Sulejow Reservoir, situated in central Poland in the middle course of the Pilica River (latitude: 51°27.2′, longitude: 19°59.2′, area:27 km^2). Water for chemical analyses was filtered through Whatman GF/F (0.45 μm) filters and stored at 4°C before use. The pH of the reservoir water was 8.11, and its concentration of dissolved organic carbon (DOC) was 6.85 mg L^{-1}. The concentrations of major ions in the reservoir water were: Ca^{2+} (34.02 mg L^{-1}), K^+ (9.79 mg L^{-1}), Na^+ (6.51 mg L^{-1}), SO_4^{2-} (21.73 mg L^{-1}), Cl^- (12.34 mg L^{-1}), NO_3^- (0.46 mg L^{-1}). Mg^{2+}, NH_4^+, Li^+, F^-, NO_2^-, Br^-, PO_4^{3-} ions were detected at concentrations of: 0.969 mg L^{-1}, 0.044 mg L^{-1}, 0.003 mg L^{-1}, 0.112 mg L^{-1}, 0.025 mg L^{-1}, 0.007 mg L^{-1}, 0.003 mg L^{-1}, respectively.

The buffer solutions (BS) were prepared at three different pH values. BS1 had similar pH to the reservoir water (RW). BS2 was used at pH 7 to compare the influences of additives in the humic acid ($SN1^{pH7}$) and nitrate (SN2) treatments. Experiments with humic acids and visible-light irradiation were done at pH 9 (BS3 and $SN1^{pH9}$).

Photodegradation experiments

The photodegradation experiments were conducted using UVC and xenon arc lamps as sources of light. The UVC direct photolysis and degradation in a H_2O_2/UV system were carried out in a rotating device with quartz test tubes (10 mL), placed between two exposure panels, each of them consisting of three 7.2 W lamps. Low pressure lamps (LP, Luzchem) emitting mainly at a wavelength $\lambda = 254$ nm were used. For more details, see [19,20]. Photon flux rate entering the reaction space, calculated on the basis of actinometric experiments with uranyl oxalate [21] was equaled 1×10^{-5} einstein L^{-1} s^{-1}.

Photosensitizing oxidation experiments were carried out with an immersion xenon arc lamp located in a quartz well with cooling jacket. A 100 W xenon arc lamp (XBO, 100 W, Osram) served as the light source to simulate solar light irradiation. The lamp was surrounded by five plate reactors (volume 0.01 L), each placed 11 cm from the light source. Each reactor consisted of two glass plates (10 cm × 6 cm) bound with silicone seal such that the distance between the inner surfaces of the plates was 0.3 cm. The tested solutions were aerated and agitated by gas bubbling. The quantity of absorbed photons was calculated using Reinecke's actinometer for wavelengths ranging from 310 to 770 nm [22], and was 3.24×10^{-4} einstein L^{-1} s^{-1}. The 2×10^{-5} M solution of $TPPS_4$ absorbed 3.77×10^{-5} einstein L^{-1} s^{-1}.

Analysis

The butylparaben decay rate was monitored by high pressure liquid chromatography (HPLC) coupled with UV detection using a Waters apparatus. Analysis was performed with a Waters Nova-Pak C18 column (3.9 mm × 150 mm) using a mobile phase consisting of a degassed mixture (70/30, v/v %) of methanol and acidified water (0.01% orthophosphoric acid) at a constant flow of 1 mL min^{-1}. The detailed description of this analysis of BP is reported elsewhere [19,23]. The spectrophotometric analysis was performed on a Unicam UV 300 spectrophotometer. The dissolved organic carbon measurements were performed on a HACH IL 550TOC-TN apparatus. The ion concentrations were determined by an ion chromatograph (Dionex model ICS) on an IonPac CS18 (for cations) and an IonPac AS18 (for anions).

The performed experiments allowed us to estimate the extent of direct reaction of butylparaben with hydrogen peroxide in the absence of radiation, ("dark reaction").

Results indicated an insignificant role of the direct reaction of studied compounds with H_2O_2. The blank reaction was carried out to investigate the hydrolysis of BP. The experiments showed no decomposition of the investigated compound in the dark after 12 h, much longer time than used during the photodegradation.

Experiments were performed in duplicate to assure accurate data acquisition.

Results and discussion

For studies on photochemical degradation of butylparaben, the comparison of BP absorbance spectra in different reaction solutions with the emission spectra of the UVC and xenon arc (XBO) lamps were performed (Figure 2). Butylparaben in neutral solutions such as the humic acid and nitrate treatments, absorbs light mainly at low wavelengths up to 300 nm so its degradation by direct photolysis is expected to be higher using UVC systems. The degradation of butylparaben occurs by direct photolysis also in reservoir water (RW) and BS1, due to overlapping of the absorption spectrum of butylparaben with UVC and xenon arc lamps. In alkaline solution the direct photolysis of BP is likely as a results of occurring with xenon arc (XBO) irradiation.

The substances commonly found in natural waters, both organic and inorganic, may affect the photodegradation process of pollutants e.g. Cl^- ions accelerate the photodegradation of selected antibiotics under UV-Visible irradiattion ($\lambda > 200$ nm), whereas simulated solar irradiation ($\lambda > 290$ nm) was incapable of antibiotics photodecomposition, irrespective of Cl^- [24].

Figure 3 shows the degradation of butyparaben in water collected from Sulejow Reservoir (RW). UVC photolysis of butylparaben proceeded slightly slower (about 6%) (Figure 3a) in natural water compared with photolysis in

(a)

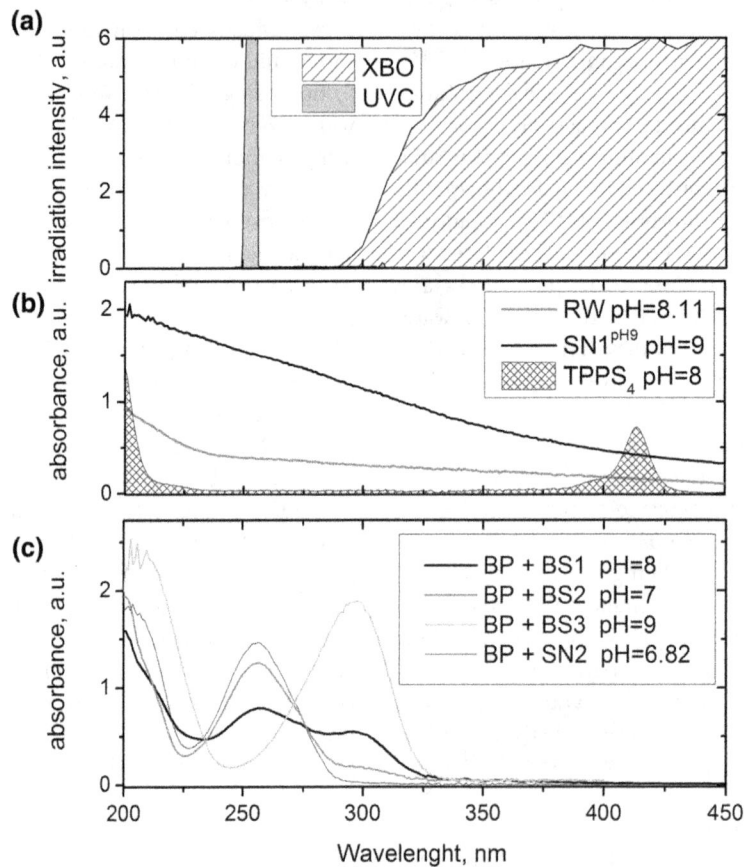

(b)

(c)

Figure 2 Emission spectra of lamps used in experiments **(a)**, absorption spectra of selected waters and TPPS$_4$ **(b)** ($C_{TPPS4} = 5 \times 10^{-6}$ M, $C_{HA} = 50$ mg L^{-1}) and BP solutions in different water matrix **(c)** ($C_{BP} = 8 \times 10^{-5}$ M).

(a)   **(b)**

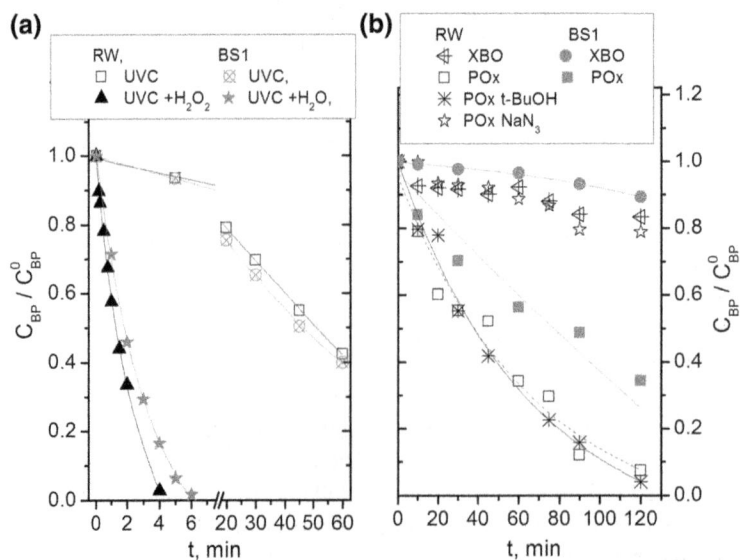

Figure 3 Changes of relative BP concentrations versus irradiation time during: UVC irradiation, and H$_2$O$_2$/UVC degradation **(a)**; VIS irradiation and photosensitized oxidation process (POx) **(b)** in natural water (pH = 8.11) and buffered solution (pH = 8.0) ($E_0^{UV} = 1 \times 10^{-5}$ einstein L^{-1} s^{-1}, $E_0^{VIS} = 3.24 \times 10^{-4}$ einstein L^{-1} s^{-1}, $C_{BP} = 8 \times 10^{-5}$ M, $C_{H2O2} = 0.02$ M, $C_{TPPS4} = 2 \times 10^{-5}$ M, $C_{t-BuOH} = 0.1$ M, $C_{NaN3} = 0.02$ M).

buffered solution (BS1), whereas, the advanced oxida-tion of butylparaben in a H_2O_2/UV system was about 18% (Figure 3a) faster in reservoir water than in BS1. Butylparaben degradation in a H_2O_2/UV system can take place through direct photolysis and reaction with hydroxyl radicals generated during hydrogen peroxide photodecomposition. Addition of H_2O_2 to the irradiated RW and BS1 shortened the BP degradation by about 40 and 30 times, respectively.

Preliminary examinations using a xenon arc lamp showed that during 2 h of irradiation of buffered solution without photosensitizer the same degree of butylparaben degradation was achieved as in an experiment without irradiation. This means that direct photolysis was not oc-curring, but decrease of butylparaben concentration dur-ing both experiments was a result of the adsorption of butylparaben on silicone tube in the reactor. After 2 h of irradiation of natural water without $TPPS_4$ the BP concen-tration decreased by 7% in comparison with experiments in buffered solutions (Figure 3b). Moreover in the dark, a decrease of butylparaben concentration was not observed if natural water was used. It is therefore possible that butylparaben decomposition occurs probably other than by adsorption. It is known that anions such as chloride, sulphate or carbonate have effects upon adsorption pro-cesses whilst phosphate and bicarbonate in the medium reduce the adsorption capacity [25,26].

Photolysis of organic compounds in water could be changed by the presence of naturally occurring organic and inorganic matter. Increasing the degree of butylpar-aben reduction by visible-light irradiation is presumably due to the natural dissolved constituents: humic acids, nitrate and chloride ions, where a strong influence on photolysis has been demonstrated [24,27]. Depending on quality and composition of water these elements can act as photosensitizers or conversely may exhibit quenching and scavenging effects [28,29].

Experiments using water from Sulejow reservoir showed that butylparaben concentration decreased by 93% after 2 h exposure in the reaction solution containing $TPPS_4$ to visible radiation.whilst in buffer solutions, butylparaben concentration fell by 66%. The large difference in the butylparaben decomposition in these experiments may be caused by reactions involving components of the Sulejow water, such as dissolved organic matter. Dissolved organic material, by absorbing light can generate reactive oxygen species [30,31], which increase the rate of butylparaben decay. The used of a free radical scavenger – tert-butanol (t-BuOH), reduced butylparaben decomposition by hy-droxyl radicals (Figure 3b). Sodium azide, a quencher of singlet oxygen, confirmed the role of singlet oxygen in the reaction mechanism. It can therefore be assumed that the DOM found in natural water, which resulted in increased reaction rate, has the ability to generate singlet oxygen. Humic substances present in low concentrations, can lead to photosensitized oxidation in the aqueous environment [32,33]. The differences in the effects of direct photolysis were observed, probably also due to the compounds con-tained in the natural water. However, experiments carried out with sodium azide resulted in inhibition of degrad-ation to the level of that of direct photolysis, and t-BuOH did not affect the reaction. But the acceleration of the

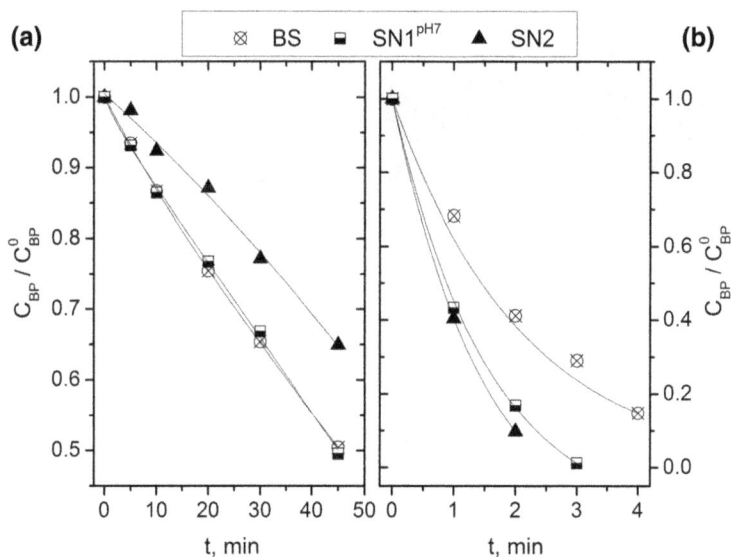

Figure 4 Changes of relative BP concentrations versus irradiation time upon degradation using UVC **(a)** and in H_2O_2/UVC system **(b)** in simulated natural water and buffered solution (SN1[pH7] pH = 6.47; SN2 pH = 6.82, BS2 pH = 7, $E_0{}^{UV}$ = 1 × 10^{-5} einstein L^{-1} s^{-1}, C_{BP} = 8 × 10^{-5} M, C_{H2O2} = 0.01 M, C_{HA} = 4.4 mg L^{-1}, C_{NO3-} = 4.0 mg L^{-1}).

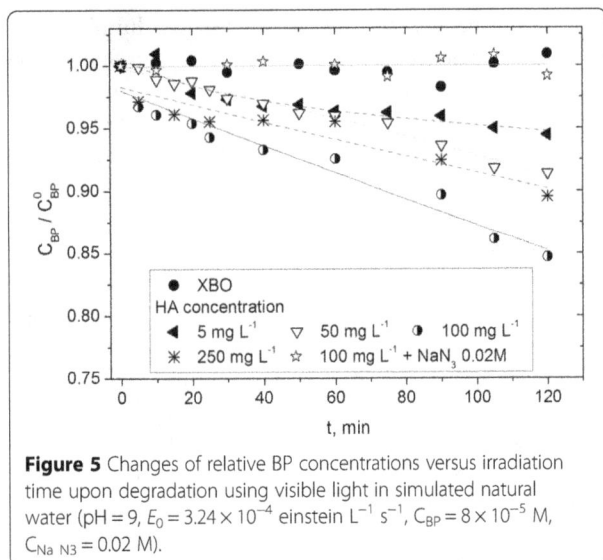

Figure 5 Changes of relative BP concentrations versus irradiation time upon degradation using visible light in simulated natural water (pH = 9, $E_0 = 3.24 \times 10^{-4}$ einstein L^{-1} s^{-1}, $C_{BP} = 8 \times 10^{-5}$ M, $C_{Na\ N3} = 0.02$ M).

direct photolysis compared with reactions in buffered solution was not related to hydroxyl radicals or singlet oxygen, but probably to the presence of organic radicals.

The influences of additives on UVC photolysis as well as H_2O_2/UV system were investigated. The presence of humic acid (SN1[pH7]) in reaction solution upon UVC irradiation did not influence on butylparaben degradation, whilst the nitrate ions (SN2) inhibited BP decay (Figure 4a). Studies by Ge et al. [24], have shown that under UV-

Visible irradiation ($\lambda > 200$ nm), NO_3^- ions are able to inhibit to some extent the photodegradation rates, mainly due to their photoshielding effects. Humic acids can behave in irradiated solution as sensitizers, light filters and quenchers of free radicals, and therefore their show distinct effects on photodegradation of pollutants under UV irradiation ($\lambda < 300$ nm) and simulated sunlight [27]. Humic acids absorb UVC radiation (Figure 2b), which should slow down the butylparaben degradation. However, studies have shown that these compounds did not undergo photolysis at a wavelength of 254 nm [34]. Photoexcitation of humic acids may lead to the generation of reactive oxygen species resulting in the intensification of the degradation process, but at the same time humic acids can quench reactive oxygen species, so their effect on butylparaben photolysis may be negligible.

In the case of the H_2O_2/UV system, the presence of humic acid (SN1[pH7]) and nitrate ions (SN2) accelerated degradation of butylparaben by about 25% and 30%, respectively (Figure 4b). The increased decay rate of organic compounds in the presence of H_2O_2 and NO_3^- was also observed by Li et al. [35] during simulated sunlight irradiation.

The effects of humic acids in visible light on the degradation of butylparaben were examined (Figure 5) in a synthetic alkaline water (pH 9; BS3 and SN1[pH9]). To avoid the adsorption of butylparaben, the silicone tubes from the reactor construction were removed. The

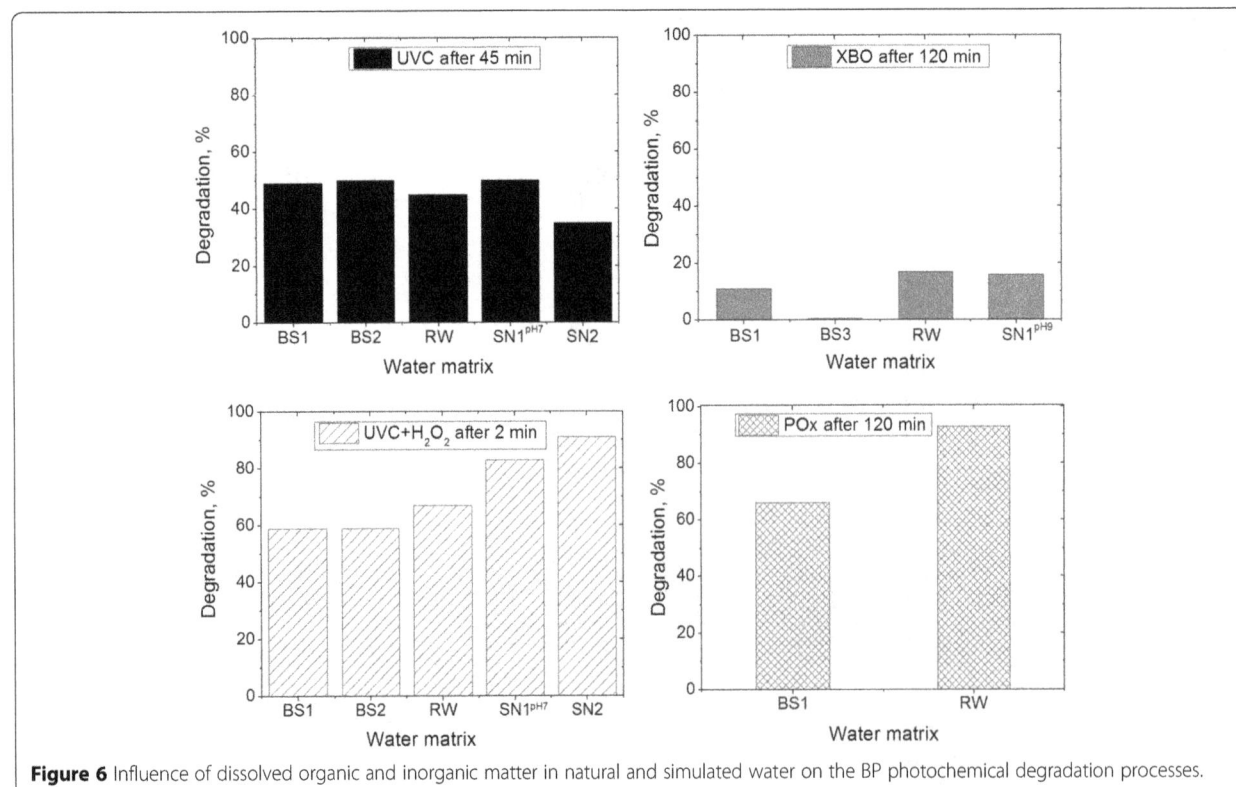

Figure 6 Influence of dissolved organic and inorganic matter in natural and simulated water on the BP photochemical degradation processes.

concentrations of humic acids (HA) were in the range: 5–250 mg L^{-1}. As can be seen in Figure 5, BP photolysis is depended on the humic acid concentration. At first, increasing humic acid concentrations resulted in acceleration of initial rate of butylparaben decay, with the highest rate at 100 mg L^{-1}. Further increase of humic acid concentrations reduced butylparaben decay primarily due to their light screening effect [36]. This process caused a decrease in efficiency of reactive oxygen species formation and butylparaben decomposition rate. It has been previously shown (Figure 1) that humic acids contain chromophores capable of sensitizing the generation of singlet oxygen. Therefore the excitation of humic acids leads to production of singlet oxygen. The experiments with azide anions, used as a singlet oxygen quencher, have indicated that singlet oxygen plays a dominant role in photodegradation of butylparaben, sensitized by humic substances. In summary, the photolysis rate of butylparaben can be enhanced by humic substances at low concentrations but could be inhibited at higher concentrations.

As can be seen from Figure 6, efficiency of the H_2O_2/ UV, direct photolysis as well as photosensitized oxidation processes is strongly dependent on composition of the water.

It can be noticed that, the impact of dissolved organic and inorganic matter on the photochemical processes is related to light sources. Photolysis under xenon arc (XBO) radiation, and photosensitized oxidation in the presence of $TPPS_4$ was fastest in reservoir water.

Conclusions

Butylparaben undergoes both direct and indirect photodegradation in aqueous solution under UVC and visible radiation. The presence of dissolved organic matter in water did not influence on UVC photolysis and increases only about 8% of BP depletion rate in H_2O_2/UV system. While during visible light photolysis and photosensitized oxidation the addition of natural water matrix causes the acceleration of reaction rate by 16% and 36%, respectively.

The results indicate that butylparaben degradation proceeds via singlet oxygen, generated from humic substances. The presence of NO_3^- lead to the highest butylparaben degradation rate in the H_2O_2/UV system. In the case of UVC light, only slight influences of water composition were observed. It could be shown that the water matrix plays an important role on the efficiency of photodegradation processes. However, the obtained results enrich the existing knowledge of the photosensitized oxidation as well as advanced oxidation processes in natural waters.

Although, the H_2O_2/UV system gave the fastest degradation of butylparaben ,the possibility of using sunlight in the photosensitized oxidation makes this method more attractive from a practical and economical point of view in the treatment of drinking water.

Abbreviations
DOM: Dissolved organic substances; ROS: Reactive oxygen species; HA: Humic acids; O_2: Singlet oxygen; OH: Hydroxyl radical; POx: Photosensitized oxidation; BP: Butylparaben; $TPPS_4$: Meso-tetra (4-sulphonatophenyl)porphin; RW: Natural water from Sulejow Reservoir; $SN1^{pH7}$: Simulated natural water with humic acids at pH 7; $SN1^{pH9}$: Simulated natural water with humic acids at pH 9; SN2: Simulated natural water with nitrate ions; BS1: Buffered solution at pH 8; BS2: Buffered solution at pH 7; BS3: Buffered solution at pH 9; XBO: Xenon arc lamp.

Competing interests
The authors declare that they have no competing interests.

Authors' contributions
MG carried out the photosensitized oxidation, UV–vis photolysis as well as experiments in the analysis of the samples and drafted the manuscript. MOK carried out the H_2O_2/UV experiments and UV photolysis, participated in the analysis of the samples. SL conceived of the study, and participated in its design and coordination and helped to draft the manuscript. All authors read and approved the final manuscript.

Acknowledgements
Authors are thankful to the Regional European Center of Ecohydrology in Lodz (Poland) performing the analysis of ion analysis in natural water from Sulejow Reservoir. Marta Gmurek is grateful for a financial support from Foundation for Polish Science within the START scholarship.

References
1. Madsen T, Boyd BH, Nylén D, Pedersen R, Petersen GI, Simonsen F. Environmental and Health Assessment of Substances in Household Detergents and Cosmetic Detergent Products. CETOX, Danish Environmental Protection Agency, Environmental Project No. 615. 2001.
2. Vethaak AD, Rijs GBJ, Shrap SM, Ruiter H, Gerritsen A, Lahr J. Estrogens and xenoestrogens in the aquatic environment of the Netherlands. Occurrence, potency and biological effects. RIZA/RIKZ report no. 2002.001. 2002.
3. Campbell CG, Borglin SE, Green FB, Grayson A, Wozei E, Stringfellow WT. Biologically directed environmental monitoring, fate, and transport of estrogenic endocrine disrupting compounds in water: a review. Chemosphere. 2006;65:1265–80.
4. Caliman FA, Gavrilescu M. Pharmaceuticals, personal care products and endocrine disrupting agents in the environment–a review. Clean. 2009;37:277–303.
5. Legrini O, Oliveros E, Braun AM. Photochemical processes for water treatment. Chem Rev. 1993;93:671–98.
6. Gogate PR, Pandit AB. A review of imperative technologies for wastewater treatment I: oxidation technologies at ambient conditions. Adv Environ Res. 2004;8:501–51.
7. Pera-Titus M, Garcia-Molina V, Baños MA, Gimenez J, Esplugas S. Degradation of chlorophenols by means of advanced oxidation processes: a general review. App Catal B. 2004;47:219–56.
8. Ning B, Graham N, Zhang Y, Nakonechny M, El-Din MG. Degradation of endocrine disrupting chemicals by ozone/AOPs. Ozone Sci Eng. 2007;29:153–76.
9. Vaughan PP, Blough NV. Photochemical formation of hydroxyl radical by constituents of natural waters. Environ Sci Technol. 1998;32:2947–53.
10. Ostroumov SA. On Some issues of maintaining water quality and self-purification. Water Resour. 2005;32:305–13.
11. Canonica S, Freiburghaus M. Electron–rich phenols for probing the photochemical reactivity of freshwaters. Environ Sci Technol. 2001;35:690–5.
12. Davies-Colley RJ, Donnison AM, Speed DJ, Ross CM, Nagels JW. Inactivation of faecal indicator microorganisms in waste stabilisation ponds: interactions of environmental factors with sunlight. Wat Res. 1999;33:1220–30.
13. Kohn T, Nelson KL. Sunlight–mediated inactivation of MS2 coliphage via exogenous singlet oxygen produced by sensitizers in natural waters. Environ Sci Technol. 2007;41:192–7.
14. Boreen AL, Arnold WA, McNeill K. Photodegradation of pharmaceuticals in the aquatic environment: a review. Aq Sci. 2003;65:320–41.

15. Zafiriou OC, Joussot-Dubien J, Zepp RG, Zika RG. Photochemistry of natural waters. Environ Sci Technol. 1984;18:358–71.

16. Blough NV, Zepp RG. Reactive oxygen species in natural waters. In: Foote CS, editor. Active Oxygen in Chemistry. New York: Chapman and Hall; 1995. p. 280–333.

17. Rodrigues A, Brito A, Janknecht P, Proença MF, Nogueira R. Quantification of humic acids in surface water: effects of divalent cations, pH, and filtration. J Environ Monit. 2009;11:377–82.

18. Gmurek M, Miller JS. Photosensitised oxidation of a water pollutant using sulphonated porphyrin. Chem Pap. 2012;66:120–8.

19. Błędzka D, Gryglik D, Miller JS. Photodegradation of butylparaben in aqueous solutions. J Photochem Photobiol A Chem. 2009;203:131–6.

20. Błędzka D, Gryglik D, Olak M, Gębicki JL, Miller JS. Degradation of butylparaben and 4–tert–octylphenol in H2O2/UV system. Radiat Phys Chem. 2010;79:409–16.

21. Murov SL, Carmichael I, Hug GL. Handbook of photochemistry. 2nd ed. New York: Basel; 1993.

22. Wegner EE, Adamson AW. Photochemistry of complex ions. III. Absolute quantum yields for the photolysis of some aqueous chromium (III) complexes. Chemical actinometry in the long wavelength visible region. J Am Chem Soc. 1966;88:394–403.

23. Gryglik D, Lach M, Miller JS. The aqueous photosensitized degradation of butylparaben. Photochem Photobiol Sci. 2009;8:549–55.

24. Ge L, Chen J, Qiao X, Lin J, Cai X. Light–source-dependent effects of main water constituents on photodegradation of phenicol antibiotics: mechanism and kinetics. Environ Sci Technol. 2009;43:3101–7.

25. Liang H, Li X, Yang Y, Sze K. Effects of dissolved oxygen, pH, and anions on the 2,3–dichlorophenol degradation by photocatalytic reaction with anodic TiO2 nanotube films. Chemosphere. 2008;73:805–12.

26. Sujana MG, Soma G, Vasumathi N, Anand S. Studies on fluoride adsorption capacities of amorphous Fe/Al mixed hydroxides from aqueous solutions. J Fluorine Chem. 2009;130:749–54.

27. Chen Y, Zhang K, Zuo Y. Direct and indirect photodegradation of estriol in the presence of humic acid, nitrate and iron complexes in water solutions. Sci Total Environ. 2013;463–464:802–9.

28. Liu S, Li QX. Photolysis of spinosyns in seawater, stream water and various aqueous solutions. Chemosphere. 2004;56:1121–7.

29. Frimmel FH. Photochemical aspects related to humic substances. Environ Int. 1994;20:373–85.

30. Canonica S, Jans U, Stemmler K, Hoigne J. Transformation kinetics of phenols in water: photosensitization by dissolved natural organic material and aromatic ketones. Environ Sci Technol. 1995;29:1822–31.

31. Chin YP, Miller PI, Zeng L, Cawley K, Weavers LK. Photosensitized degradation of Bisphenol A by dissolved organic matter. Environ Sci Technol. 2004;38:5888–94.

32. Manjun Z, Xi Y, Hongshen Y, Lingren K. Effect of natural aquatic humic substances on the photodegradation of bisphenol A. Front Environ Sci Eng Chin. 2007;1:311–5.

33. Sakkas VA, Lambropoulou DA, Albanis TA. Study of chlorothalonil photodegradation in natural waters and in the presence of humic substances. Chemosphere. 2002;48:939–45.

34. Veselinović A, Bojić A, Purenović M, Bojić D, Andjelković T. Photodegradation of humic acids in the presence of hydrogen peroxide. Zbornik radova Tehnološkog fakulteta u Leskovcu. 2009;19:220–6.

35. Li Y, Duan X, Li X, Tang X. Mechanism study on photodegradation of nonylphenol in water by intermediate products analysis. Acta Chim Sin. 2012;70:1819–26.

36. Milne PJ, Zika RG. Luminescence quenching of dissolved organic matter in seawater. Mar Chem. 1989;27:147–64.

6

Removal of inorganic mercury from aquatic environments by multi-walled carbon nanotubes

Kamyar Yaghmaeian[1], Reza Khosravi Mashizi[1*], Simin Nasseri[1,2], Amir Hossein Mahvi[1], Mahmood Alimohammadi[1] and Shahrokh Nazmara[1]

Abstract

Background: Mercury is considered as a toxic heavy metal in aquatic environments due to accumulation in bodies of living organisms. Exposure to mercury may lead to different toxic effects in humans including damages to kidneys and nervous system.

Materials and methods: Multi-walled carbon nanotubes (MWCNTs) were selected as sorbent to remove mercury from aqueous solution using batch technique. ICP instrument was used to determine the amount of mercury in solution. Moreover, pH, contact time and initial concentration of mercury were studied to determine the influence of these parameters on the adsorption conditions.

Results: Results indicate that the adsorption strongly depended on pH and the best pH for adsorption is about 7. The rate of adsorption process initially was rapid but it was gradually reduced with increasing of contact time and reached the equilibrium after 120 min. In addition, more than 85 % of initial concentration of 0.1 mg/l was removed at 0.5 g/l concentration of sorbent and contact time of 120 min. Meanwhile, the adsorption process followed the pseudo second-order model and the adsorption isotherms could be described by both the Freundlich and the Langmuir models.

Conclusion: This study showed that MWCNTs can effectively remove inorganic mercury from aqueous solutions as adsorbent.

Keywords: Adsorption, Aqueous solutions, Heavy metals, Nano material

Background

Water resource pollution with industrial effluents is known as a serious environmental problem, nowadays [1].Of these, scientists have focused mostly on the presence of mercury due to its bioaccumulation in organisms, toxic effects and persistence in environment [2, 3]. In addition, it should be noted that mercury has been widely used in various industrial fields as an element including chlor-alkali, pharmaceutical, producing barometer and thermometer, mining, dental practices. It is proved that high concentrations of mercury are released constantly from the mentioned industries to the environment [4, 5].

From the toxicological point of view, the level of toxicity of mercury is highly related to its chemical form [6]. To put it another way, mercury transforms biologically, physically, and chemically through its cycle in nature, which results in the formation of various forms of mercury. Organic mercury is the most toxic among these forms [7]. Mercury is mostly in its inorganic (Hg^{+2}) or methylmercury forms in aquatic environments [6]. However, at the presence of specific kind of bacteria the inorganic form of mercury transforms to methylmercury which is highly toxic for human and other organisms at food chain [8]. Exposing to mercury results in neurological disorders, damage to central nervous systems, and also negatively affects the kidney and liver [9]. Considering these facts, there should be a proper way to handle this element and remove it from the environment.

Various methods have been used to remove mercury from water and wastewater, including chemical precipitation, ion exchange and membrane methods [10].

* Correspondence: khosravireza60@yahoo.com
[1]Department of Environmental Health Engineering, School of Public Health, Tehran University of Medical Sciences, Tehran, Iran
Full list of author information is available at the end of the article

Nevertheless, these methods have their own weaknesses including high level of either energy or chemical compound is needed, and most importantly these methods are not able to remove low concentration of mercury from the environment. However, adsorption due mostly to its high performance, recoverability, and reactive ability of adsorbent can be considered as a suitable method in terms of economy [11, 12].

In adsorption process, it is needed to have an adsorbent with wide specific surface. The surface results from the existence of tiny pores at it which the chemical property, area, size and distribution of these pores influence the level of an adsorbent's specific surface. Different materials, such as fruit shell [13], chitosan [14], marine macroalga [15], bagasse pith [16], furfural [17], and rubber [18] have been applied as adsorbents to remove mercury from aqueous environments.

After the discovery of carbon nanotubes, scientists paid specific attention to them because of their particular efficiency in construction, electricity, chemistry, and physics. Also, these materials have widely been used to produce nano-structured materials, nanocomposites, sensors, and gas adsorption. In 2004, when EPA proposed a research into the environmental application of these

materials [19], wide ranges of studies were conducted on these nanotubes which have tiny pores with uniform size and also wide specific surface [20]. Usage of carbon nanotubes was studied to remove pollutants, such as fluoride [21], dichlorobenzene [22], trihalomethanes [23], zinc [24], chromium [25], nickel [26, 27], and cadmium [28] from water and waste water. In this paper, Multi-walled carbon nanotubes) $MWCNT_S$ (were used to remove inorganic mercury from aqueous solutions.

Materials and methods

Mercury solution was prepared by using $HgCl_2$ (Merk) and deionized water. Carbon nanotubes were obtained from research division of Iranian Petroleum Industry. The characteristics of applied nanotubes in this study were as follow: the BET surface area of 270 m^2/g; diameters of 10–30 nm, respectively; length of 10 µm, and over 95 % purity. Furthermore, the morphology and size of carbon nanotubes were characterized by transmission electron microscope (TEM) and scanning electron microscope (SEM). The surface functional groups of multi-walled carbon nanotubes were detected by Fourier transform infrared spectroscopy (FTIR). Figure 1 shows the TEM, SEM and FTIR images of carbon nanotubes.

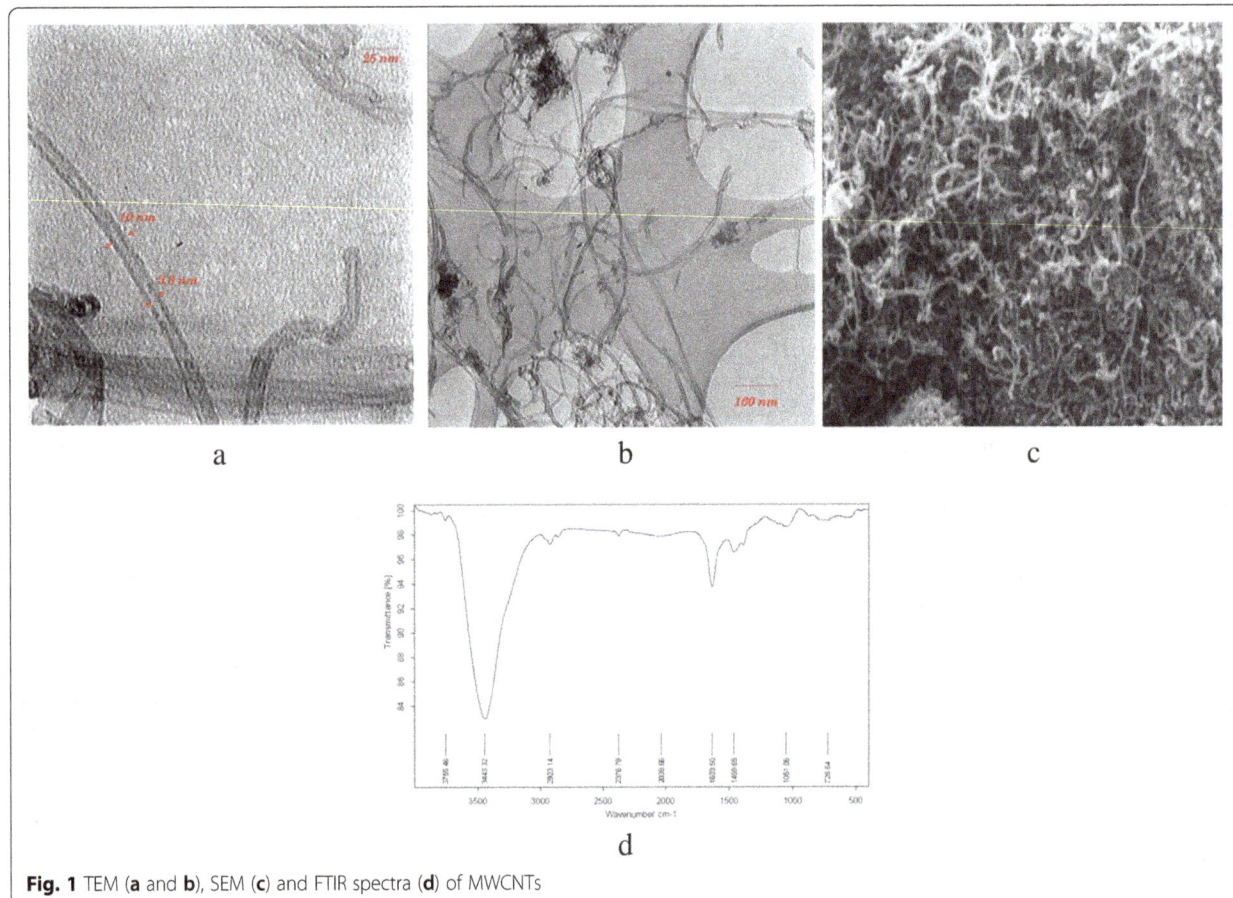

Fig. 1 TEM (**a** and **b**), SEM (**c**) and FTIR spectra (**d**) of MWCNTs

Table. 1 considered variables in previous studies the removal of metal ions from aquatic environments by carbon nanotubes

Metal ion	Variables	Ref	Metal ion	Variables	Ref
zinc	pH, contact time, initial metal ion concentration. Isotherm models	[24]	Zinc	adsorption kinetic, Isotherm models	[30]
Chromium	pH, contact time, agitation speed	[25]	copper	pH, ionic concentration, Isotherm models	[31]
nickel	pH, contact time, initial metal ion concentration, adsorbent's concentration, Isotherm models	[26]	Lead	pH, contact time, agitation speed, adsorbent's concentration, adsorption kinetic, Isotherm models	[32]
cadmium	pH, contact time, initial metal ion concentration, temperature, adsorption kinetic, Isotherm models	[28]	Lead	contact time, pH, ionic strength, foreign ions	[33]

Batch reactors with the same volumes of 250 ml were used. The reactors were filled with 100 ml of mercury solution with the concentrations of 0.1, 1 and 10 mg/l. The pH of the solutions were adjusted by nitric acid 0.1 N and NaOH 0.1 N (Merk). After adjusting the pH, the solutions were agitated under the temperature of 25 °C and 150 rpm on an Incubator Shaker (Innova 4340, USA). Thereafter, the solution was passed through 0.2 μm Millipore filter in order to separate the adsorbents from the aqueous solutions. Then, the pH was adjusted and lowered to <2 by using nitric acid. It should be noted that the prepared solutions were kept in glass containers at 4 °C. Besides, Cold vapour inductively coupled plasma optical emission spectrometry (Spectro, Germany) was applied to measure the concentration of mercury. This method has high sensitivity, excellent detection limits, rapid analysis and Easy to use. Also chemical and spectral interference is less than other methods [29]. For the reliable determination of mercury all variables were measured at least twice.

Some variables were considered in previous studies the removal of metal ions from aquatic environments by carbon nano tubes (Table 1). The effects of contact time, primary concentration of the solution, adsorbent's concentration, ionic strength of the solution, and pH on the process of adsorption were assessed in this research.

Freundlich and Langmuir adsorption isotherms were used to study the adsorption isotherm. Also pseudo-first- order and pseudo-second-order kinetics models were applied to determine the adsorption kinetic model.

The amount of adsorbed Hg^{2+} was calculated through the following equation [24–30]:

$$q = \frac{(C_o - C_t) \times V}{m} \tag{1}$$

where q denotes the amount of Hg^{2+} adsorbed on adsorbent at any time (mg/g), C_0 denotes the initial Hg^{2+} concentration (mg/l), C_t denotes the concentration of Hg^{2+} at any time (mg/l), V denotes the volume of the solution (l) and *m* denotes the adsorbent mass (g).

Results and discussion
The surface functional groups of MWCNTs
Figure 1(d) depicts the FTIR spectra of multi-walled carbon nanotubes. This spectrum displays major peaks at 3755, 3443, 2923, 1633, 1459 and 1051 cm^{-1}. The peak at 3755 cm^{-1} is associated with free hydroxyl groups. The peak at 3443 cm^{-1} is attributed to the O–H stretch from carboxyl groups (O = C – OH and C – OH). The peak at 2923 cm^{-1} is associated with (CH2). The peak at 1633 cm^{-1} is related to carbonyl groups. The peak at

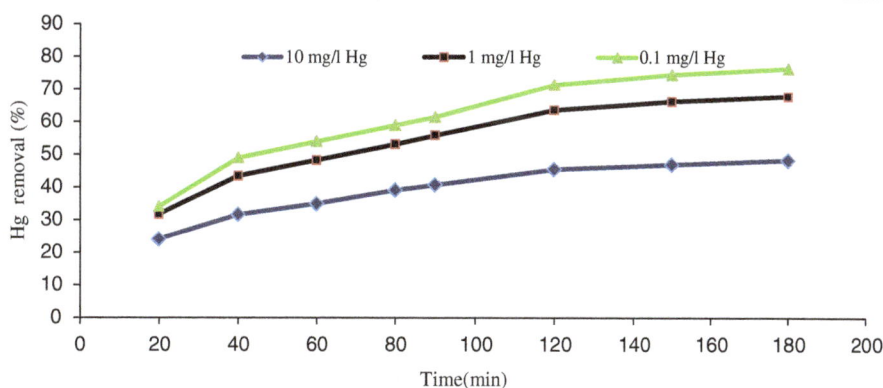

Fig. 2 Effect of contact time on the removal of mercury by MWCNT$_S$ at different mercury concentrations (adsorbent dose = 0.5 g/l, T = 25 °C, pH =5, agitation speed = 150 rpm)

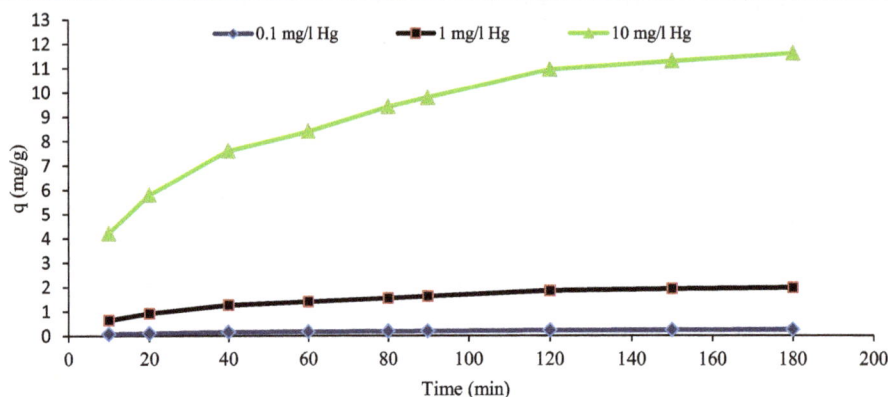

Fig. 3 Effect of contact time on the q at different mercury concentrations adsorbed on to MWCNTs (adsorbent dose = 0.5 g/l, T = 25 °C, pH =5, agitation speed = 150 rpm)

$1459 \ cm^{-1}$ is associated with carboxylic acids and phenolic groups (O–H). The peak at $1051 \ cm^{-1}$ is attributed to the (C–O). These functional groups can prepare numerous chemical sorption sites on surface of the MWCNTs [24, 28, 30, 32, 33].

Effect of contact time

As it is indicated in the Fig. 2, the effect of contact time on the level of adsorption of inorganic mercury was evaluated on the surface of multi-walled carbon nanotubes. In this test, the primary concentrations of inorganic mercury were 0.1, 1, and 10 mg/l and the other factors including temperature, pH, concentration of adsorbent, and agitation speed were kept constant. The adsorption of mercury increased with time until it reached equilibrium. The contact time for attainment to equilibrium was 120 min. Also Fig. 2 shows that the increase in the initial concentrations of mercury ions did not affect the equilibrium time. Similar findings have been reported for sorption of Ni^{+2} on carbon nanotubes [26] In addition, the rate of adsorption was reduced with the increase of contact time, which this reduction can be due

to the saturation of adsorption points on the carbon nanotubes. 'q' which is the amount of adsorbed mercury per the weight of carbon nanotubes raised by increase of contact time (Fig. 3).

Effect of primary concentration

The level of mercury removal in the pH of 5, contact time of 120 min, temperature of 25 °C, agitation speed of 150 rpm, and primary concentration of 0.1, 1, and 10 were 71.4, 63.6 and 45.6 % respectively (Fig. 4). the percentage of mercury removal was declined when the concentration of mercury increased from 0.1 to 10 mg/l. Primary concentration also affects the removal of chrome from aqueous solutions by carbon nanotubes [34]. When the concentration of adsorbent in the solution is constant, the number of adsorbent places is also constant for metallic ions. In this condition, the increase in the number of such ions results in the increase of competence among these ions to be adsorbed and it leads to the decrease of metallic ions' removal due mainly to electrical repulsive force [34]. It is noteworthy that the increase of primary concentration do not

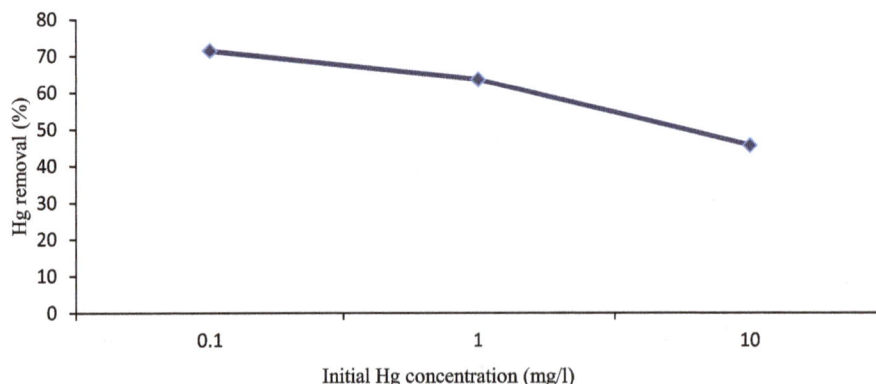

Fig. 4 Effect of primary Hg concentration on the removal of mercury by MWCNTs (adsorbent dose = 0.5 g/L, T = 25 °C, pH =5, agitation speed = 150 rpm, contact time = 120 min)

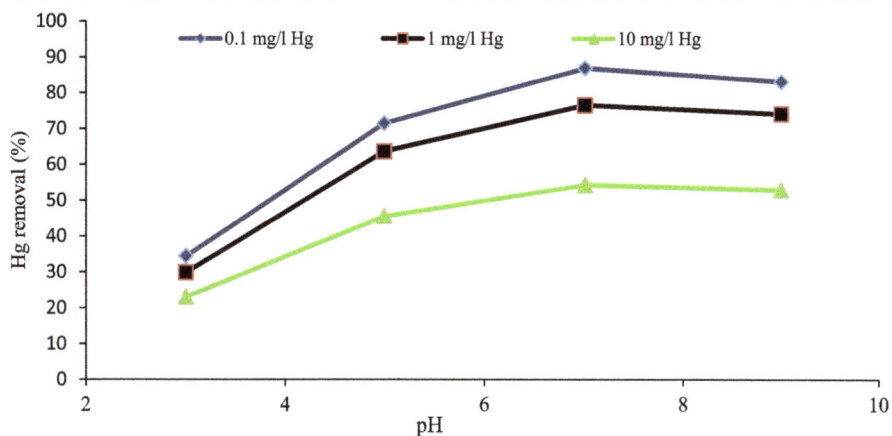

Fig. 5 Effect of pH on the removal of mercury by MWCNT$_S$ at different mercury concentrations (adsorbent dose = 0.5 g/l, T = 25 °C, contact time = 120 min, agitation speed = 150 rpm)

influence the balance time but it significantly affects 'q' (Fig. 3). 'q' raised with increasing of zinc primary concentration in the removal of zinc from aqueous solutions by carbon nanotubes [24].

Effect of solution's pH

pH is known as an important and effective parameter on adsorption process. The pH effects not only include the type of ion but also include the properties of adsorbent surface such as surface active groups [35]. As it is shown in the Fig. 5, the effects of pH on the level of mercury's adsorption on carbon nanotubes in the *conditions* of mercury's concentrations of 0.1, 1, and 10 mg/l and temperature of 25 °C, contact time of 120 min, adsorbent dosage of 0.5 g/l and agitation speed of 150 rpm were assessed. Based on the results, pH has a significant effect on the process of adsorption and the increase of it

raises the capacity of mercury's adsorption. Besides, the appropriate pH achieved in this study is 7, and below the 5, mercury transforms to Hg^{+2} which competes with H^+ for the adsorbent places. Hence, at the low pH, the level of adsorption decreases. It is mainly due to this fact that the H^+ level declines when the pH increases. The results of the present study are in the line with the studies conducted by Gao et al. [36] on the removal of Cu, Cd, Ni, and Zn by carbon nanotubes, and Atieh et al. [32] on the removal of Pb by the same adsorbent.

After the test, the pH was measured again and it was cleared that in the primary pH of upper than 5, the final pH decreases. The release of H^{2+} from the surface of carbon nanotubes into the solution could be the reason of this decrease. In the same contact time, the decrease of final pH was higher for the high concentration of mercury then the low concentration of mercury. When

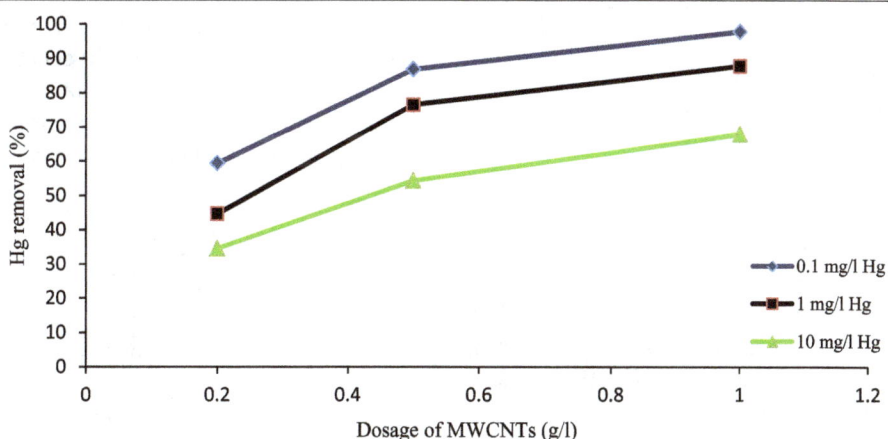

Fig. 6 Effect of adsorbent dosage on the removal of mercury by MWCNT$_S$ at different mercury concentrations (T = 25 °C, contact time = 120 min, agitation speed = 150 rpm, pH = 7)

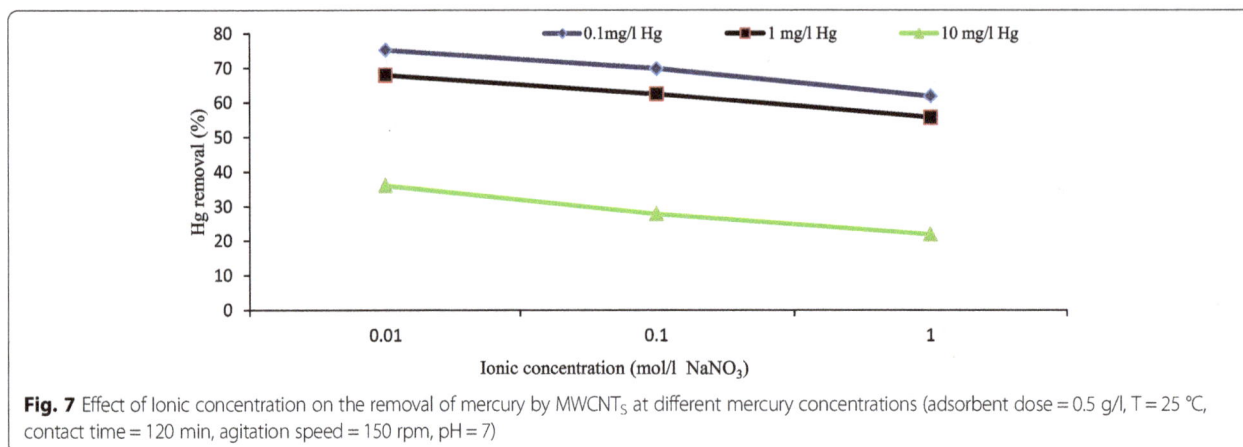

Fig. 7 Effect of Ionic concentration on the removal of mercury by MWCNT$_S$ at different mercury concentrations (adsorbent dose = 0.5 g/l, T = 25 °C, contact time = 120 min, agitation speed = 150 rpm, pH = 7)

the primary concentration of mercury increases, the amount of adsorption of Hg^{+2} also gets higher by which the rate of H^{2+} release from the adsorbent surface increases and causes to the drop of the pH level of the solution [37]. This process can be the indicator of chemical adsorption of inorganic mercury on the surface of the carbon nanotubes in pH of over 5.

Effect of adsorbent's concentration

The results indicated that the percentage of mercury's removal increased when the concentration of carbon nanotubes increased from 0.2 to 0.5 and finally to 1 mg/l (Fig. 6). This is mostly due to this reason that the increased concentration of adsorbent makes more active surface [33]. Similar findings have been reported for sorption of Ni^{+2} on carbon nanotubes [27].

Effect of ionic concentration

Based on the results, high ionic concentration has negative effect on the removal of mercury (Fig. 7). The reason is probably the impact of ionic concentration on the transfer of Hg^{+2} from the solution to the surface of adsorbent. In a study conducted by Lu and Liu [27], the authors found that the rate of Ni^{2+} adsorption on the surface of carbon nanotubes decreases when the ionic concentration increase.

Adsorption isotherm

Adsorption isotherms describe the distribution of metal ions between the liquid and solid phase at equilibrium. Adsorption balance of metallic ions is usually studied by the Freundlich and Langmuir adsorption isotherm. Langmuir isotherm is the indicator of active surface adsorption on the homogenous surface, while the Freundlich isotherm is used for heterogeneous surfaces [11]. The linear form of Langmuir and Freundlich equations are as below.

Langmuir

$$\frac{C_e}{q_e} = \frac{1}{kq_m} + \frac{C_e}{q_m} \tag{2}$$

Freundlich

$$\log q_e = \log K_F + 1/n \log C_e \tag{3}$$

Where C_e denotes the equilibrium concentration of Hg^{+2} (mg/l), q_e denotes the amount adsorbed (mg/g), q_m denotes the theoretical saturated adsorption capacity (mg/g), K denotes the Langmuir constant (l/mg). The values of K and q_m were calculated by plotting C_e/q_e versus C_e. K_F and n are the Freundlich constants related to adsorption capacity and adsorption intensity, The Freundlich constants n and K_F were calculated by plotting $\log q_e$ versus $\log C_e$.

The information of the two adsorption isotherms of inorganic mercury on the surface of carbon nanotubes are indicated in Table 2. As can be inferred from the information, the process of adsorption follows both Freundlich and Langmuir isotherms. Li et al. [38] depicted that adsorption of lead on to carbon nanotubes follows both Langmuir and Freundlich equations. Besides, the maximum adsorption capacity obtained from the Langmuir isotherm is 25.641 mg/g. Numerous low cost organic and inorganic adsorbents (e.g. activated carbon) have been

Table. 2 The parameters of Langmuir and Freundlich isotherm models for the removal of Hg^{+2} by MWCNT$_S$

Isotherm models	Parameters	value
Langmuir	q_m (mg/g)	25.641
	k (l/mg)	0.565
	R^2	0.948
Freundlich	K_F (l/ g)	0.077
	N	0.7173
	R^2	0.999

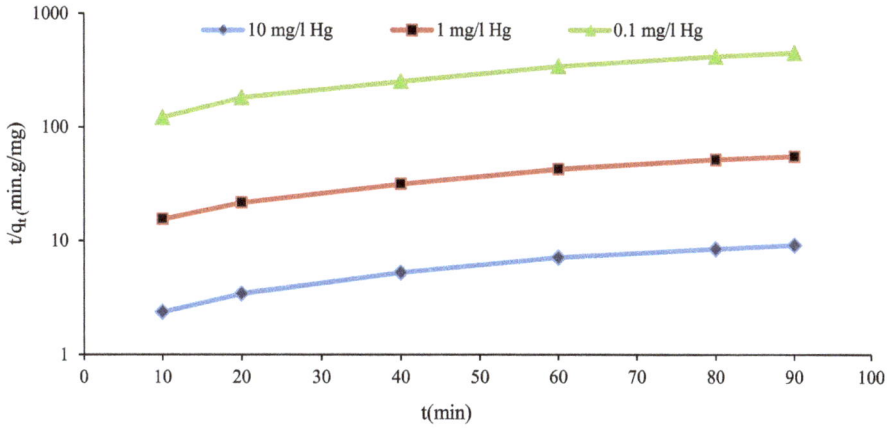

Fig. 8 Pseudo- second -order kinetics plot for the removal of Hg^{+2} by $MWCNT_S$ at different mercury concentrations (adsorbent dose = 0.5 g/l, T = 25 °C, agitation speed = 150 rpm, pH = 5)

explored for mercury removal [39, 40]. Zabihi et al. [41], Di Natale et al. [42], Asasian et al. [43] used different activated carbons to remove mercury that adsorption capacities of activated carbons were different. These amounts were strongly dependent on adsorption conditions such as solution's pH, adsorbent's concentration, temperature, ionic concentration and especially initial concentration of mercury [42]. However, activated carbons have disadvantages like weak physical stability, low selectivity for mercury, poor reactive ability of adsorbent and the release of mercury vapor into the atmosphere [39].

Adsorption kinetic

The adsorption kinetic model can provide suitable information for designing the removal of pollutants from water and wastewater. In order to assess the adsorption kinetic of inorganic mercury on the surface of carbon nanotubes, the pseudo-first and pseudo-second orders of kinetic equations were applied.

The pseudo first-order equation is written as:

$$\frac{dq_t}{dt} = k_l(q_e\text{-}q_t) \tag{4}$$

Where q_t is the amount of Hg^{+2} adsorbed at any time (mg/g), q_e is the amount of Hg^{+2} adsorbed at equilibrium (mg/g), K_1 is the adsorption rate constant (1/min).

The integrating for the boundary conditions t = 0 to t = t and q_t = 0 to q_t = q_e, gives the linear relationship between the amount of Hg^{+2} adsorbed (q_t) and time (t).

$$\text{Log}(q_e\text{-}q_t) = \text{Log}\,q_e\text{-}\frac{k_l}{2.303}t \tag{5}$$

A straight line log (q_e -q_t) versus t indicates the applicability of the pseudo-first order kinetic model.

The pseudo second-order equation can be expressed as:

$$\frac{dq_t}{dt} = k_2(q_e\text{-}q_t)^2 \tag{6}$$

Where k_2 denotes the second-order sorption rate constant (g/mg min). Integrating for the boundary conditions, t = 0 to t = t and q_t = 0 to q_t = q_e, gives the following equation:

$$\frac{t}{q_t} = \frac{1}{k_{2qe^2}} + \frac{1}{q_e}t \tag{7}$$

The information regarding these two equations is provided in Fig. 8 and Table 3. The achieved q_e from the pseudo-first order adsorption kinetic for the concentration of 0.1, 1, and 10 mg/l are 0.17, 1.40, and 8.1 mg/g, respectively. The corresponding scores for the pseudo-second order adsorption kinetic are 0.24, 1.99, and 11.76 mg/g, respectively, while the experimental q_e for the primary concentrations of mercury are 0.23, 1.84,

Table. 3 Kinetic parameters for the removal of Hg^{+2} by $MWCNT_S$ at different mercury concentrations

Kinetic model	primary Hg^{+2} concentration (mg/l)		
	0.1	1	10
Pseudo-first order			
q_e, exp (mg/g)	0.23	1.84	10.94
q_e, cal (mg/g)	0.17	1.40	8.1
K_1 (1/min)	0.0184	0.0184	0.0207
R^2	0.99	0.991	0.992
Pseudo-second-order			
q_e, cal (mg/g)	0.24	1.99	11.76
k_2 (g/mg min)	.0187	0.022	0.0042
R^2	0.996	0.997	0.995

and 10.94, respectively. The process of adsorption of inorganic mercury on the surface of multi-walled carbon nanotubes mostly fits with the second order adsorption kinetic and it is because of this that the q_e achieved from the pseudo-second order equations is closer to q_e of the base experiments, compared to the pseudo-first order equations. Also, R^2 conceived from the pseudo-second order adsorption kinetic graph is higher than that of pseudo-first order graph. Previous studies also showed that absorbed cadmium [28], zinc [30], lead [33] and chromium [34] on carbon nanotubes follow pseudo-second order adsorption kinetic.

Qu et al. [44] reported that nanomaterials have been widely used to remove heavy metals from water due to their large surface area, high reactivity, short intra particle diffusion distance and low temperature modification. In spite of that Tang et al. [45] suggested the reuse and management of the used nanomaterials is an important issue and has not been considered seriously. Only a few relevant studies are available in literature. It would be worthwhile to investigate the reusability of the used nanomaterials.

Conclusion

Multi-walled carbon nanotubes were assessed as adsorbent to remove inorganic mercury from aqueous solutions. The adsorption rate of Hg^{2+} on the surface of adsorbent is highly affected by pH, and the increase of pH from 3 to 7 increases the percentage of removal. The best contact time is 120 min. Also, the increases of primary concentration of inorganic mercury and ionic concentration solution have negative effect on adsorption process. Finally, the process of adsorption follows both Freundlich and Langmuir isotherms, and the pseudo-second order adsorption kinetic can well describe adsorption process. The present study indicated that Carbon nanotubes have high efficiency in adsorbing of mercury.

Abbreviations
ICP-OES-EOP: Inductively coupled plasma optical emission spectrometry (End-on-Plasma); MWCNTS: Multi-walled carbon nanotubes.

Competing interests
The authors declare that they have no competing interests.

Authors' contribution
Authors participated in this research including design, experiments and data analysis, and manuscript preparation. All authors read and approved the final manuscript.

Acknowledgements
This research was supported by Tehran University of Medical Sciences & health Services (Project No. 17314).
Our aim of this research is Removal of mercury from water and waste water that it plays an important role to maintain environmental and human health. Usage of nano materials as a new technology has been increased in different research. Hence, efficiency of carbon nanotubes to remove mercury from water and waste water was studied in this paper.

Author details
[1]Department of Environmental Health Engineering, School of Public Health, Tehran University of Medical Sciences, Tehran, Iran. [2]Center for Water Quality Research (CWQR), Institute for Environmental Research (IER), Tehran University of Medical Sciences, Tehran, Iran.

References
1. Jaafari J, Mesdaghinia AR, Nabizadeh R, Hoseini M, Kamani H, Mahvi AH. Influence of upflow velocity on performance and biofilm characteristics of Anaerobic Fluidized Bed Reactor (AFBR) in treating high-strength wastewater. J Environ Health Sci Engineer. 2014;12:1–10.
2. Li Y, Wu CY. Role of moisture in adsorption, photocatalytic oxidation, and reemission of elemental mercury on a SiO_2–TiO_2 nanocomposite. Environ Sci Technol. 2006;40:6444–8.
3. Girginova PI, Daniel-da-Silva AL, Lopes CB, Figueira P, Otero M, Amaral VS, et al. Silica coated magnetite particles for magnetic, removal of Hg^{2+} from water. J Colloid Interface Sci. 2010;345:234–40.
4. Bailey SE, Olin TJ, Bricka RM, Adrian DD. A review of potentially low-cost sorbents for heavy metals. Water Res. 1999;33:2469–79.
5. Gibicar D, Horvat M, Logar M, Fajon V, Falnoga I, Ferrara R, et al. Human exposure to mercury in the vicinity of chlor-alkali plant. Environ Res. 2009;109:355–67.
6. Balarama KMV, Karunasagar D, Rao SV, Arunachalam J. Preconcentration and speciation of inorganic and methyl mercury in waters using polyaniline and gold trap-CVAAS. Talanta. 2005;68:329–35.
7. Zhang L, Chang X, Hu Z, Zhang L, Shi J, Gao R. Selective solid phase extraction and preconcentration of mercury (II) from environmental and biological samples using nanometer silica functionalized by 2, 6-pyridine dicarboxylic acid. Microchim Act. 2010;168:79–85.
8. Wang X, Pehkonen SO, Ray AK. Photocatalytic reduction of Hg (II) on two commercial TiO_2 catalysts. Electrochim Acta. 2004;49:1435–44.
9. Morel FMM, Kraepiel AML, Amyot M. The chemical cycle and bioaccumulation of mercury. Annual Review of Ecology and Systematics. 1998;29:543-66. doi:10.1146/annurev.ecolsys.29.1.543.
10. Miretzky P, Cirelli AF. Hg(II) removal from water by chitosan and chitosan derivatives: a review. Hazard Mater. 2009;167:10–23.
11. Rao PD, Lu C, Su F. Sorption of divalent metal ions from aqueous solution by carbon nanotubes: a review. Sep Purif Technol. 2007;58:224–31.
12. Skodras G, Diamantopoulou IR, Zabaniotou A, Stavropoulos G, Sakellaropoulos GP. Enhanced mercury adsorption in activated carbons from biomass materials and waste tires. Fuel Process Technol. 2007;88:749–58.
13. Inbaraj SB, Sulochana N. Mercury adsorption on a carbon sorbent derived from fruit shell of terminalia catappa. Hazard Mater. 2006;133:283–90.
14. Jeon C, Holl WH. Chemical modification of chitosan and quilibrium study for mercury ion removal. Water Res. 2003;37:4770–80.
15. Herrero R, Lodeiro P, Castro CR, Vilarin T, Vicente MESD. Removal of inorganic mercury from aqueous solutions by biomass of the marine macroalga Cystoseira baccata. Water Res. 2005;39:3199–210.
16. Krishnan KA, Anirudhan TS. Removal of mercury (II) from aqueous solutions and chlor-alkali industry effluent by steam activated and sulphurised activated carbons prepared from bagasse pith: ineticsand equilibrium studies. Hazard Mater. 2002;92:161–83.
17. Budinova T, Savova D, Petrov N, Razvigorova M, Minkova V, Ciliz N, et al. Mercury adsorption by different modifications of furfural adsorbent. Ind Eng Chem Res. 2003;42:2223–9.
18. Vizuete ME, Garc AM, NGisbert A, Gonz CF, Serrano VG. Adsorption of mercury by carbonaceous adsorbents prepared from rubber of tyre wastes. Hazard Mater. 2005;119:231–8.
19. Feng X: Application of single walled carbon nanotubes in environmental engineering: adsorption and desorption of environmentally relevant species studied by infrared spectroscopy and temperature programmed desorption. Doctoral Dissertation 2006, University of Pittsburgh.
20. Mohan D, Pittman CU. Activated carbons and low cost adsorbents for the remediation of tri- and hexavalent chromium fromwater. Hazard Mater. 2006;137:762–811.
21. Li YH, Wang S, Zhang X, Wei J, Xu C, Luan Z, et al. Adsorption of fluoride from water by aligned carbon nanotubes. Mater Res Bull. 2003;38:469–76.
22. Peng X, Li Y, Luan Z, Di Z, Wang H, Tian B, et al. Adsorption of 1,2-

dichlorobenzene from water to carbon nanotubes. Chem Phys Lett. 2003;376:154–8.

23. Lu C, Chung YL, Chang KF. Adsorption of trihalomethanes from water with carbon nanotubes. Water Res. 2005;39(6):1183–9.

24. Lu C, Chiu H. Adsorption of zinc (II) from water with purified carbon nanotubes. Chem Eng Sci. 2006;61:1138–45.

25. Gupta VK, Agarwal S, Saleh TA. Chromium removal by combining the magnetic properties of iron oxide with adsorption properties of carbon nanotubes. Water Res. 2011;45(6):2207–12.

26. Kandah MI, Meunier JL. Removal of nickel ions from water by multi-walled carbon nanotubes. Hazard Mater. 2007;146:283–8.

27. Lu C, Liu C. Removal of nickel (II) from aqueous solution by carbon nanotubes. Chem Technol Biotechnol. 2006;81:1932–40.

28. Vukovi GD, Marinkovi AD, Coli M, Risti MD, Aleksi R, Peri'c-Gruji AA, et al. Removal of cadmium from aqueous solutions by oxidized and ethylenediamine-functionalized multi-walled carbon nanotubes. Chem Eng J. 2010;157:238–48.

29. Hellings J, Adeloju S B, Verheyen T V: Rapid determination of ultra-trace concentrations of mercury in plants and soils by cold vapour inductively coupled plasma-optical emission spectrometry. Microchem Journal 2013, http://dx.doi.org/10.1016/j.microc.2013.02.007.

30. Lu C, Chiu H. Chemical modification of multiwalled carbon nanotubes for sorption of Zn^{2+} from aqueous solution. Chem Eng. 2008;2008(139):462–8.

31. Sheng G, Li J, Shao D, Hu J, Chen CH, Chen Y, et al. Adsorption of copper(II) on multiwalled carbon nanotubes in the absence and presence of humic or fulvic acids. J Hazard Mater. 2010;178:333–40.

32. Atieh MA, Bakather OY, Al-Tawbini B, Bukhari AA, Abuilaiwi FA, Fettouhi MB. Effect of carboxylic functional group functionalized on carbon nanotubes surface on the removal of lead from water. Bioinorg Chem Appl. 2010;2010:1–9.

33. Xu D, Tan X, Chen C, Wang X. Removal of Pb(II) from aqueous solution by oxidized ultiwalled carbon nanotubes. Hazardous Material. 2008;154:407–16.

34. Hu J, Chen C, Zhu X, Wang X. Removal of chromium from aqueous solution by using oxidized multiwalled carbon nanotubes. Hazard Mater. 2009;162:1542–50.

35. Qu R, Sun C, Ma F, Zhang Y, Ji Ch XQ, Wang C, et al. Removal and recovery of Hg (II) from aqueous solution using chitosan-coated cotton fibers. Hazard Mater. 2009;167:717–27.

36. Gao Z, Bandosz TJ, Zhao Z, Han M, Qiu J. Investigation of factors affecting adsorption of transition metals on oxidized carbon nanotubes. Hazard Mater. 2009;167:357–65.

37. Wang XK, Chen CL, Hu WP, Ding AP, Xu D, Zhou X. Sorption of 243Am(III) to multi-walled carbon nanotubes. Environ Sci Technol. 2005;39:2856–60.

38. Li YH, Wang S, Wei J, Zhang X, Xu C, Luan Z, et al. Lead adsorption on carbon nanotubes. Chem Phys Lett. 2002;357:263–6.

39. Presto AA, Granite EJ. Impact of sulphur oxides on mercury capture by activated carbon. Environ Sci Technol. 2007;41:6579–84.

40. Anirudhan TS, Divya L, Ramachandran M. Mercury (II) removal from aqueous solutions and wastewaters using a novel cation exchanger derived from coconut coir pith and its recovery. Hazardous Material. 2008;157:620–62.

41. Zabihi M, Haghighi Asl A, Ahmadpour A. Studies on adsorption of mercury from aqueous solution on activated carbons prepared from walnut shell. Hazard Mater. 2010;174:251–6.

42. Di Natale F, Erto A, Lancia A, Musmarra D. Mercury adsorption on granular activated carbon in aqueous solutions containing nitrates and chlorides. Hazard Mater. 2011;192:1842–50.

43. Asasian N, Kaghazchi T, Soleimani M. Elimination of mercury by adsorption onto activated carbon prepared from the biomass material. Ind Eng Chem. 2012;18:283–9.

44. Qu XL, Alvarez PJJ, Li QL. Applications of nanotechnology in water and wastewater treatment. Water Res. 2013;47(12):3931–46.

45. Tang WW, Zeng GM, Gong JL, Liang J, Xu P, Zhang C, et al. Impact of humic/fulvic acid on the removal of heavy metals fromaqueous solutions using nanomaterials: A review. Sci Total Environ. 2014;468–469:1014–27.

7

Fabrication and characterization of a polysulfone-graphene oxide nanocomposite membrane for arsenate rejection from water

Reza Rezaee[1], Simin Nasseri[2,1*], Amir Hossein Mahvi[1,3,4], Ramin Nabizadeh[1], Seyyed Abbas Mousavi[5], Alimorad Rashidi[6], Ali Jafari[7] and Shahrokh Nazmara[1]

Abstract

Background: Nowadays, study and application of modified membranes for water treatment have been considered significantly. The aim of this study was to prepare and characterize a polysulfone (PSF)/graphene oxide (GO) nanocomposite membrane and to evaluate for arsenate rejection from water.

Materials and methods: The nanocomposite PSF/GO membrane was fabricated using wet phase inversion method. The effect of GO on the synthesized membrane morphology and hydrophilicity was studied by using FE-SEM, AFM, contact angle, zeta potential, porosity and pore size tests. The membrane performance was also evaluated in terms of pure water flux and arsenate rejection.

Results: ATR-FTIR confirmed the presence of hydrophilic functional groups on the surface of the prepared GO. FE-SEM micrographs showed that with increasing GO content in the casting solution, the sub-layer structure was enhanced and the drop like voids in the pure PSF membrane changed to macrovoids in PSF/GO membrane along with increase in porosity. AFM images indicated lower roughness of modified membrane compared to pure PSF membrane. Furthermore, contact angle measurement and permeation experiment showed that by increasing GO up to 1 wt%, membrane hydrophilicity and pure water flux were increased. For PSF/GO-1, pure water flux was calculated about 50 L/m^{2}h at 4 bar. The maximum rejection was obtained by PSF/GO-2 about 83.65 % at 4 bar. Moreover, it was revealed that arsenate rejection depended on solution pH values. It was showed that with increasing pH, the rejection increased.

Conclusions: This study showed that application of GO as an additive to PSF casting solution could enhance the membrane hydrophilicity, porosity, flux and arsenate rejection.

Keywords: Mixed matrix membrane, Polysulfone, Graphene oxide, Hydrophilicity, Arsenate

Background

In recent years, a growing public concern has arisen over release of toxic pollutants such as inorganic ions, metals and synthetic organic matters into the water due to increasingly industrial and agricultural activities. Among these toxicants, arsenic is a serious threat in water resources of some regions [1, 2]. Toxicological and epidemiological studies proven that inorganic arsenic could cause carcinogenic and non-carcinogenic effects in human [3]. World health organization (WHO) and united state protection agency (USEPA) had classified arsenic as class A carcinogens list [4]. International agency for research on cancer (IARC) also classified inorganic form of arsenic in class I carcinogens list [5]. With regard to strict regulations for control and removal of arsenic in drinking water, and limitations of conventional water treatment processes (e.g. generation of toxic intermediates and low efficiencies) looking for new technologies is of great interest [3, 6]. Membrane process can be considered as a promising technology for arsenic removal due to its several advantages such as no need to add

* Correspondence: naserise@tums.ac.ir
[2]Center for Water Quality Research, Institute for Environmental Research, Tehran University of Medical Sciences, Tehran, Iran
[1]Department of Environmental Health Engineering, School of Public Health, Tehran University of Medical Sciences, Tehran, Iran
Full list of author information is available at the end of the article

chemicals, no generation of sludge, ease of system capacity development, separation in continuous mode, ease of integration with other processes, minimum dependency to environmental conditions and capable of microorganisms and solutes removal [7, 8]. However, certain drawbacks associated with common membranes are low water recovery, fouling problem and high-energy consumption [9, 10]. In recent years to overcome these drawbacks, different studies have been conducted in order to modify polymeric membranes to enhance the permeability, rejection and decreasing fouling problem and reduce the investment and operational costs [11]. Accordingly, various works such as physical blending, plasma treatment, polymer grafting and chemical reactions have been carried out to modify the membranes [12, 13]. Among these methods physical blending is preferred due to the simplicity procedure using phase inversion technique [14]. Physical blending consist of mixing of polymeric materials with inorganic nanoparticles (e.g. TiO_2 [15], ZnO [16], silica [17]) and recently carbon allotropes [11, 18, 19]. Adding inorganic nanoparticles to membrane matrix can enhance the membrane hydrophilicity, strength, permeability and antifouling characteristics [18, 20]. Graphene and its derivatives due to unique two-dimensional structure, one-atom-layer-thick, high theoretical surface area (2630 m^2/g), good mechanical properties, non-harmful effects, low cost production have attracted interest for different application especially polymeric membrane modification [21, 22]. Graphene oxide (GO) is also highly hydrophilic due to presence of oxygen containing functional groups (e.g. hydroxyl, carboxyl, carbonyl and epoxy) [12, 23]. When thin sheets of carbon atoms (GO) are added to a polymer matrix at low content and proper procedure, it could significantly improve the physical properties of the base polymer [21, 24]. Among different synthetic polymers, polysulfone (PSF) is the one that is widely used for various membranes fabrication such as filtration, ultrafiltration, hemodialysis and bioreactor technologies [13, 25]. The reasons for wide use of this type of polymer are good characteristics such as desire mechanical and thermal properties, high chemical stability, high resistance in a wide range of pH and high solubility in a broad range of polar solvents (dimethylformamide, dimethylacetamide, dimethylsulfoxide) [13, 21, 25]. One of the main drawbacks of PSF membrane is fouling problem and consequently reduction of the membrane lifetime. Actually, this type of membrane is influenced by fouling problem more than other membrane materials because of the hydrophobic nature of the membrane and interactions between the membrane surface charges and the foulants [13, 21]. A few studies have used GO in casting solution to improve the water permeability, antifouling properties and mechanical strength characteristics of the mixed matrix membrane. Zhao et,al showed that synthesized

PVDF/GO ultrafiltration membrane had higher pure water flux compared to PVDF due to improvement of the surface hydrophilicity [12]. Wang et, al also reported that GO nanosheet as a hydrophilic modifier could enhance the water flux of the fabricated ultrafiltration membrane with an improvement in the antifouling property [14]. In another study, Zinadini et.al showed that water permeability, hydrophilicity and antifouling properties of the PES/GO membrane were enhanced compared to pure PES membrane [26]. Xia et, al also revealed that employment of a certain amount of GO in the matrix could improve the water flux, hydrophilicity and antifouling characteristics of a type of synthesized PVDF/GO membrane used for natural organic matter removal [27]. The aim of this study is to synthesis and characterizes a PSF/GO nanocomposite membrane in order to reject arsenic from water. In this work, GO was applied to PSF matrix as a hydrophilic agent. The performance of the synthesized membranes was evaluated by pure water flux measurement and arsenate rejection.

Materials and methods
Materials
All chemicals used in the experiments were of reagent grade. Graphite fine powder extra pure (with a mean particle size of <50 μm) was purchased from Merck-Germany. PSF (with average M_w = 22,000 g/mol) was obtained from Sigma-Aldrich Co-Germany. N,N-Dimethylformamide >(DMF) was purchased from Sigma-Aldrich and used without purification as a solvent to prepare cast solution. Analytical grade H_2SO_4 (98 %-Merck), $NaNO_3$ (99 %, Sigma−Aldrich), $KMnO_4$ (99 %, Sigma−Aldrich) and H_2O_2 (30 % solution, stabilized-Merck) were used as received. Sodium arsenate dibasic heptahydrate ($Na_2HAsO_4 \cdot 7H_2O$) was obtained from Sigma-Aldrich. The deionized (DI) water was used in the sample preparation and for pure water flux measurements.

Preparation and characterization of graphene oxide (GO)
In this study GO was prepared using modified Hummer's method [28]. Firstly, 5 g graphite powder and 2.5 g sodium nitrate were added to a 500-ml neck flask containing 120 ml concentrated sulfuric acid in ice bath and thoroughly mixed for 30 min. Then under vigorous mixing, 15 g $KMnO_4$ was slowly added to the suspension and mixing was continued for 30 min. The rate of adding was controlled to maintain temperature of the reaction below 20 °C. After that, ice bath was removed and the mixture was stirred overnight at room temperature. By elapsing the time, the mixture changed in to sticky and the color changed to brown. Then under mixing condition, 150 ml distilled water was slowly added to the mixture. The temperature was rapidly increased to 98 °C and the color turned to yellow. This aqueous suspension

was stirred at 98 °C for 24 h. In order to remove KMnO$_4$, 50 ml H$_2$O$_2$ (30 %) was added to the liquid mixture. For more purification, the liquid mixture of GO was washed by HCL (5 %) and DI water and centrifuged for several times to reach the pH to natural range. Finally, for exfoliating the product, sonication was conducted for 1 h. Then it was filtered and dried in a vacuum oven (at 40 °C for 24 h) to obtain a grey color GO nanoplate powder. Raman spectra of the GO was obtained in the spectral range of 100-4200 cm^{-1} and with 532 nm wavelength incident laser light (Almega Thermo Nicolet Dispersive Raman Spectrometer, Germany). The measurements of the attenuated total reflectance fourier transform infrared spectroscopy (ATR-FTIR) of the GO was performed using a ATR-FTIR spectroscopy in the range between 600 cm – 1 and 4000 cm^{-1} (Tensor 27, Bruker Inc., Germany).

Fabrication of PSF/GO nanocomposite membrane

In present work, PSF/GO nanocomposite membrane was fabricated via common phase inversion method [14, 29]. For this purpose, PSF was used as bulk material, DMF as solvent, GO nanoplate as the additive and hydrophilic modifier, DMF as the solvent and DI water as the nonsolvent in coagulation bath. The casting solution consist of PSF = 15 %wt, DMF = 85 wt% and GO = (0-0.5-1-2 wt% PSF). PSF and GO powder were dried in vacuum oven at 60 °C for 4 h. At first, four different amounts of GO were dispersed in DMF and was sonicated for 1 h to obtain a homogenous casting solution. Then, under continuous stirring condition, PSF was added to GO/DMF mixture and was allowed to stir for 24 h. Then the casting solution was maintained in room temperature for 24 h without stirring. Finally, casting solution was sonicated to remove remaining air bubbles. The prepared casting solution was casted uniformly onto a smooth and clean glass plate using a casting knife at a thickness of 200 μm. The casted film on the glass was left for air exposure (20 s) followed by immersing into the nonsolvent coagulation bath (DI water at 25 °C). The glass plate was kept in the coagulation bath for 10 min to guarantee complete phase inversion process. Finally the peeled off synthesized membrane was washed with DI water for several times until all the residual solvent removed. The membranes were kept in DI water for characterization and experiments. The synthesized membrane based on GO content named pure PSF, PSF/GO-0.5, PSF/GO-1 and PSF/GO-2.

Characterization of the prepared membranes

The structure and surface morphology of the membranes were evaluated using a field-emission scanning electron microscope (FE-SEM, S-4160, Hitachi, Japan). For sample preparation, membrane were cut into small pieces and washed with distilled water. For obtaining a

good cross section image, the wet pieces were immersed in liquid nitrogen for 1 min to freeze. The frozen pieces of the membranes were fractured and kept in air to dry. The dried samples were coated with a thin layer of gold to increase the electric conductivity before FE-SEM imaging. Atomic force microscopy (AFM) was applied for top surface morphology and roughness analysis. Thermo microscopes Auto probe CP Research (Veeco Instruments, Sunnyvale, CA, USA) was used for AFM analysis. The samples were cut into small pieces (1 cm × 1 cm), washed with distilled water and dried in room temperature. In this study, surface hydrophilicity changes of different fabricated membranes were determined via the contact angle and Zeta potential. The contact angle was analyzed using a water contact angle measurement (OCA 15 Plus, Dataphsycs, Germany). Before contact angle measurement, the samples were dried in oven at 50 °C for 4 h. For more accuracy in the determination of contact angle, 5 different top surface points were measured and the average was reported. The zeta potentials of fabricated membrane were measured by streaming potential method using Electro kinetic Analyzer (EKA, Anton Paar GmbH, Austria) equipped with plated sample cell. For this purpose membrane were cut in 5 cm × 5 cm pieces and zeta potentials were measured at 26 °C and pH of 7. In this measurement method, 0.001 M KCl solution was applied as electrolyte and zeta potential were measured in triplicate for each membrane.

Membrane porosity and pore size

To evaluate the effect of GO on the membrane structure, the porosity, as the ratio of the volume of voids to the total volume of membrane, was measured using a gravimetry method. For this, the membranes were dried in an air-circulating oven at 50 °C for 24 h. Then the samples were cut into small pieces (1 cm × 1 cm) (5 pieces for each membrane) and weighted. The pieces were immersed in distilled water for 24 h at 25 °C. After removing the droplets on the surface of membrane by a paper filter, the membrane was weighted again. The average of dry and wet weights for each membrane was recorded and the porosity (ε) was calculated using the gravimetry equation 1 [18, 30, 31].

$$\varepsilon = \frac{\frac{W_1 - W_2}{\rho_w}}{\frac{W_1 - W_2}{\rho_w} + \frac{W_2}{\rho_m}} \times 100\% \qquad (1)$$

Where, W_1 and W_2 are wet and dry weights of membrane respectively (g), ρ_w is the density of distilled water (0.998 g/mL) and ρ_m is the density of polymer (PSF = 1.24 g/mL at 25 °C). The average pore radius (r_m) of the membranes was calculated by following equation known as Guerout–Elford–Ferry equation (Eq. 2) [18, 32].

$$rm = \sqrt{\frac{(2.9-1.75\varepsilon) \times 8\eta lQ}{\varepsilon \times A \times \Delta P}} \qquad (2)$$

where rm is the mean pore radius (m), η is the water viscosity (8.9×10^{-4} Pa·s), l is the membrane thickness (m), Q is the volume of the permeate water per unit time (m^3/s), A is the effective area of the membrane (m^2) and ΔP is the operational pressure (Pa).

Permeation tests and arsenate rejection experiments
In this study, to evaluate the permeation flux and arsenate rejection by fabricated nanocomposite membrane, a lab-scale filtration system was used at dead end mode operation. The main components of the filtration system include a 2-L feed tank (equipped with mixer and temperature control), low and high pressure feed pumps (1 to 15 bar), stainless steel flat membrane module (effective area of 9.6 cm^2), valves and pressure gauges (Fig. 1). For flux measurements, the membranes were first immersed in distilled water for 24 h. Then the membranes were compacted under 7 bar of distilled water at 25 ± 0.5 °C for 30 min until a constant flux was achieved. Immediately the pressure was reduced to 4 bar and pure water flux test was conducted for 1 h with collecting and measuring the filtrate volume at 5 min intervals. Finally, the flux was calculated using equation 3 [11, 33].

$$J_w = \frac{V}{A\Delta t} \qquad (3)$$

Where J_w is the pure water flux (L/m^{2}h), V is the volume of permeated pure water (L), A is the effective area of membrane (m^2) and Δt is the sampling time (h).

after measuring the pure water flux, arsenate sodium solution with initial concentration of 300 ± 10 µg/L was prepared based on a standard procedure [34]. The rejection of arsenate was evaluated at 4 bar. The permeate was collected each 20 min for arsenate analysis, finally the average was reported. Arsenic concentrations was measured by inductively coupled plasma optical emission spectroscopy (HG-ICP/OES) (Model Spectro arcos, Specro Inc, Germany) connected to a hydride generator. The percentage of rejection was calculated using equation 4 [26].

$$\%R = \left(1-\frac{C_p}{C_f}\right) * 100 \qquad (4)$$

Where R is the rejection of arsenate (%), and C_p and C_f are the concentrations of arsenate in the permeation and feed solution, respectively (µg/L). All pure water flux and rejection experiments were performed in triplicate.

Results and discussion
Characterization of graphene oxide
Figure 2 depicts the Raman spectrum of synthesized GO. From the Figures D, G and 2D appeared at 1348, 1585 and 2700 cm^{-1} as GO known peaks [35]. Generally, in graphene Raman spectrum, D band indicates the disordered and defect in graphene structure, G band shows that normal structure of graphene and 2D band is related to number of layers. Graphene Raman spectrum from a single layer and a few layer graphene consist of peak G at around 1580 cm^{-1} and peak 2D at around 2700 cm^{-1}. In GO Raman spectrum, as well as G and 2D peaks, typical D band is obvious which appears at around 1350 cm^{-1}. This peak (D) is absent in ordered graphene, while in GO, presence of D band is assigned to the developed defect structure in due to oxygen containing functional groups(e.g. hydroxyl) at the edge of

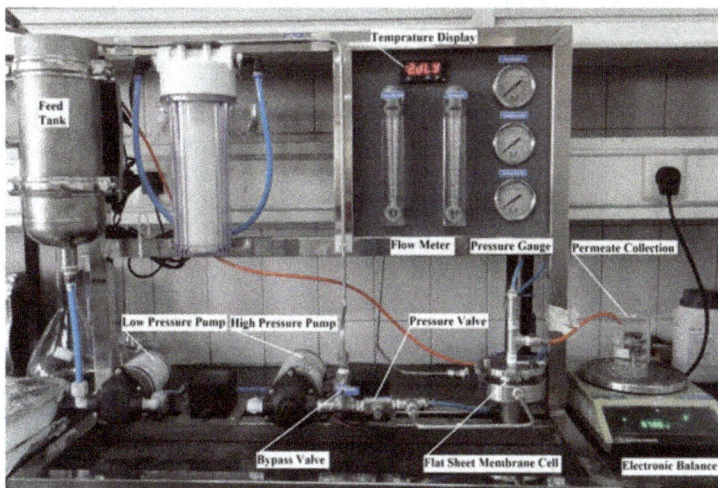

Fig. 1 Membrane filtration system used in experiment

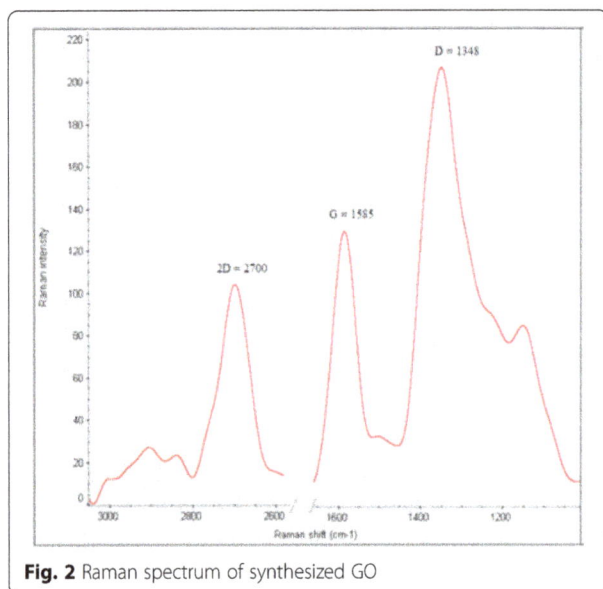

Fig. 2 Raman spectrum of synthesized GO

hydroxyl groups. O-H bond may exist in forms of alcoholic, phenolic, carboxylic and so on. This peak also confirms the hydrophilic properties of GO. The band 1395.86 cm^{-1} can be attributed to O-H deformation vibration [21]. The absorption peak in the 1713.63 cm^{-1} shows the carbonyl stretching vibration (C = O) and indicate the presence of carboxyl functional group. In addition, appearance of an adsorption peak at 1110.65 cm^{-1} could be assigned to the C-O bond stretching vibration [38]. With regard to presence of oxygen containing functional groups it is proved that the synthesized GO is highly hydrophilic. These observations are consistent with the results reported in other works [12, 14].

Characterization of the PSF/GO membrane
Effect of GO addition on membrane morphology
In next stage of the study, the effect of loaded GO on the micro-structure of the PSF membrane was analyzed. Cross-sectional FE-SEM micrographs of the prepared membranes are presented in Fig. 4. General structure of the membrane consists of a dense skin layer on the top and a porous support sub layer. Pure PSF membrane with mainly sponge structure and few separated closed end drop-like pores shown in Fig. 4a. With the addition of GO, the main characteristics of a asymmetric structure appears composing of a dense skin layer on top and a thick porous layer with finger like pores in the bottom (Fig. 4b-d). From Fig. 4b, in membrane with 0.5 wt% GO, drop like pores have been replaced by finger-like pores in the pure PSF membrane but the walls of the pores are thick and with closed ends, and the sponge parts are still exist as a significant part of the membrane. With further increase in loaded GO, finger like channels turned into a large, open-end macrovoids and the spongy portion decreased significantly. Furthermore, from the figure, the

graphene plates. As the intensity of peak D is higher, the sample has higher disordered structure. In addition, Raman spectrum can be used for analysis of graphene quality and determination of the layers (up to 5 layers) through the 2D peak shape, width and position. With increase in number of the layers, 2D peak shifts to higher wavelengths and will broaden [36, 37]. In this study based on appeared peaks from Raman spectrum analysis confirmed a few-layer structure of GO.

In addition to Raman spectra, IR spectrum is also used for GO characterization. Figure 3 shows the GO ATR-FTIR spectrum. From the figure, a prominent adsorption peak appeared at 3411.94 cm^{-1} that reveals the typical GO characteristics. This strong peak assigns to stretching vibration O-H bond and indicates the presence of

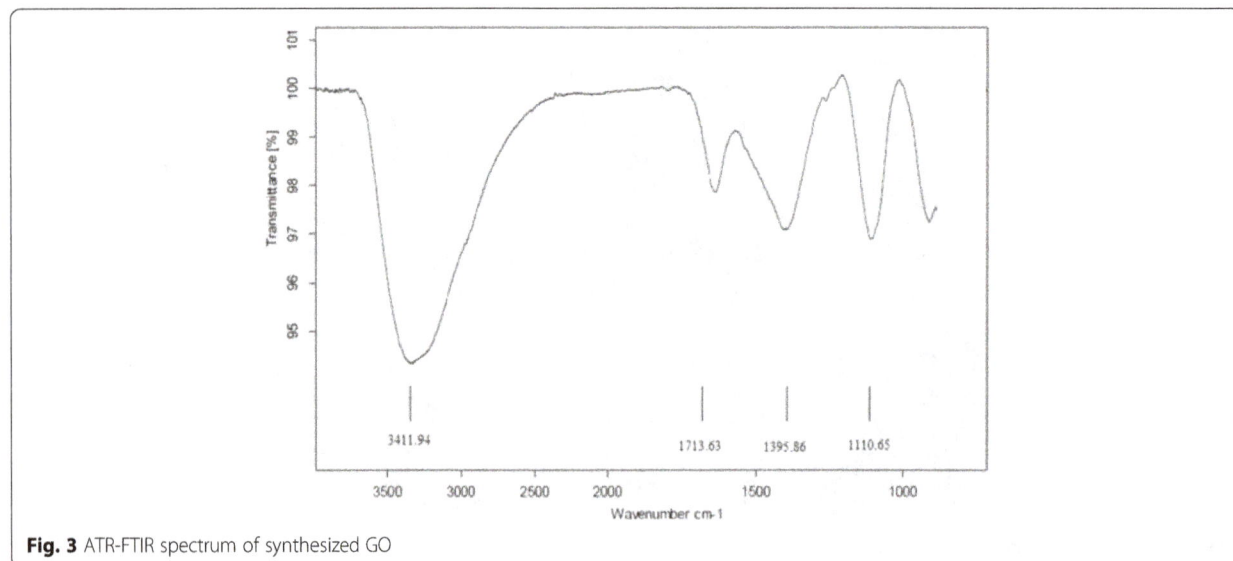

Fig. 3 ATR-FTIR spectrum of synthesized GO

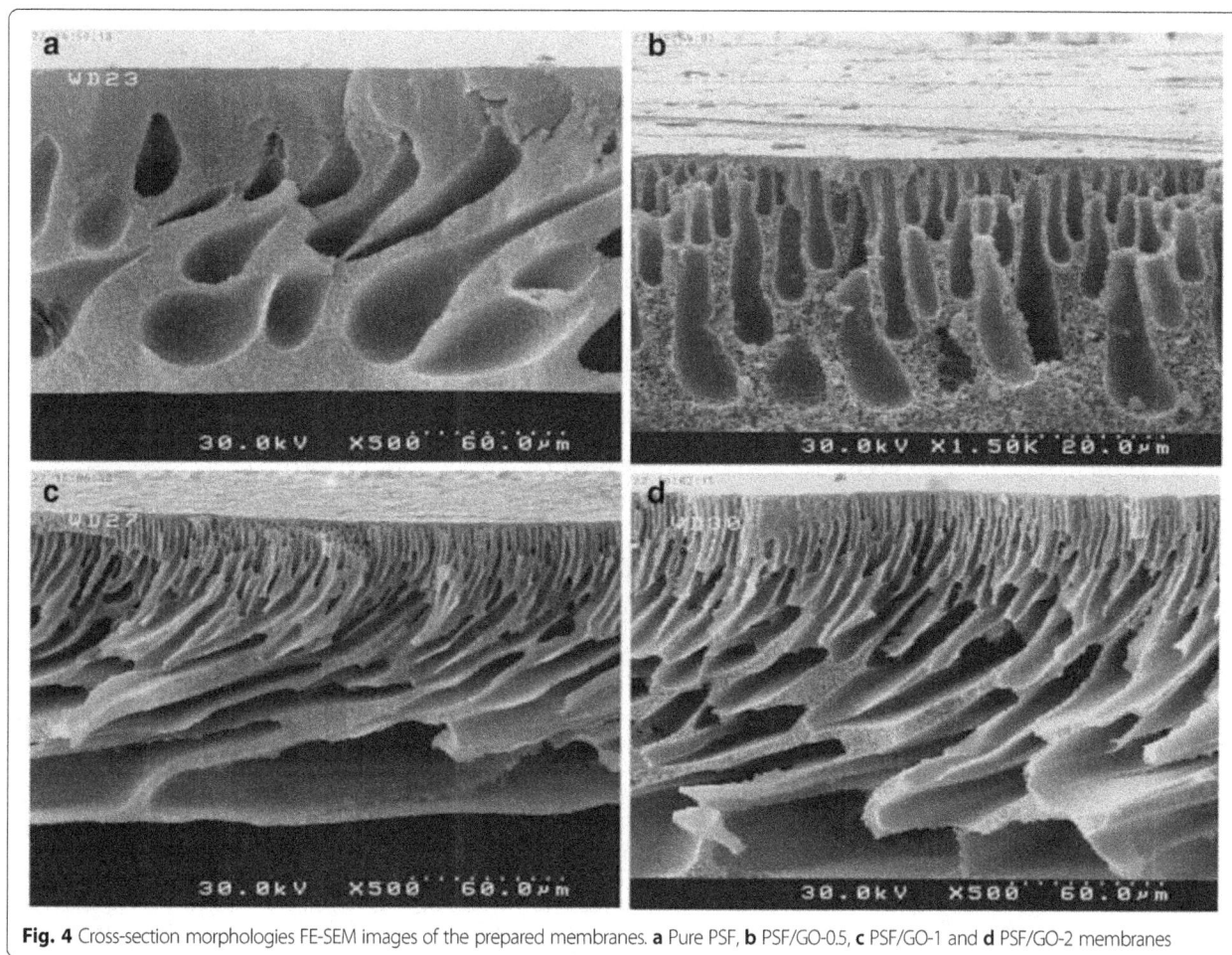

Fig. 4 Cross-section morphologies FE-SEM images of the prepared membranes. **a** Pure PSF, **b** PSF/GO-0.5, **c** PSF/GO-1 and **d** PSF/GO-2 membranes

number of their pores increased and walls thickness decreased compare to pristine PSF membrane (Fig. 4c-d). Generally, these structures have a low resistance to water permeation [39]. In addition, in the membranes with 1 and 2 wt% GO, horizontal channels appeared that can improve the water permeability. This issue is confirmed by other similar studies [14, 26]. The rate of pores production is the directly related to the exchange rate of solvent and non-solvent in the coagulation bath of phase inversion process. However, the faster the exchange rate of solvent and non-solvent in the coagulation process, the larger pores, more finger like pores and more channels. In contrast, the slower the exchange rate of solvent and non-solvent in the coagulation process, the smaller pores, more drop like pores and a spongy or non-void structure is resulted which finally alter the membrane permeability [40, 41]. By adding GO to the matrix of membrane the sub layer is effectively modified. This capability is attributed to GO hydrophilicity which results in thermodynamic instability in the casting solution, consequently rapid mass transfer between the solvent and nonsolvent is occurred. As a result, large pores are formed in the sub layer of

membrane [5]. In this study, to evaluate the surface morphology of the synthesized membranes, AFM was used. In Fig. 5, three-dimensional images of the four types of synthesized membranes are illustrated. As it is obvious, the bright areas exhibit the highest points and dark areas depict the valleys or pores of the fabricated membranes. It seems that the direction of the dents is pointed to direction of applied coagulation bath. In addition, Table 1 presents the different roughness parameters of the membranes. From the Table, surface roughness of pure PSF membrane is greater than the modified membrane with 0.5 and 1 wt% GO, but it is less than membrane with 2 wt% GO. Adding a certain amount of GO changes the large peaks and valleys of the membrane to a large number of small peaks and valleys [26]. Actually, in low loading of carbon modifiers such as carbon nanotube and grapheme oxide, due to low electrostatic interaction and good compatibility with the membrane matrix, these nanomaterials could develop a suitable structure in the membrane, reducing the membrane roughness and thus create a smooth surface [42]. Similar behavior has been reported in previous studies [11, 43].

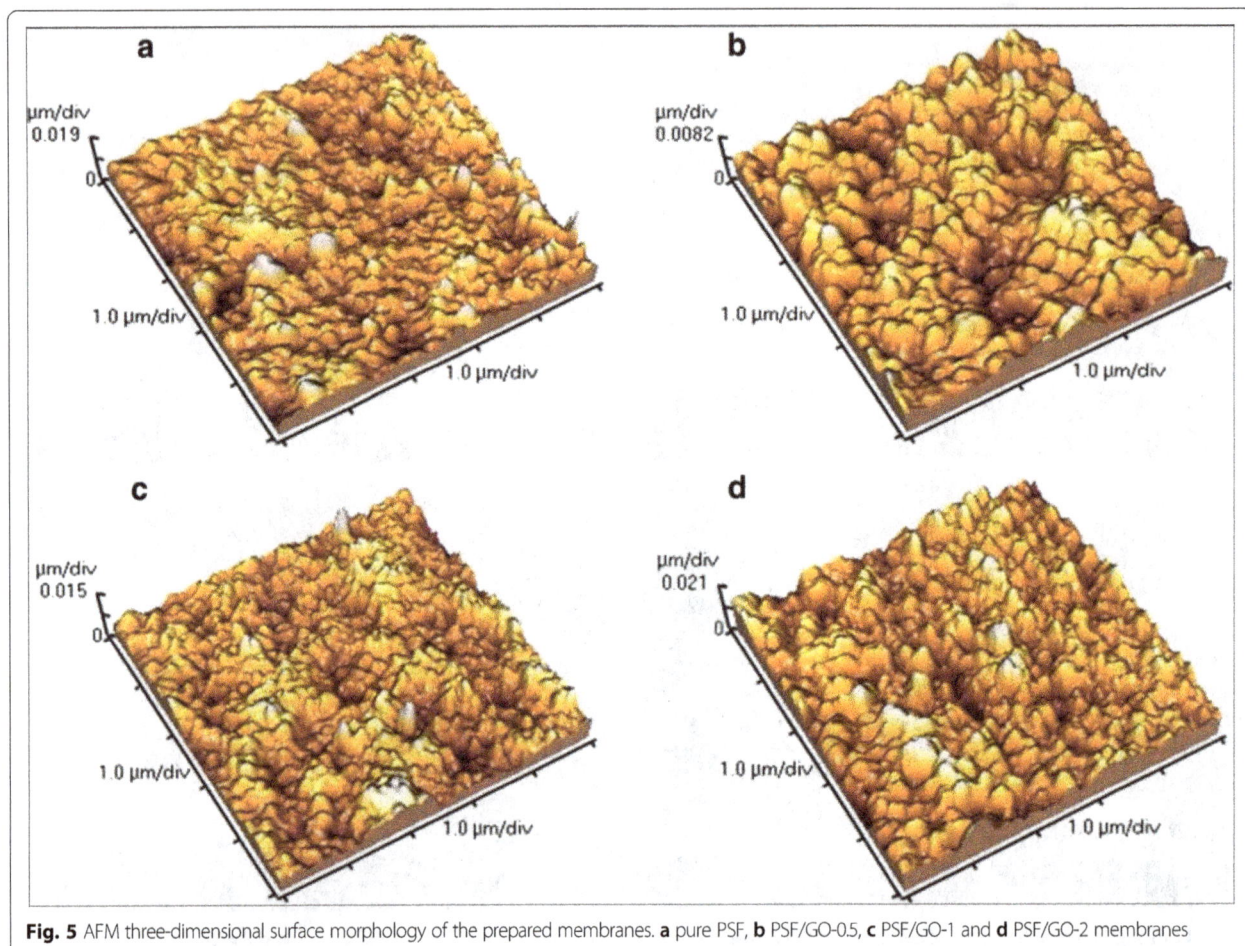

Fig. 5 AFM three-dimensional surface morphology of the prepared membranes. **a** pure PSF, **b** PSF/GO-0.5, **c** PSF/GO-1 and **d** PSF/GO-2 membranes

Membrane hydrophobicity, Water permeation flux and pore structure parameters

Zeta potential values for synthesized membranes are presented in Fig. 6. As shown, all given values are negative. With increasing the amount of GO to 1 wt%, negatively charge and zeta potential increased. The results of contact angle measurements, porosity, pore size and water flux are given in Table 2. As shown in the table with an increase of 1 % GO nanoparticles to the polymer matrix, water contact angle decreased, in contrast the porosity, pore size and water flux increased. Accordingly, net

Table 1 Surface roughness parameters of the prepared membranes obtained from analyzing six randomly chosen surface AFM images

Membranes	Roughness parameters	
	Mean surface roughness (Ra-nm)	Root mean square roughness (Rq-nm)
Pure PSF	2.9 ± 0.23	3.9 ± 0.47
PSF/GO-0.5	2 ± 0.14	2.5 ± 0.15
PSF/GO-1	2.5 ± 0.30	3.4 ± 0.36
PSF/GO-2	4.4 ± 0.32	5.8 ± 0.50

PSF membrane has the highest contact angle, lowest values of porosity, pore size and flux. Among the membranes, PSF/GO-1 has the lowest contact angle and maximum porosity, pore size and water flux. From the table, PSF/GO-2 has slightly higher contact angle and lower pure water flux compare to PSF/GO-1 membrane. Generally, zeta potential plays an important role in flux and anti-fouling properties of membranes [44]. The surface charge is an indication of presence of charged functional groups on the membrane surface. Inducing of hydroxylic and carboxylic functional groups can produce negative charges on the membrane surface [20]. During the phase inversion process hydrophilic functional groups in GO migrate to the surface resulting in negatively charged surface. Blended nano particles in the membrane casting solution migrate to the top of the membrane that is initially exposed to the non-solvent (water) liquid. Increase of hydrophilic groups density on the membrane surface results in decrease of intermediate energy (interface energy) with water. As a result, with increasing the surface hydrophilicity, contact angle decreased [12]. The hydrophilic nature of GO speed up the exchange process of solvent and non solvent in the phase inversion method which increases the

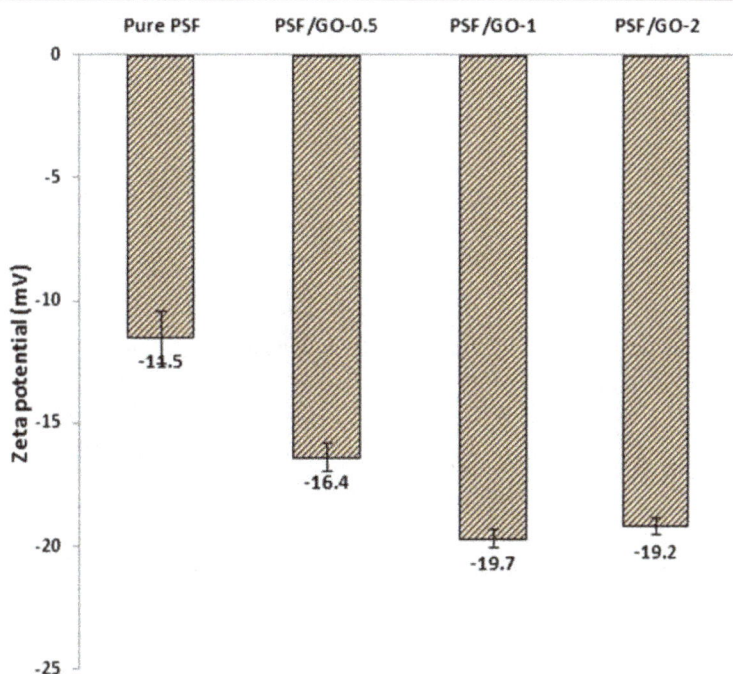

Fig. 6 Surface zeta potential of the prepared membranes with various GO contents

porosity and pore size of the membrane as clearly seen in FE-SEM micrographs (Fig. 4). These changes in the membrane properties enhance the membrane permeability [14, 45]. At GO contents of more than 1 wt%, (namely PSF/GO-2), the hydrophilicity of the membrane relatively reduced. This phenomenon may be due to accumulation and irregular positioning of GO nanoplates and decrease of the functional groups on the membrane surface. In addition, the reduction of water flux through membrane with GO content more than 1 wt%, is attributed to decrease of membrane porosity and pore size due to high viscosity of casting solution and delay of solvent and non-solvent exchange. In this situation, the pores are blocked by high concentrations of GO, resulting in flux reduction [11, 14]. The results of GO effects on the membrane characteristics are consistent with the similar works [12, 13, 21, 26].

Arsenate rejection performance evaluation

The results of arsenate rejection and membrane flux for different fabricated membranes are presented in Fig. 7. In the mentioned operating condition arsenate rejection

for the pure PSF membrane and the modified membrane with 0.5, 1 and 2 wt% GO were 25.87 %, 65.80 %, 82.30 %, and 83.65 % respectively. The rejection by modified membranes are substantially higher than that of the pure PSF membrane. Moreover, by increasing the weight of GO in the casting solution the arsenate ions rejection increased. The reasons for this increase are described below. Negative hydrophilic functional groups such as hydroxyl and carboxyl groups on surface of GO can build up a high zeta potential by inducing negative charges on surface of the membrane. Negative charge of arsenate and negative charge on the membrane surface result in increase of Donnan repulsion, resulting in an increase in the arsenate rejection [46]. This feature does not exist in pure PSF membrane. Fundamentally, the charge repulsion of ions depends on the membrane charge, ionic strength and ions capacity [11]. Lohokare et al. reported that dominant removal mechanism of arsenate was Donnan exclusion using a modified hydrophilic UF membrane [46]. Moreover some researchers have previously proposed that modified hydrophilic membranes due to strong bonds with water can

Table 2 Effect of GO content on water contact angle, pure water flux and pore structure parameters of the prepared membranes

Membranes	Contact angle (deg)	Porosity (%)	Mean pore radius (nm)	Pure water flux (L/m²h)
Pure PSF	73.5 ± 2.1	48.3 ± 2.6	6.9 ± 0.56	19.7 ± 3.2
PSF/GO-0.5	66.7 ± 1.6	77.9 ± 2.2	8.3 ± 0.31	32.3 ± 3.5
PSF/GO-1	51.3 ± 1.2	86.5 ± 1.8	9.1 ± 0.63	49.9 ± 2.6
PSF/GO-2	54.8 ± 1.4	82.1 ± 1.3	8.7 ± 0.42	46.4 ± 2.0

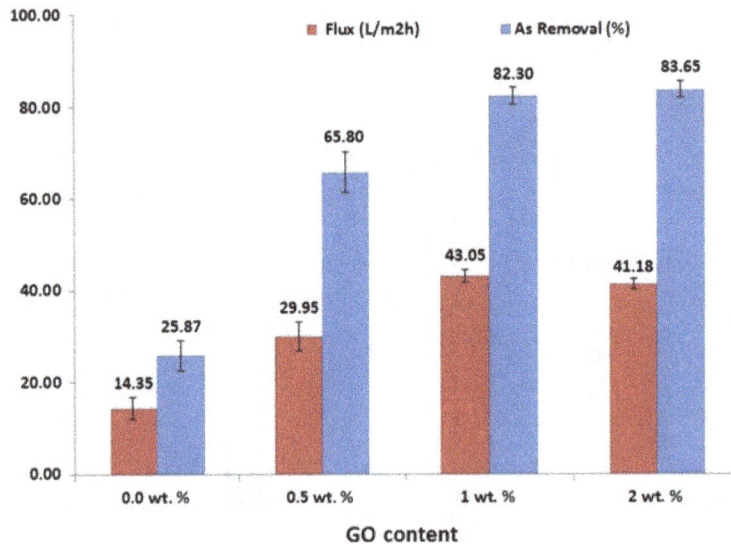

Fig. 7 As(V) rejection and flux of the prepared membranes with various GO contents. (Operating pressure =4 bar, pH = 8.5 ± 0.2, Initial As(V) concentration = 300 ± 10 µg/L, feed temperature =25 ± 0.5 °C)

effectively prevent the passage of molecules [27]. However, it has been expressed that carbon nano-materials, could absorb foulants by surface reactions and consequently increase the rejection rate [32]. From Fig. 7, with increase in GO to 1 wt%, the increase in arsenate rejection is obvious, which is justified by the previously mentioned reasons. However, with increase to more than 1 wt% GO (PSF/GO-2) the removal efficiency is not very significant compared to the PSF/GO-1 membrane. This could be because of high density of irregular GO on membrane structure, reducing the functional groups on the membrane surface, resulting in decrease in membrane hydrophilicity [26]. Consequently with reduction

of functional groups, negatively charged on membrane surface is reduced, thus the removal of arsenate does not increase proportion to loaded GO [11, 14]. In addition, a slight increase of arsenate rejection in PSF/GO-2 compare to PSF/GO-1 can be assigned to lower flux and pore size value. From Fig. 7, the results of flux of arsenate solution filtration for modified membranes showed an approximate 10 % reduction compared to pure water flux (Table 2), while the flux reduction is more in pure PSF membrane (about 27 %). This difference can be attributed to the nature of the hydrophilic and anti-fouling properties of the modified membranes. The rejection of arsenic by membrane could be affected by various

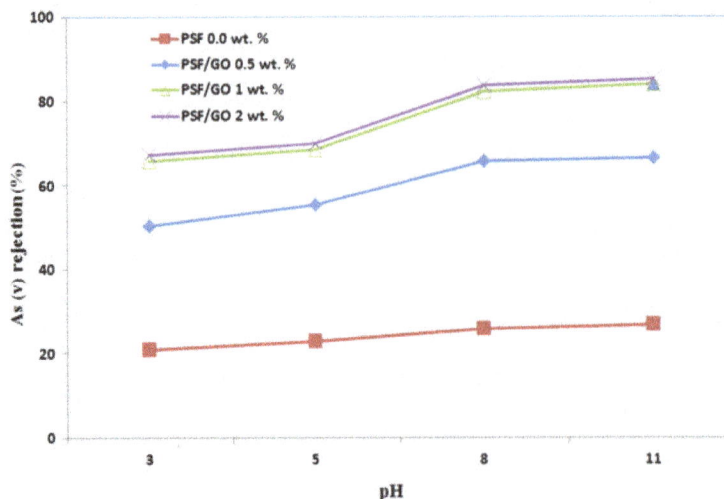

Figure 8 Percentage rejection of As (V) at different pH by prepared membranes with various GO contents. (Operating pressure =4 bar, Initial As (V) concentration = 300 ± 10 µg/L, feed temperature =25 ± 0.5 °C)

parameters such as operating pressure, initial concentration, pH, ionic strength [46]. However, one of the most influential parameters is solution pH, which plays a major role in the rejection of arsenic by membrane systems [21, 47]. The effect of pH on the rate of arsenate rejection by synthesized membranes has been presented in Fig. 8. With increasing pH, the rejection increased due to some reasons. First, by increasing the pH, zeta potential of membrane increases and membrane surface charge becomes more negative [46, 48]. Moreover, arsenic charge which is controlled by pH, becomes more negative with increasing the pH [49]. Second, changing the pH values will change the predominant species of arsenic in the environment. So that at pH <6.9 monovalent ions ($H_2AsO_4^-$) are dominant, while at pH > 6.9 divalent ions ($HAsO_4^{2-}$) are dominant. Therefore, with increasing pH, monovalent ions converted into divalent ions. Since the repulsive effect of Donnan is more dominant for divalent ions than monovalant ions, thus the rejection increased in higher pH values [47, 49, 50]. Accordingly Seidel et al. showed that the removal of arsenic by nanofiltration membranes was reduced from 85 % at pH = 8.5 to 8 % in pH = 4.5 [50]. Based on the good results of As (v) rejection obtained from the synthesized membrane, it seems that with determining the optimum operating parameters, the proposed standards for arsenic, especially in surface waters with the dominant species of arsenate, is achievable.

Conclusion

In present study, GO nanoplate were directly added to PSF casting solution to fabricate a mixed matrix membrane via phase inversion method. The results showed that presence of abundant containing hydrophilic functional groups on GO, strongly enhance the hydrophilicity and permeability of the synthesized membrane. Graphene oxide also could modify the morphology of the membrane so that the spongy structure and closed-end drop like pores of the pure PSF membrane could change to finger like pores and larger open-end channels in PSF/GO membrane. Adding GO up to 1 wt% in casting solution resulted in enhancement of membrane morphology so that the contact angle reduced and the porosity and pure water flux increased due to the improvement of the membrane surface hydrophilicity. The results also showed that the rejection of arsenic in the PSF/GO membranes has substantially increased compared to pure PSF membrane. In addition, with increase in GO weight in the casting solution the rejection of arsenate ions increased. The experiments also showed that the predominant mechanism of arsenate rejection is, Donnan repulsion due to the negative charges induced by GO on the membrane surface. The results of this study revealed that due to unique properties of GO especially hydrophilicity, it can be considered as a promising nanomaterial for membrane fabrication and modification.

Competing interests
The authors declare that they have no competing interests.

Authors' contributions
RR and SN has participated in all stages of the study (design of the study, conducting the experiment, analyzing of data and manuscript preparation). AHM and RN participated in the design of the study, final revised of manuscript and intellectual helping for analyzing of data. SAM and AMR carried out statistical and technical analysis of data and intellectual helping for analyzing of data. AJ carried out statistical and technical analysis of data, participated in design of study and manuscript preparation. SN performed data collection and carried out technical analysis. All authors read and approved the final manuscript.

Acknowledgements
This research was part of a PhD dissertation of the first author and has been financially supported by a grant (NO, 22716-46-02-92) from Center for Water Quality Research, Institute for Environmental Research, Tehran University of Medical Sciences, Tehran, Iran. The authors would like to express their thanks to the Department of Environmental Health Engineering, School of Public Health, Tehran University of Medical Sciences for their collaboration.

Author details
[1]Department of Environmental Health Engineering, School of Public Health, Tehran University of Medical Sciences, Tehran, Iran. [2]Center for Water Quality Research, Institute for Environmental Research, Tehran University of Medical Sciences, Tehran, Iran. [3]Center for Solid Waste Research, Institute for Environmental Research, Tehran University of Medical Sciences, Tehran, Iran. [4]National Institute of Health Research, Tehran University of Medical Sciences, Tehran, Iran. [5]Department of Chemical and Petroleum Engineering, Sharif University of Technology, Tehran, Iran. [6]Nanotechnology Research Center, Research Institute of Petroleum Industry (RIPI), Tehran, Iran. [7]Department of Environmental Health Engineering, School of Public Health, Lorestan University of Medical Sciences, Khorramabad, Iran.

References
1. Bhatnagar A, Sillanpää M. A review of emerging adsorbents for nitrate removal from water. Chem Eng J. 2011;168:493–504.
2. Maleki A, Amini H, Nazmara S, Zandi S, Mahvi AH. Spatial distribution of heavy metals in soil, water, and vegetables of farms in Sanandaj, Kurdistan, Iran. J Environ Health Sci Eng. 2014;29:33.
3. Jain CK, Singh RD. Technological options for the removal of arsenic with special reference to South East Asia. J Environ Manage. 2012;107:1–18.
4. Saitua H, Gil R, Padilla AP. Experimental investigation on arsenic removal with a nanofiltration pilot plant from naturally contaminated groundwater. Desalination. 2011;274:1–6.
5. Addo Ntim S, Mitra S. Adsorption of arsenic on multiwall carbon nanotube–zirconia nanohybrid for potential drinking water purification. J Colloid Interface Sci. 2012;375:154–9.
6. Ebrahimi R, Maleki A, Shahmoradi B, Daraei H, Mahvi AH, Barati AH, et al. Elimination of arsenic contamination from water using chemically modified wheat straw. Desalination Water Treat. 2013;51:2306–16.
7. Dutta T, Bhattacherjee C, Bhattacherjee S. Removal Of Arsenic Using Membrane Technology–A Review. International Journal of Engineering Research and Technology. 2012;1:1–23.
8. Jafari A, Mahvi AH, Nasseri S, Rashidi A, Nabizadeh R, Rezaee R, et al. Ultrafiltration of natural organic matter from water by vertically aligned carbon nanotube membrane. J Environ Health Sci Eng. 2015;13:51.
9. Shi X, Tal G, Hankins NP, Gitis V. Fouling and cleaning of ultrafiltration membranes: a review. J Water Process Eng. 2014;1:121–38.
10. Zazouli M, Nasseri S, Mahvi A, Gholami M, Mesdaghinia A, Younesian M. Retention of humic acid from water by nanofiltration membrane and influence of solution chemistry on membrane performance. Iran J Environ Health Sci Eng. 2008;5:11–8.
11. Vatanpour V, Madaeni SS, Moradian R, Zinadini S, Astinchap B. Fabrication and characterization of novel antifouling nanofiltration membrane prepared from oxidized multiwalled carbon nanotube/polyethersulfone nanocomposite. J Membr Sci. 2011;375:284–94.

12. Zhao C, Xu X, Chen J, Yang F. Effect of graphene oxide concentration on the morphologies and antifouling properties of PVDF ultrafiltration membranes. J Environ Chem Eng. 2013;1:349–54.

13. Lee J, Chae H-R, Won YJ, Lee K, Lee C-H, Lee HH, et al. Graphene oxide nanoplatelets composite membrane with hydrophilic and antifouling properties for wastewater treatment. J Membr Sci. 2013;448:223–30.

14. Wang Z, Yu H, Xia J, Zhang F, Li F, Xia Y, et al. Novel GO-blended PVDF ultrafiltration membranes. Desalination. 2012;299:50–4.

15. Rajaeian B, Heitz A, Tade MO, Liu S. Improved separation and antifouling performance of PVA thin film nanocomposite membranes incorporated with carboxylated TiO 2 nanoparticles. J Membr Sci. 2015;485:48–59.

16. Liang S, Xiao K, Mo Y, Huang X. A novel ZnO nanoparticle blended polyvinylidene fluoride membrane for anti-irreversible fouling. J Membr Sci. 2012;394:184–92.

17. Hassanajili S, Khademi M, Keshavarz P. Influence of various types of silica nanoparticles on permeation properties of polyurethane/silica mixed matrix membranes. J Membr Sci. 2014;453:369–83.

18. Zhao C, Xu X, Chen J, Yang F. Optimization of preparation conditions of poly(vinylidene fluoride)/graphene oxide microfiltration membranes by the Taguchi experimental design. Desalination. 2014;334:17–22.

19. Crock CA, Rogensues AR, Shan W, Tarabara W. Polymer nanocomposites with graphene-based hierarchical fillers as materials for multifunctional water treatment membranes. Water Res. 2013;47:3984–96.

20. Yin J, Deng B. Polymer-matrix nanocomposite membranes for water treatment. J Membr Sci. 2015;479:256–75.

21. Ganesh BM, Isloor AM, Ismail AF. Enhanced hydrophilicity and salt rejection study of graphene oxide-polysulfone mixed matrix membrane. Desalination. 2013;313:199–207.

22. Yari M, Norouzi M, Mahvi AH, Rajabi M, Yari A, Moradi O, et al. Removal of Pb (II) ion from aqueous solution by graphene oxide and functionalized graphene oxide-thiol: effect of cysteamine concentration on the bonding constant. Desalination Water Treat. 2015; In press, doi: 10.1080/19443994.2015.1043953.

23. Hegab HM, Zou L. Graphene oxide-assisted membranes: fabrication and potential applications in desalination and water purification. J Membr Sci. 2015;484:95–106.

24. Hu K, Kulkarni DD, Choi I, Tsukruk VV. Graphene-polymer nanocomposites for structural and functional applications. Prog Polym Sci. 2014;39:1934–72.

25. Shah P, Murthy CN. Studies on the porosity control of MWCNT/polysulfone composite membrane and its effect on metal removal. J Membr Sci. 2013;437:90–8.

26. Zinadini S, Zinatizadeh AA, Rahimi M, Vatanpour V, Zangeneh H. Preparation of a novel antifouling mixed matrix PES membrane by embedding graphene oxide nanoplates. J Membr Sci. 2014;453:292–301.

27. Xia S, Ni M. Preparation of poly (vinylidene fluoride) membranes with graphene oxide addition for natural organic matter removal. J Membr Sci. 2014;473:54-62.

28. Hummers Jr WS, Offeman RE. Preparation of graphitic oxide. J Am Chem Soc. 1958;80:1339–9.

29. Blanco J-F, Sublet J, Nguyen QT, Schaetzel P. Formation and morphology studies of different polysulfones-based membranes made by wet phase inversion process. J Membr Sci. 2006;283:27–37.

30. Dong C, He G, Li H, Zhao R, Han Y, Deng Y. Antifouling enhancement of poly(vinylidene fluoride) microfiltration membrane by adding Mg(OH)2 nanoparticles. J Membr Sci. 2012;387–388:40–7.

31. Wang Y, Ou R, Ge Q, Wang H, Xu T. Preparation of polyethersulfone/carbon nanotube substrate for high-performance forward osmosis membrane. Desalination. 2013;330:70–8.

32. Xu Z, Zhang J, Shan M, Li Y, Li B, Niu J, et al. Organosilane-functionalized graphene oxide for enhanced antifouling and mechanical properties of polyvinylidene fluoride ultrafiltration membranes. J Membr Sci. 2014;458:1–13.

33. Zhao S, Wang Z, Wang J, Wang S. The effect of pH of coagulation bath on tailoring the morphology and separation performance of polysulfone/polyaniline ultrafiltration membrane. J Membr Sci. 2014;469:316–25.

34. APHA, AWWA, WEF. Standard methods for the examination of water and wastewater. 21st ed. Washington DC: APHA, AWWA, WEF; 2005.

35. Wu J, Tang Q, Sun H, Lin J, Ao H, Huang M, et al. Conducting film from graphite oxide nanoplatelets and poly (acrylic acid) by layer-by-layer self-assembly. Langmuir. 2008;24:4800–5.

36. Zhu Y, Murali S, Cai W, Li X, Suk JW, Potts JR, et al. Graphene and graphene oxide: synthesis, properties, and applications. Adv Mater. 2010;22:3906–24.

37. Yang D, Velamakanni A, Bozoklu G, Park S, Stoller M, Piner RD, et al. Chemical analysis of graphene oxide films after heat and chemical treatments by X-ray photoelectron and Micro-Raman spectroscopy. Carbon. 2009;47:145–52.

38. Lin Y, Jin J, Song M. Preparation and characterisation of covalent polymer functionalized graphene oxide. J Mater Chem. 2011;21:3455–61.

39. Lin D-J, Chang C-L, Huang F-M, Cheng L-P. Effect of salt additive on the formation of microporous poly(vinylidene fluoride) membranes by phase inversion from LiClO4/Water/DMF/PVDF system. Polymer. 2003;44:413–22.

40. Li YQ, Xi DL, Fan SL. Preparation and characterization of novel hollow fiber membrane with multicomponent polymeric materials. Adv Mater Res. 2012;534:8–12.

41. Kuilla T, Bhadra S, Yao D, Kim NH, Bose S, Lee JH. Recent advances in graphene based polymer composites. Prog Polym Sci. 2010;35:1350–75.

42. Qiu S, Wu L, Pan X, Zhang L, Chen H, Gao C. Preparation and properties of functionalized carbon nanotube/PSF blend ultrafiltration membranes. J Membr Sci. 2009;342:165–72.

43. Zhao H, Wu L, Zhou Z, Zhang L, Chen H. Improving the antifouling property of polysulfone ultrafiltration membrane by incorporation of isocyanate-treated graphene oxide. Phys Chem Chem Phys. 2013;15:9084–92.

44. Zazouli MA, Nasseri S, Ulbricht M. Fouling effects of humic and alginic acids in nanofiltration and influence of solution composition. Desalination. 2010;250:688–92.

45. Sun M, Su Y, Mu C, Jiang Z. Improved antifouling property of PES ultrafiltration membranes using additive of silica – PVP nanocomposite. Ind Eng Chem Res. 2009;49:790–6.

46. Lohokare H, Muthu M, Agarwal G, Kharul U. Effective arsenic removal using polyacrylonitrile-based ultrafiltration (UF) membrane. J Membr Sci. 2008;320:159–66.

47. Fang J, Deng B. Rejection and modeling of arsenate by nanofiltration: contributions of convection, zdiffusion and electromigration to arsenic transport. J Membr Sci. 2014;453:42–51.

48. Zhao G, Li J, Ren X, Chen C, Wang X. Few-layered graphene oxide nanosheets as superior sorbents for heavy metal ion pollution management. Environ Sci Technol. 2011;45:10454–62.

49. Akbari H, Mehrabadi AR, Torabian A. Determination of nanofiltration efficency in arsenic removal from drinking water. Iran J Environ Health Sci Eng. 2010;7:273–8.

50. Seidel A, Waypa JJ, Elimelech M. Role of charge (Donnan) exclusion in removal of arsenic from water by a negatively charged porous nanofiltration membrane. Environ Eng Sci. 2001;18:105–13.

8

Sonoelectrochemical mineralization of perfluorooctanoic acid using Ti/PbO_2 anode assessed by response surface methodology

Gholamreza Bonyadinejad[1,2], Mohsen Khosravi[3], Afshin Ebrahimi[1*], Roya Nateghi[1], Seyed Mahmood Taghavi-Shahri[4] and Hamed Mohammadi[1,2]

Abstract

Background: Perfluorocarboxylic acids (PFCAs) are emerging pollutant and classified as fully fluorinated hydrocarbons containing a carboxylic group. PFCAs show intensively resistance against chemical and biological degradation due to their strong C–F bond. The Sonoelectrochemical mineralization of the synthetic aqueous solution of the perfluorooctanoic acid (PFOA) on Ti/PbO_2 anode was investigated using the response surface methodology based on a central composite design with three variables: current density, pH, and supporting electrolyte concentration.

Methods: The defluorination ratio of PFOA was determined as an indicator of PFOA mineralization. Fluoride ion concentration was measured with an ion chromatograph unit. The Ti/PbO2 electrode was prepared using the electrochemical deposition method. The ultrasonic frequency was 20 kHz.

Results: The optimum conditions for PFOA mineralization in synthetic solution were electrolyte concentration, pH, and current density of 94 mM, 2, and 83.64 mA/cm^2, respectively. The results indicated that the most effective factor for PFOA mineralization was current density. Furthermore, the PFOA defluorination efficiency significantly enhanced with increasing current density. Under optimum conditions, the maximum mineralization of PFOA was 95.48 % after 90 min of sonoelectrolysis.

Conclusions: Sonoelectrolysis was found to be a more effective technique for mineralization of an environmentally persistent compound.

Keywords: Ultrasonics, Lead dioxide, Perfluorooctanoic acid, Central composite design

Introduction

The Perfluorocarboxylic acids are classified as fully fluorinated hydrocarbons materials that contain a carboxylic group. Due to their high surface active nature, high thermal and chemical stabilities, PFCAs are extensively using in fire retardants, industrial surfactants and in fluoropolymers manufacturing processes [1]. This extensive utilization of PFCAs results in releasing of these compounds into the environment, which was estimated at 3200–7300 tons for the period of 1950–2004 by direct and indirect emissions [2]. PFCAs show intensive resistance against chemical and biological degradation. Conventional treatment methods are ineffectual for the degradation of PFCAs because it is intrinsically recalcitrant to biological and conventional chemical treatment [3]. The marvelous stability of perfluorinated compounds (PFCs) is ascribed to their strong C–F bond which makes them very persistent to most natural conditions [1, 4]. Perfluorooctanoic acid (PFOA) is one of the PFCAs' family, which is categorized as a likely potential carcinogen by The US EPA's Science Advisory Board in 2006. Recently, PFOA and its precursors have been universally detected in wildlife, water and human body [5]. For instance, the river Elbe in Germany was polluted by PFOA due to effluents discharged from wastewater treatment plant [6]. Since, PFOA has

* Correspondence: a_ebrahimi@hlth.mui.ac.ir
[1]Environment Research Center and Department of Environmental Health Engineering, School of Health, Isfahan University of Medical Sciences (IUMS), Isfahan 81676-36954, Iran
Full list of author information is available at the end of the article

toxicity to organisms and humans and persistence in the environment, a lot attention has been paid to treatment of PFOA from water which uses effective methods under moderate conditions. Lately, some chemical technologies for PFOA degradation have been reported, such as photocatalytic oxidation, direct photolysis, photochemical reduction, photochemical oxidation, thermally-induced reduction and sonochemical pyrolysis. Yet, most of the techniques could not effectively decompose PFOA [7]. The goal of destructive methods is cleaving the C–F bonds to form F-ions [2]. The electrochemical oxidation is a technology that has presented its capacity to degrade refractory organic pollutants such as emerging contaminants contained in the secondary effluents of wastewater treatment plants and also removal of various pollutants from aqueous environments [8, 9]. Electrochemical techniques generally carried out oxidatively, have the advantage of contaminant elimination without the addition of chemicals. Although electrochemical methods are easy to use and have high removal efficiency, energy consumption is the main disadvantage of these methods. Nevertheless, this disadvantage has been suppressed by the development of new anode materials [10, 11]. PbO$_2$ is a low-cost electrode material that can be quickly and easily prepared, and is being used by many researchers for electro-oxidation of refractory pollutants in aquatic solutions [12–15]. During the process of oxidation of polluted waters, OH$^{\bullet}$ specimen is generated on the PbO$_2$ electrode surface, and a mechanism for the electrochemical processes occurring in the gel-crystal structure of the PbO$_2$ layer of the electrode has been proposed [16]. The Pb*O(OH)$_2$ active centers placed in the hydrated PbO$_2$ layer on the surface of the crystalline PbO$_2$ anode provide electrons to the crystal district, becoming positively charged (Pb*O(OH)$^{+}$ (OH)$^{\bullet}$). This electric charge is neutralized according to the following reaction:

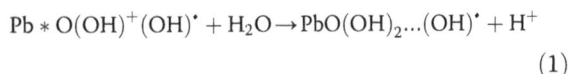

$$\text{Pb} * \text{O(OH)}^{+}(\text{OH})^{\bullet} + \text{H}_2\text{O} \rightarrow \text{PbO(OH)}_2...(\text{OH})^{\bullet} + \text{H}^{+}$$

$$(1)$$

in which hydroxyl radicals are produced in the active centers. The OH$\bullet$ can depart from the active centers and react with the pollutant in the aqueous solution. Thus, the PbO$_2$ anode is expected to perform quite well in organic pollutant mineralization. However, the main problem of PbO$_2$ anode is the release of poisonous ions, Pb^{2+} [17]. On the other hand, sonochemical treatment applies pyrolytic cleavages for organic pollutants degradation and is an emerging and impressive method which can be used efficiently to eliminate PFOA particularly [18]. Ultrasonic (US) treatment acts as cavitation, which not only generates plasma in water, and degrading molecules by pyrolysis, but also produces free radicals and other reactive types that can improve the amount of collisions between free radicals and contaminants. Recently, using

the combination of ultrasonic method and other techniques for the treatment of organic wastewater has been largely studied. (e.g. phenol and pharmaceutical compounds) [19–21]. Optimization of the operating conditions of an experimental system and recognition of the way in which the experimental parameters affect the final output of the system are realized by using modeling techniques [22]. In addition, it is possible to determine the relationships and interactions between the variables through these techniques. In this regard, statistical methodologies, such as the response surface methodology (RSM), are suitable for studying and modeling a particular system [23]. The aim of the present study was to investigate the electrochemical mineralization of PFOA as an emerging contaminant using Ti/PbO$_2$ anode coupled with ultrasonic irradiation (sonoelectrochemical) and determine the optimum conditions by RSM. The effects of current density (CD), pH of the solution, and supporting electrolyte (EL) concentration were evaluated in terms of PFOA defluorination.

Materials and methods
Chemicals
Analytical-grade PFOA was purchased from Sigma Aldrich co., and used without further purification. Pb(NO$_3$)$_2$ (Sigma Aldrich co.), Triton X-100 (Merck co.), and CuSO$_4$.5H$_2$O (Merck co.) were used for electrode preparation. Other chemicals were purchased from Merck co. The initial pH of the solutions was adjusted using sodium hydroxide and sulfuric acid. Sodium sulfate was used as the supporting electrolyte. All the solutions were prepared using deionized water.

Preparation of Ti/PbO$_2$ electrode
The Ti substrate with 2 mm thickness was cut into a strip (4.8 cm × 4 cm, 99.7 % Aldrich) and pre-treated according to the following procedure: the substrate was polished on 320-grit paper strips [24] to eliminate the superficial layer of TiO$_2$ (an electric semiconductor) and increase surface roughness (for efficient adherence of PbO$_2$). Then, the substrate was degreased in an ultrasonic bath of acetone for 10 min and then in distilled water for 10 min. Afterwards, the substrate was etched for 1 h in a boiling solution of oxalic acid (10 %) and rinsed with ultrapure water [25]. Finally, the cleaned Ti substrate was transferred to an electrochemical deposition cell, which contained 12 % (w/v) Pb(NO$_3$)$_2$ solution comprising 5 % (w/v) CuSO$_4$.5H$_2$O and 3 % (w/v) surfactant (Triton X-100). The role of the surfactant was to minimize the surface tension of the solution for better wetting of the substrate and also to increase the adhesion of PbO$_2$ to the Ti substrate. The electrodeposition of PbO$_2$ was performed at a constant anodic current of 20 mA/cm^2 for 60 min at 80 °C with continuous stirring [26]. The X-ray diffraction (XRD) tests were performed

using a Bruker, D8 Advance, Germany. The samples were scanned under Co Kα radiation (wavelength: 1.7890 Å) at 40 kV and 40 mA. Scanning electron microscope (SEM; Philips XI30, Netherlands) was employed to observe the surface morphology of the electrodes, which presented a typical pyramid shape similar to that reported in the literature [27].

Sonoelectrochemical mineralization of PFOA

Synthetic wastewater was prepared by dissolving PFOA in distilled water at a concentration of 50 mg/L. The sonoelectrochemical mineralization of PFOA was performed in a temperature-controlled water batch reactor (0.45 L) equipped with an ultrasonic probe (Bandelin SONOPULS,UW 3200, TT 13, Germany), a 41.12 cm^2 Ti/PbO$_2$ as the anode and a 80.32 cm^2 stainless steel plate as the cathode in conjunction with an adjustable power supply unit (HANI, Iran). The gap between the anode and cathode was 1 cm. The temperature of the reaction solution was kept constant at 25 ± 1 °C. The duration of all the electrolysis experiments was 90 min. The reactor was placed on a magnetic stirrer to mix its content during the experiment in order to maximize mass transport (Fig. 1).

Analytical procedure

Defluorination ratio of PFOA was determined based on the release of fluoride into the solutions. The defluorination ratio, is the indicator of PFOA mineralization [28]. Fluoride ions concentration were measured with an Ion chromatograph system (IC762, Metrohm, USA) equipped with an automatic sample injector, a degasser, a pump, a guard column (Metrosep RP guard, Metrohm), a separation column (Metrosep Anion Dual 2, Metrohm), and a conductivity detector with a suppressor device. The mobile phase was an aqueous solution containing NaHCO$_3$ (2 mM) and Na$_2$CO$_3$ (1.3 mM). The flow rate was 0.8 mL/min. The defluorination ratio (R) was calculated as Eq. 2:

$$R = \frac{C_{F^-}}{C_0 \times 15} \times 100 \tag{2}$$

where C_{F^-} is the concentration of fluoride in mM, C_0 is the initial concentration of PFOA in mM, and the value of 15 represents the number of fluorine atoms contained in one PFOA molecule.

Experimental design

The central composite design (CCD) technique coupled with RSM has been opted for modeling and design of experimental tests [29]. The CD, EL, and pH parameters were selected as input variables. The rotatable experimental plan was performed with the three variables at five levels (−1.68, −1, 0, 1, 1.68). Table 1 shows the values and levels of the variables. Five replications were done at the center point of the design to evaluate the pure error and consequently the lack of fit. Statistica ver. 10, and R ver. 3.1.2 software were used to design and analyze the experiments. Table 2 shows the CCD matrix of the mineralization experiments.

The relationship between response Y and the three independent variables X_1, X_2, and X_3 could be approximated by quadratic polynomial Eq. 3 as follows:

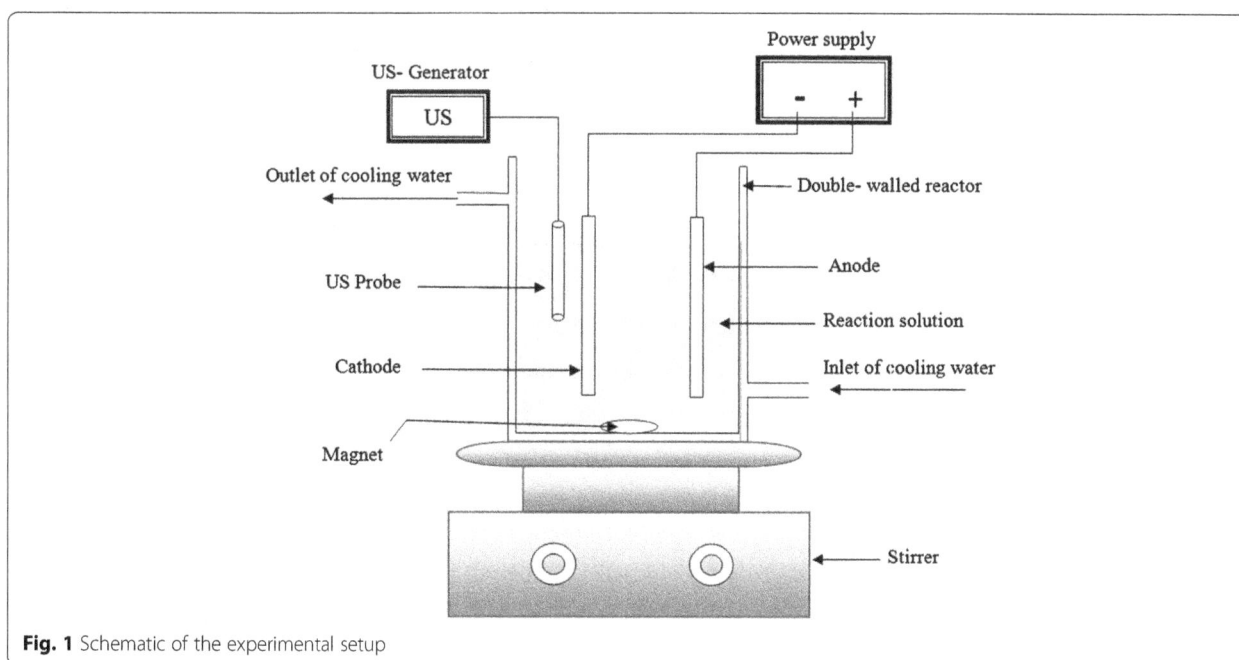

Fig. 1 Schematic of the experimental setup

Table 1 The range and codification of the independent variables (X_i) used in the experimental design

Variables	Actual values of the coded values				
	−1.68	−1	0	1	1.68
pH (X_1)	1.95	4	7	10	12.05
EL (mM) (X_2)	32.96	50	75	100	117.04
CD (mA/cm^2) (X_3)	16.36	30	50	70	83.64

$$Y = b_0 + b_1X_1 + b_2X_2 + b_3X_3 + b_{11}X_1{}^2 + b_{22}X_2{}^2 + b_{33}X_3{}^2 + b_{12}X_1X_2 + b_{13}X_1X_3 + b_{23}X_2X_3 \tag{3}$$

where Y is the predicted response; b_0 is a constant; b_1, b_2, and b_3 are the linear coefficients, b_{12}, b_{13}, and b_{23} are the cross-product coefficients; and b_{11}, b_{22}, and b_{33} are the quadratic coefficients. In the present study, backward variable selection was used for multiple regression modeling [30]. The assumption of final regression model was verified using the Anderson–Darling test for normality of residuals [31], Breusch–Pagan test for constant variance of residuals [32], and Durbin–Watson test for independence of residuals [33, 34]. Lack of fit test was performed to assess the fit of the final model. Validation of the final model was established using predicted R-squares (R^2), which estimates the prediction power of the model with new observations based on the leave-one-out technique [35]. The optimum values of the final model were calculated using numerical methods. In this regard, the experimental range predictors were divided into a grid and then the final model was calculated for all possible combinations of predictors in the grid.

Result and discussion
Characterization of the Ti/PbO$_2$ electrode
XRD pattern of the Ti/PbO$_2$ electrode has been represented in Fig. 2, it can be observed that PbO$_2$ was deposited in the form of two known polymorphs, namely, orthorhombic α-PbO$_2$ and tetragonal β-PbO$_2$, which occur naturally as scrutinyite and plattnerite, respectively. Figure 3 shows the SEM image of the surface microstructure of the Ti/PbO$_2$ electrode. It can be observed that the PbO$_2$ layer is crack free and composed of packed faceted micro crystallites. Such morphology verifies that only PbO$_2$ is involved in the electrochemical mineralization of the PFOA and protects the surface of the Ti substrate. Furthermore, energy-dispersive X-ray spectroscopy (EDS) analysis (data not shown) confirmed the presence of lead and oxygen atoms on the surface of the Ti/PbO$_2$ electrode.

CCD analysis and modeling
The interaction effect of input variables, which were statistically designed by using CCD method were studied through different combination of experimental parameters. The CCD design matrix, the predicted results and observed defluorination ratio values for the sonoelectrochemical mineralization of PFOA are shown in Table 2. Equation 4 represents the first model that was developed with all linear, quadratic, and two-way interaction of predictors:

Table 2 CCD matrix of sonoelectrochemical mineralization of PFOA

Exp. No.	pH	EL	CD	Defluorination ratio (%)	
				Observed	Predicted
1	4.00	50.00	30.00	54.16	41.90
2	4.00	50.00	70.00	87.45	81.34
3	4.00	100.00	30.00	57.47	44.16
4	4.00	100.00	70.00	89.49	83.61
5	10.00	50.00	30.00	51.89	38.57
6	10.00	50.00	70.00	85.55	78.03
7	10.00	100.00	30.00	51.46	40.84
8	10.00	100.00	70.00	87.46	80.29
9	1.95	75.00	50.00	75.71	67.15
10	12.05	75.00	50.00	69.6	61.57
11	7.00	32.96	50.00	69.12	61.36
12	7.00	117.04	50.00	74.26	65.18
13	7.00	75.00	16.36	37.73	23.20
14	7.00	75.00	83.64	91.86	89.55
15	7.00	75.00	50.00	74.32	65.36
16	7.00	75.00	50.00	75.51	65.36
17	7.00	75.00	50.00	74.66	65.36
18	7.00	75.00	50.00	73.82	65.36
19	7.00	75.00	50.00	74.32	65.36

$$Y = 6.04096 - 0.045 \times pH + 0.267 \times EL + 1.6625 \times CD - 0.00565 \times pH \times EL + 0.01108 \times pH \times CD + 0.000106 \times EL \times CD - 0.04639 \times pH^2 - 0.001247 \times EL^2 - 0.008789 \times CD^2 \tag{4}$$

The predicted R^2 of this initial model was 98.0 %. Backward elimination method was used to achieve a parsimonious model with significant predictors. In the first step, linear form of pH was removed from the initial model ($p = 0.88$). In three following steps, interactions of EL*CD ($p = 0.73$), pH*EL ($p = 0.17$), and pH*CD ($p = 0.33$) were removed from the model. Finally, linear and quadratic form of EL and CD, with quadratic form of pH, EL, and CD remain significant in the prediction model. The final model was as Eq. 5:

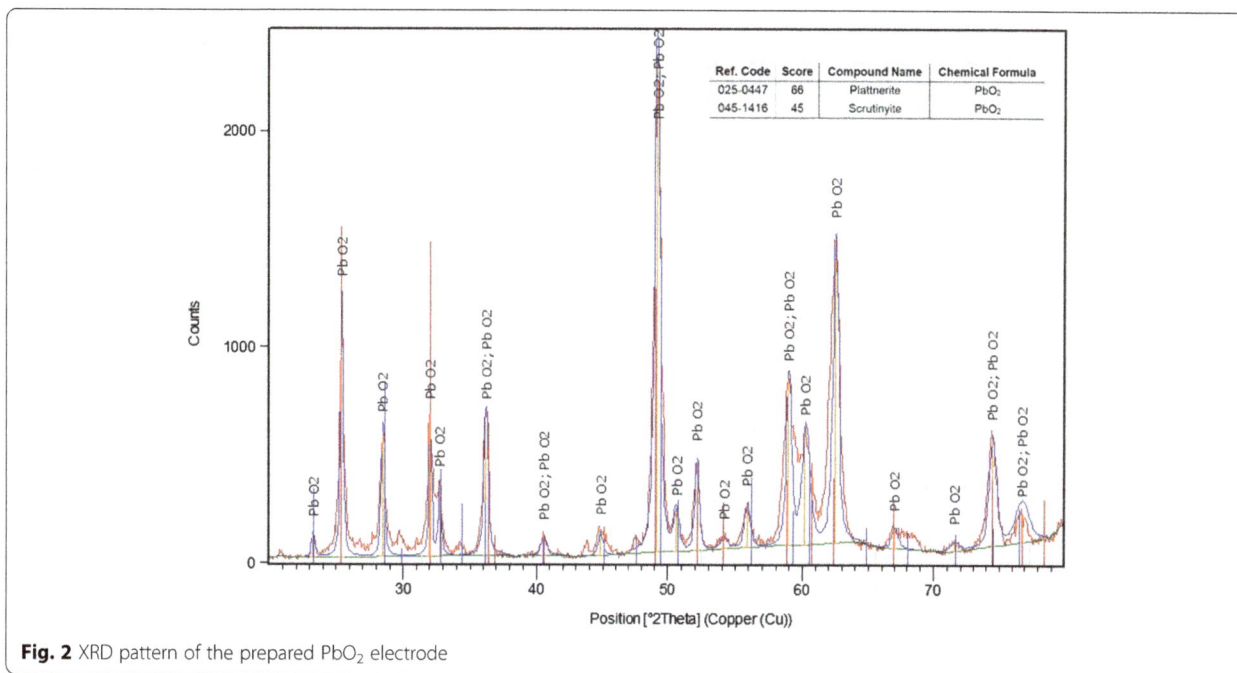

Fig. 2 XRD pattern of the prepared PbO$_2$ electrode

$$Y = 5.069 + 0.22311 \times EL + 1.62182 \\ \times CD - 0.039453 \times pH^2 - 0.0011852 \\ \times EL^2 - 0.0079442 \times CD^2 \quad (5)$$

A partial F test was performed for comparison between first model and the final model. The value of F was 0.86 with 4 and 9° of freedom, and a P-value of 0.52. Hence, the difference of these two models was not significant, although the final regression model (Eq. 5) had four predictors less than the initial model (Eq. 4). Table 3, represent the final regression model for PFOA mineralization. The P-value of Anderson-Darling, Breusch-Pagan, and Durbin-Watson tests were 0.34, 0.82 and 0.75, respectively, which confirmed the assumptions of regression model. Lack of fit test also was non-significant with P-value of 0.09 which means the final model fitted experimental data satisfactorily. (F = 4.08, degree of freedom for lack of fit = 9, degree of freedom for pure error = 4) (Fig. 4).

The final model was validated using leave-one-out technique. Predicted R-squared was 98.8 % which confirmed

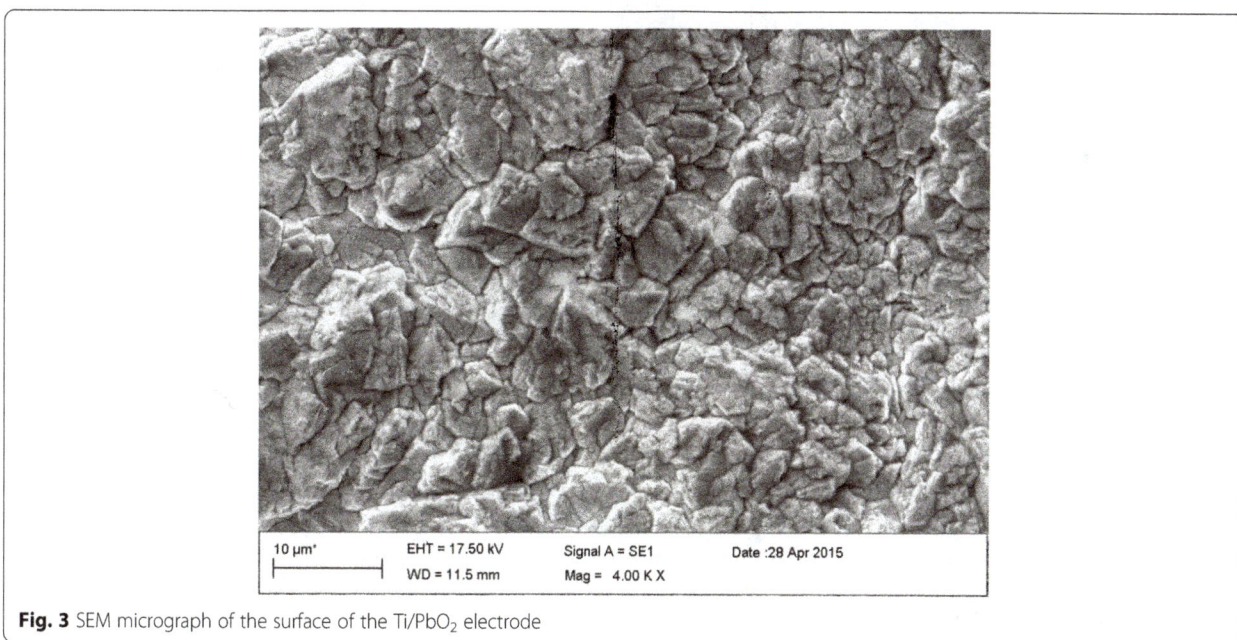

Fig. 3 SEM micrograph of the surface of the Ti/PbO$_2$ electrode

Table 3 Final regression model for PFOA mineralization

Variable	Coefficients	Std. Error	T statistic	P-value
(Constant)	5.07	3.36	1.51	0.155
EL	2.23×10^{-1}	7.23×10^{-2}	3.09	0.009
CD	1.62	7.57×10^{-2}	21.42	<0.001
pH^2	-3.95×10^{-2}	6.98×10^{-3}	-5.65	<0.001
EL^2	-1.18×10^{-3}	4.75×10^{-4}	-2.49	0.027
CD^2	-7.94×10^{-3}	7.42×10^{-4}	-10.71	<0.001

R-squared = 99.6 %, Adjusted R-squared = 99.4 %, Predicted R-squared = 98.8 %

external validity of the final model. Leave-one-out prediction of the final model vs. observed values of response represented in Fig. 5.

The final model was evaluated through the range of experimental predictors with a numerical method. The grid combination of predictors was ranged between 2 to 12 with 0.1 increments for pH, from 33 to 117 with 1 increments for EL and from 16 to 85 with 1 increments for CD. Model prediction was calculated for 579,600 different combination of predictors and optimum values of input parameters were pH = 2, EL = 94, CD = 83.64 which resulted in 95.48 % PFOA mineralization. Since, there are no interaction effects in the final model, effects of pH, EL, and CD on PFOA mineralization can be investigated independently. Figures 6, 7, and 8 show effects of pH, CD, and EL on PFOA mineralization, respectively, when other factors were at optimum values.

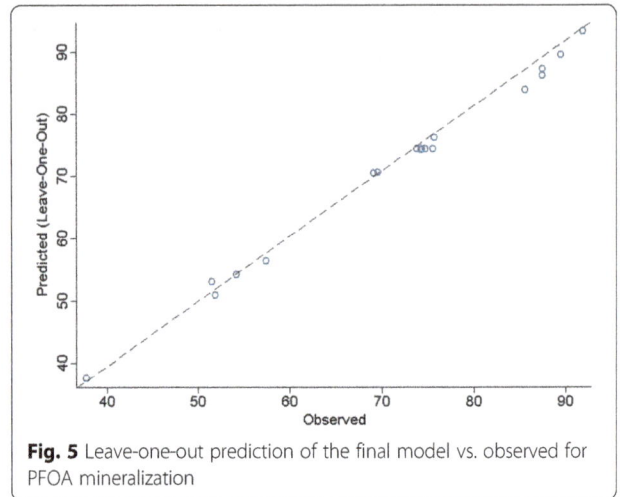

Fig. 5 Leave-one-out prediction of the final model vs. observed for PFOA mineralization

Synergistic effect

In order to evaluate the simultaneous effect of ultrasonic and electrochemical processes on PFOA mineralization, three different pretests for PFOA mineralization were done as follow: sonolysis, electrochemical and sonoelectrochemical. In all three experiments, reaction time, PFOA concentration and initial pH value were 90 min, 50 mg/L, and 7, respectively. The mineralization results were 8 %, 21 % and 73.9 % for sonolysis (frequency = 20 khz), electrochemical ($CD = 50$, mA/cm^2, and EL = 75, mM) and sonoelectrochemical (frequency = 20 khz, $CD = 50$ mA/cm^2, and EL = 75, mM) processes, respectively. This indicates

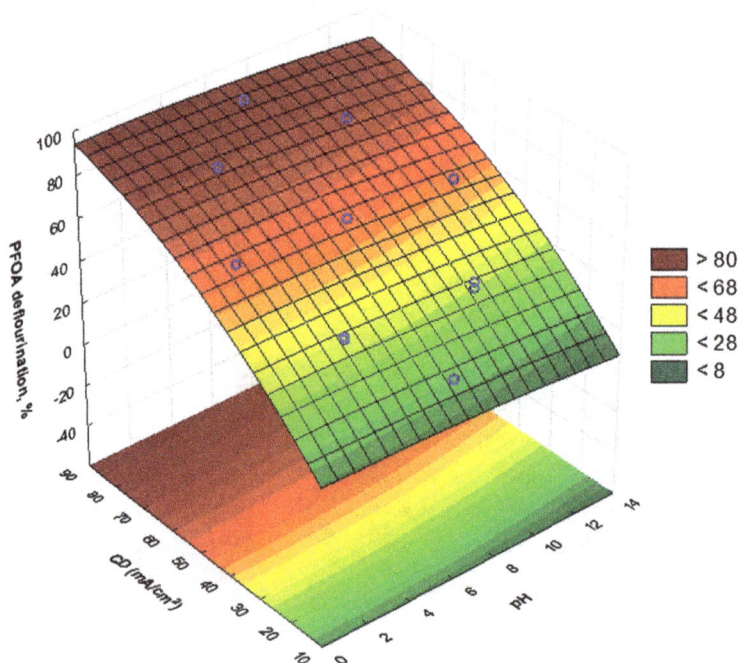

Fig. 4 3D surface plot for the PFOA deflourination as a function of CD and pH

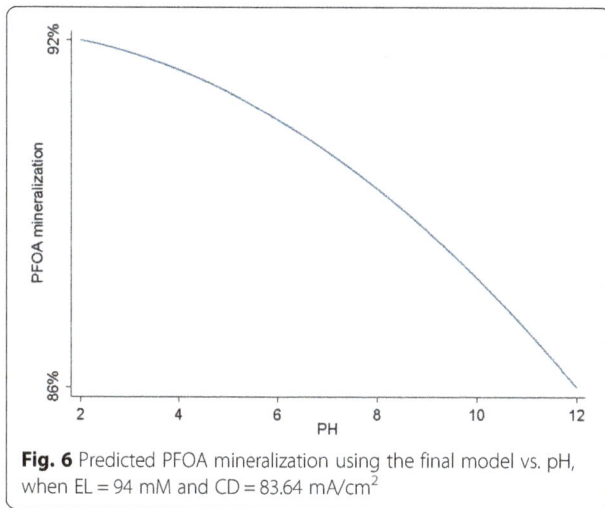

Fig. 6 Predicted PFOA mineralization using the final model vs. pH, when EL = 94 mM and CD = 83.64 mA/cm^2

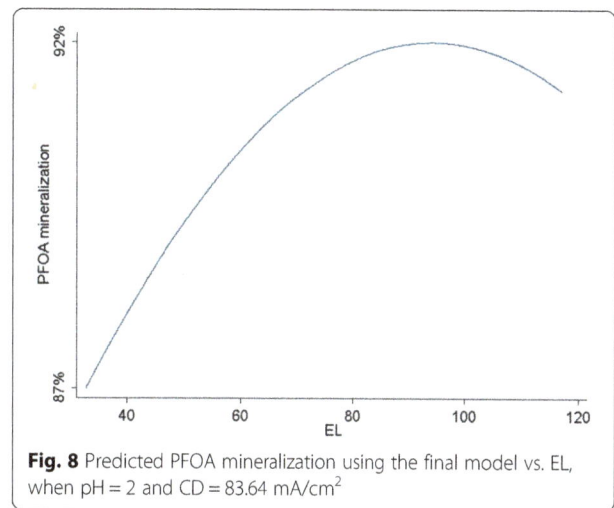

Fig. 8 Predicted PFOA mineralization using the final model vs. EL, when pH = 2 and CD = 83.64 mA/cm^2

that combination of sonolysis and electrochemical, have a remarkable synergistic effect on PFOA mineralization. Some of the basic concepts are mentioned here to clarify the sonoeletochemical, results in a higher mineralization efficiency than sonolysis and electrochemical. In the sonolysis process, the propagation of ultrasound waves through the bulk of liquid cause cavitational bubbles, which can oxidize organic substances either directly by formation of •OH by the sonolysis of water, or as a result of thermolytic reactions taking place inside [36]. The ultrasonic technique is powerful, however with respect to total input energy, there is no economical justification for using this method without the help of other techniques [37].

The synergism observed between sonolysis and electrochemical oxidation can be associated with following reasons:

1- Formation of sulfate radical by ultrasonic irradiation in the presence of sulfate ions when sodium sulfate used as supporting electrolyte [38].

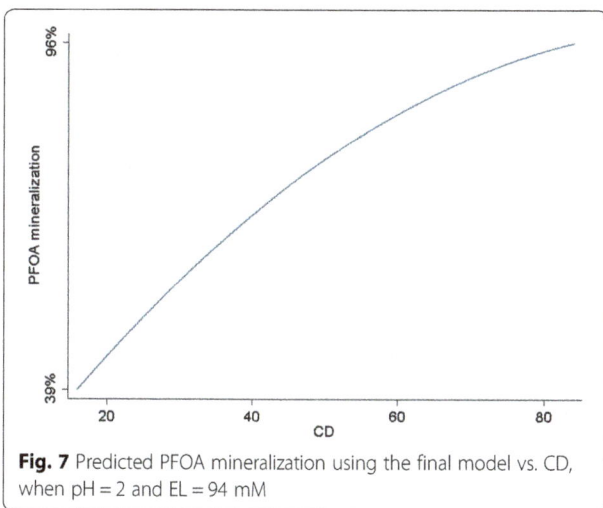

Fig. 7 Predicted PFOA mineralization using the final model vs. CD, when pH = 2 and EL = 94 mM

2- Ultrasonic waves facilitated the mass-transfer on the electrode surface, resulted in increasing diffusion of the produced hydroxyl radicals,which increased the OH radical concentration in the solution [36, 39, 40].

3- Cleaning of the electrode surface by cavitational bubbles. The mechanical effects of cavitation lead to cleaning of the electrode surface and inhibit any passive layer formation. This effect guarantee the normal electrochemical operation process with a stable electric current in the period of the treatment time [36, 40]

Effect of initial pH

Figure 4 shows the effect of initial pH of the solutions on PFOA mineralization which has been adjusted to the following values: 1.95, 4, 7, 10, and 12.05. There are many disagreements about the mechanism of influence of pH in the literature which is due to diversity of the organic structures and electrode materials, however the initial pH is one of the main factors in the oxidation process [27].

As shown in Fig. 4 and by Eq. 5, the PFOA mineralization efficiency was increased with decrease of pH. From Eq. (3), the highest and the lowest level of PFOA mineralization efficiency can be reached at an acidic (pH = 2) and alkaline (pH = 12) conditions, respectively. An increase in the PFOA mineralization with the decreasing pH from 12 to 2 can be explained as follows; first, in the alkaline conditions, sulfate radicals might react with OH resulting O$^-$ which has a lower oxidation potential lead to PFOA mineralization rate decrease [38]. Second, the enhancement of the PFOA mineralization efficiency at pH values lower than neutral is due to the increase in oxygen over-potential that abate the anodic oxygen evolution reaction and favors the production of more potent oxidizers such as OH radicals that are appropriate for the oxidation of organic compounds [41–43]. Third, many micro bubbles formed at alkaline pH in the aqueous solution in contrast with the amounts of bubbles formed at acidic pH. This situation leads to adherence of bubbles to the sonicator's probe and therefore restrict ultrasound energy distribution through the bulk of solution [37].

However, in the strong acidic solutions, the life of anode decreases [44]. As a result, in the present study, a pH of 2 was found as the optimum pH value for maximum PFOA mineralization. Other researchers also reported similar results [45]. Equation 5 and Fig. 6 show that the difference between the minimum and maximum PFOA mineralization efficiency related to pH is 6 %, which indicates sonoelectrochemical mineralization of PFOA using PbO_2 anode is not very sensitive to the initial pH and PFOA could be mineralized under a broad range of pH. Therefore, within the scope of the present study, it can be suggested that pre-adjustment of pH with the addition of chemicals is not necessary for sonoelectrochemical mineralization of the studied compound in the full scale treatment, unless a little increase in the mineralization efficiency is logical.

Effect of CD

In the present study, the effect of CD was investigated at five levels (16.36, 30, 50, 70, and 83.64 mA/cm^2) in combination with pH and EL (Fig. 4). As shown in this figure, which is the output of the CCD, the PFOA mineralization efficiency significantly increased with the increasing CD. Respect to Eq. (5), the maximum PFOA mineralization efficiency was obtained at the CD of 83.64 mA/cm^2 in the range of investigation. Equation (5) and Fig. 7 show that the difference between the minimum and maximum PFOA mineralization efficiency related to CD is 57 %, indicating that, the most important variable for the enhancement of PFOA mineralization was CD. The main reason for the increase in efficiency with increasing current density can be attributed to increasing the number of OH specimen produced which is in well agreement with other studies [42, 44, 46]. As a result, in the present study, a CD of 83.64 mA/cm^2 was chosen as the optimum CD value for maximum PFOA mineralization.

Effect of EL

The effect of EL was investigated at five levels (32.96, 50, 75, 100, and 117.04 mM) in combination with pH and CD. It can be concluded from Fig. 8 that despite of increasing PFOA mineralization with increasing the concentration of the EL, its impact is not significant. Many researchers which have used electrochemical process for wastewater treatment, believe that the degradation rate of pollutants is not affected by electrolyte concentration [46–48], however this negligible raise in the efficiency with increasing the concentration of the electrolyte can be explained by increasing the concentration of sulphate

radicals, which are generated through the irradiation of sulfate ions by ultrasonic that was mentioned before. Change the electrolyte concentration in the range of tests, leading up to a 5 % change in the PFOA mineralization efficiency.

Conclusions

In the present study, sonoelectrochemical mineralization of the PFOA was investigated and modeled by employing CCD coupled with RSM for the prediction and optimization of the PFOA mineralization in synthetic wastewater using Ti/PbO_2 as the anode and stainless steel as the cathode. The use of RSM based on CCD allowed determination of the behavior of the sonoelectrolysis on mineralization, without requiring large number of experiments, and provided sufficient information. Moreover, the CCD facilitates the process of selecting optimum conditions for defluorination. The final model was validated using the leave-one-out technique, and the predicted R^2 was 98.8, which confirmed the external validity of the model. In addition, lack of fit test was nonsignificant with a P-value of 0.09, which confirmed the fit of the final model. The results of the present study demonstrated that sonolysis and electrochemical (using Ti/PbO_2 anode) processes are not able to mineralize PFOA significantly and combination of them as sonoelectrochemical process is a suitable and an environment-friendly method for the mineralization of refractory PFOA in aqueous solution.

Abbreviations

PFOA: Perfluorooctanoic Acid; PFCA: Perfluorocarboxylic acid; AOP: Advanced Oxidation Process; EAOP: Electrochemical Advanced Oxidation Process; EO: Electrochemical Oxidation; CD: Current Density; EL: Supporting Electrolyte; RSM: Response Surface Methodology; CCD: Central Composite Design; SEM: Scanning Electron Microscope; XRD: X-Ray Diffraction.

Competing interests

The authors declare that they have no competing interests.

Authors' contributions

GB designed the study, carried out the experiments, analyzed the data and drafted the manuscript. MK participated in design of the study, electrochemical setup and contributed in the result interpretation. AE participated in the design of the study, reviewed the article and drafted the manuscript. RN helped to draft manuscript and participated in modeling process. SMTS participated in the statistical analysis and interpretation. HM helped to carry out the experiments. All authors read and approved the final manuscript.

Acknowledgments

This study is a part of a PhD approved Thesis (No. 393267) performed at Isfahan University of Medical Sciences (IUMS), Iran. The authors are thankful for the funding provided by the Department of Environmental Health Engineering and Environment Research Center, IUMS.

Author details

[1]Environment Research Center and Department of Environmental Health Engineering, School of Health, Isfahan University of Medical Sciences (IUMS), Isfahan 81676-36954, Iran. [2]Student Research Center, School of Health, IUMS, Isfahan, Iran. [3]Nanotechnology Department, University of Isfahan, Isfahan 81744-73441, Iran. [4]Research Center for Environmental Pollutants, Qom University of Medical Sciences, Qom 37136-49373, Iran.

References

1. Panchangam SC, Lin AY-C, Tsai J-H, Lin C-F. Sonication-assisted photocatalytic decomposition of perfluorooctanoic acid. Chemosphere. 2009;75(5):654–60.

2. Urtiaga A, Fernández-González C, Gómez-Lavín S, Ortiz I. Kinetics of the electrochemical mineralization of perfluorooctanoic acid on ultrananocrystalline boron doped conductive diamond electrodes. Chemosphere. 2015;129:20–6.

3. Qu Y, Zhang C-J, Chen P, Zhou Q, Zhang W-X. Effect of initial solution pH on photo-induced reductive decomposition of perfluorooctanoic acid. Chemosphere. 2014;107:218–23.

4. Liu D, Xiu Z, Liu F, Wu G, Adamson D, Newell C, et al. Perfluorooctanoic acid degradation in the presence of Fe (III) under natural sunlight. J Hazard Mater. 2013;262:456–63.

5. Zhao B, Lv M, Zhou L. Photocatalytic degradation of perfluorooctanoic acid with β-Ga 2 O 3 in anoxic aqueous solution. J Environ Sci. 2012;24(4):774–80.

6. Ahrens L, Felizeter S, Sturm R, Xie Z, Ebinghaus R. Polyfluorinated compounds in waste water treatment plant effluents and surface waters along the River Elbe. Germany Mar Pollut Bull. 2009;58(9):1326–33.

7. Song C, Chen P, Wang C, Zhu L. Photodegradation of perfluorooctanoic acid by synthesized TiO 2–MWCNT composites under 365nm UV irradiation. Chemosphere. 2012;86(8):853–9.

8. Bazrafshan E, Mahvi A, Nasseri S, Shaieghi M. Performance evaluation of electrocoagulation process for diazinon removal from aqueous environments by using iron electrodes. Iran J Environ Health Sci. 2007;4(2):127–32.

9. Bazrafshan E, Mahvi A, Nasseri S, Mesdaghinia A, Vaezi F, Nazmara S. Removal of cadmium from industrial effluents by electrocoagulation process using iron electrodes. Iran J Environ Health Sci. 2006;3(4):261–6.

10. Mondal S. Methods of dye removal from dye house effluent-an overview. Environ Eng Sci. 2008;25(3):383–96.

11. Martínez-Huitle CA, Brillas E. Decontamination of wastewaters containing synthetic organic dyes by electrochemical methods: a general review. Appl Catal B-Environmental. 2009;87(3):105–45.

12. Ju P, Fan H, Guo D, Meng X, Xu M, Ai S. Electrocatalytic degradation of bisphenol A in water on a Ti-based PbO₂–ionic liquids (ILs) electrode. Chem Eng J. 2012;179:99–106.

13. Hmani E, Elaoud SC, Samet Y, Abdelhédi R. Electrochemical degradation of waters containing O-Toluidine on PbO₂ and BDD anodes. J Hazard Mater. 2009;170(2):928–33.

14. Samet Y, Elaoud SC, Ammar S, Abdelhedi R. Electrochemical degradation of 4-chloroguaiacol for wastewater treatment using PbO₂ anodes. J Hazard Mater. 2006;138(3):614–9.

15. Abaci S, Yildiz A. Electropolymerization of thiophene and 3-methylthiophene on PbO₂ electrodes. J Electroanal Chem. 2004;569(2):161–8.

16. Panizza M, Cerisola G. Direct and mediated anodic oxidation of organic pollutants. Chem Rev. 2009;109(12):6541–69.

17. Siedlecka EM, Stolte S, Gołębiowski M, Nienstedt A, Stepnowski P, Thöming J. Advanced oxidation process for the removal of ionic liquids from water: The influence of functionalized side chains on the electrochemical degradability of imidazolium cations. Sep Purif Technol. 2012;101:26–33.

18. Moriwaki H, Takagi Y, Tanaka M, Tsuruho K, Okitsu K, Maeda Y. Sonochemical decomposition of perfluorooctane sulfonate and perfluorooctanoic acid. Environ Sci Technol. 2005;39(9):3388–92.

19. Rokhina EV, Repo E, Virkutyte J. Comparative kinetic analysis of silent and ultrasound-assisted catalytic wet peroxide oxidation of phenol. Ultrason Sonochem. 2010;17(3):541–6.

20. Merouani S, Hamdaoui O, Saoudi F, Chiha M, Pétrier C. Influence of bicarbonate and carbonate ions on sonochemical degradation of Rhodamine B in aqueous phase. J Hazard Mater. 2010;175(1):593–9.

21. Su S, Guo W, Yi C, Leng Y, Ma Z. Degradation of amoxicillin in aqueous solution using sulphate radicals under ultrasound irradiation. Ultrason Sonochem. 2012;19(3):469–74.

22. Montgomery DC. Design and analysis of experiments. 7th ed. New York: John Wiley & Sons; 2009.

23. Aquino JM, Rocha-Filho RC, Bocchi N, Biaggio SR. Electrochemical degradation of the Disperse Orange 29 dye on a β-PbO₂ anode assessed by the response surface methodology. J Environ Chem Eng. 2013;1(4):954–61.

24. Polcaro A, Palmas S, Renoldi F, Mascia M. On the performance of Ti/SnO₂ and Ti/PbO₂ anodesin electrochemical degradation of 2-chlorophenolfor wastewater treatment. J Appl Electrochem. 1999;29(2):147–51.

25. Del Río A, Benimeli M, Molina J, Bonastre J, Cases F. Electrochemical Treatment of CI Reactive Black 5 Solutions on Stabilized Doped Ti/SnO₂ Electrodes. Int J Electrochem Sci. 2012;7:13074–92.

26. Ghalwa NA, Gaber M, Khedr AM, Salem MF. Comparative Study of Commercial Oxide Electrodes Performance in Electrochemical Degradation of Reactive Orange 7 Dye in Aqueous Solutions. Int J Electrochem Sci. 2012;7:6044–58.

27. Xu L, Guo Z, Du L, He J. Decolourization and degradation of C.I. Acid Red 73 by anodic oxidation and the synergy technology of anodic oxidation coupling nanofiltration. Electrochim Acta. 2013;97:150–9.

28. Lee Y, Lo S, Kuo J, Hsieh C. Decomposition of perfluorooctanoic acid by microwaveactivated persulfate: Effects of temperature, pH, and chloride ions. Front Environ Sci Eng. 2012;6(1):17–25.

29. Myers RH, Montgomery DC, Anderson-Cook CM. Response surface methodology: process and product optimization using designed experiments. USA: John Wiley & Sons; 2009.

30. Kutner MH, Nachtsheim C, Neter J. Applied linear regression models. USA: McGraw-Hill/Irwin; 2004.

31. Anderson TW, Darling DA. A test of goodness of fit. J Am Stat Assoc. 1954;49(268):765–9.

32. Breusch TS, Pagan AR. A simple test for heteroscedasticity and random coefficient variation. Econometrica. 1979;47(5):1287-94.

33. Durbin J, Watson GS. Testing for serial correlation in least squares regression. I. Biometrika. 1950;37(3-4):409–28.

34. Durbin J, Watson GS. Testing for serial correlation in least squares regression. II. Biometrika. 1951:38(1-2):159–77.

35. Efron B, Gong G. A leisurely look at the bootstrap, the jackknife, and cross-validation. Am Stat. 1983;37(1):36–48.

36. Barros WR, Steter JR, Lanza MR, Motheo AJ. Degradation of amaranth dye in alkaline medium by ultrasonic cavitation coupled with electrochemical oxidation using a boron-doped diamond anode. Electrochim Acta. 2014;143:180–7.

37. Thi L-AP, Do H-T, Lo S-L. Enhancing decomposition rate of perfluorooctanoic acid by carbonate radical assisted sonochemical treatment. Ultrason Sonochem. 2014;21(5):1875–80.

38. Lin J-C, Lo S-L, Hu C-Y, Lee Y-C, Kuo J. Enhanced sonochemical degradation of perfluorooctanoic acid by sulfate ions. Ultrason Sonochem. 2015;22:542–7.

39. Yang B, Zuo J, Tang X, Liu F, Yu X, Tang X, et al. Effective ultrasound electrochemical degradation of methylene blue wastewater using a nanocoated electrode. Ultrason Sonochem. 2014;21(4):1310–7.

40. Ren Y-Z, Wu Z-L, Franke M, Braeutigam P, Ondruschka B, Comeskey DJ, et al. Sonoelectrochemical degradation of phenol in aqueous solutions. Ultrason Sonochem. 2013;20(2):715–21.

41. Dai Q, Shen H, Xia Y, Chen F, Wang J, Chen J. The application of a novel Ti/SnO₂-Sb₂O₃/PTFE-La-Ce-β-PbO₂ anode on the degradation of cationic gold yellow X-GL in sono-electrochemical oxidation system. Sep Purif Technol. 2013;104:9–16.

42. Niu J, Maharana D, Xu J, Chai Z, Bao Y. A high activity of Ti/SnO₂-Sb electrode in the electrochemical degradation of 2,4-dichlorophenol in aqueous solution. J Environ Sci. 2013;25(7):1424–30.

43. Neto SA, De Andrade A. Electrooxidation of glyphosate herbicide at different DSA® compositions: pH, concentration and supporting electrolyte effect. Electrochim Acta. 2009;54(7):2039–45.

44. Zhong C, Wei K, Han W, Wang L, Sun X, Li J. Electrochemical degradation of tricyclazole in aqueous solution using Ti/SnO₂-Sb/PbO₂ anode. J Electroanal Chem. 2013;705:68–74.

45. Lin H, Niu J, Ding S, Zhang L. Electrochemical degradation of perfluorooctanoic acid (PFOA) by Ti/SnO₂-Sb, Ti/SnO₂-Sb/PbO₂ and Ti/SnO₂-Sb/MnO₂ anodes. Water res. 2012;46(7):2281–9.

46. Zhou M, He J. Degradation of cationic red X-GRL by electrochemical oxidation on modified PbO₂ electrode. J Hazard Mater. 2008;153(1):357–63.

47. Niu J, Bao Y, Li Y, Chai Z. Electrochemical mineralization of pentachlorophenol (PCP) by Ti/SnO₂-Sb electrodes. Chemosphere. 2013;92(11):1571–7.

48. Shao D, Liang J, Cui X, Xu H, Yan W. Electrochemical oxidation of lignin by two typical electrodes: Ti/Sb-SnO₂ and Ti/PbO₂. Chem Eng J. 2014;244:288–95.

9

Investigation of potential genotoxic activity using the SOS Chromotest for real paracetamol wastewater and the wastewater treated by the Fenton process

Emel Kocak

Abstract

Background: The potential genotoxic activity associated with high strength real paracetamol (PCT) wastewater (COD = 40,000 mg/L, TOC = 12,000 mg/L, BOD_5 = 19,320 mg/L) from a large-scale drug-producing plant in the Marmara Region, was investigated in pre- and post- treated wastewater by the Fenton process (COD = 2,920 mg/L, TOC = 880 mg/L; BOD_5 = 870 mg/L).

Methods: The SOS Chromotest, which is based on *Escherichia coli* PQ37 activities, was used for the assessment of genotoxicity. The corrected induction factors (CIF) values used as quantitative measurements of the genotoxic activity were obtained from a total of four different dilutions (100, 50, 6.25, and 0.078 % v/v.) for two samples, in triplicate, to detect potentially genotoxic activities with the SOS Chromotest.

Results: The results of the SOS Chromotest demonstrated CIF_{max} value of 1.24, indicating that the PCT effluent (non-treated) is genotoxic. The results of the SOS Chromotest showed an CIF_{max} value of 1.72, indicating that the wastewater treated by Fenton process is genotoxic.

Conclusions: The findings of this study clearly reveal that the PCT wastewater (non-treated) samples have a potentially hazardous impact on the aquatic environment before treatment, and in the wastewater that was treated by the Fenton process, genotoxicity generally increased.

Keywords: Induction factor, *Escherichia coli* PQ37, Fenton process, Real paracetamol wastewater, SOS Chromotest

Introduction

Pharmaceutical drugs can reach the aquatic environment from domestic waste or industrial wastewater, hospitals, and health care centers [1]. Considering the complex nature of a large variety of pharmaceutical drugs, genotoxicologic studies of real paracetamol wastewater remained very superficial, and therefore it is necessary to further examine assay systems that have the ability to evaluate the substantial impact of some of the more persistent pollution sources. Unused drugs, manufacturing waste, and sewage sludge can also be introduced to the environment by way of landfill leachates [2].

Over the past few years, advanced oxidation processes (AOPs) are used to reduce contamination based on the presence of stable pharmaceuticals [3–6]. Complex organic chemicals are formed during the production of pharmaceuticals and it is not easy to remove these compounds biologically. As a result, AOPs are more appropriate than conventional methods to treat pharmaceutical wastewater [7]. AOPs comprises Fenton, photo-Fenton, and ozonation combined with UV-light and/or H_2O_2, mainly TiO_2-mediated photocatalysis [6], electrolysis, wet air oxidation, ultrasound and ionizing radiation, microwaves, pulsed plasma, and the ferrate reagent [5]. Different reviews of the literature have reported the fate of some pharmaceutical compounds as well as their occurrence and effects in the aquatic environment [8, 9]. Some of the substances found in wastewater are genotoxic and

Correspondence: ekocak@yildiz.edu.tr
Department of Environmental Engineering, Faculty of Civil Engineering, Yildiz Technical University, Istanbul, Turkey

are suspected to be a possible cause of the cancers observed in previous decades. Water genotoxicity studies are of interest because epidemiologic investigations have shown a link between genotoxic drinking water intake and a rise in cancer [10–12]. The results of these studies must however, be interpreted with caution because the exposure to genotoxic water was only estimated and not truly measured. However, these results emphasized the importance of the determination of water genotoxicity with an aim of controlling the population exposure. Thus, the monitoring of water contamination for potentially carcinogenic compounds represents a major concern for human health [13]. It is extremely difficult to quantify the risk associated with these chemical pollutants because they usually occur in concentrations too low to allow analytical determination, and putative mutagens, with few exceptions, have never ever been identified. Moreover, the composite effects of mixtures cannot be readily assessed via analytical methods. Thus, toxicity is often evaluated by means of biological tests, as well as by bacterial genotoxicity tests which do not require a priori knowledge of toxicant identity and/or physical-chemical properties [14].

To the best of the authors' knowledge, there are no studies dealing with the high strength real paracetamol (PCT) wastewater genotoxicity by way of the SOS Chromotest. However, all hospital wastewater studies show that this kind of wastewater including drugs and antibiotics could have a genotoxic potential [15–18]. Genotoxicity was studied with the SOS Chromotest, which allows for the detection of primary DNA damaging agents in *Escherichia coli*. Based on the above-mentioned facts, the specific objectives of the present study were as follows: (1) to evaluate the main characteristics of the real PCT wastewater, (2) to use the SOS Chromotest microplate assay to investigate the genotoxic activities of the non-treated PCT wastewater, (3) to use the SOS Chromotest microplate assay to investigate the genotoxic activities of the wastewater treated by the Fenton process.

Materials and methods
Characteristics of the PCT wastewater
High COD values of real PCT wastewater are related to the high concentrations of PCT, PAP, and aniline. Pollutant concentrations of the wastewater can change day by day according to the process operations [19]. In this study, wastewater with a COD concentration of 40,000 mg/L was used for the investigation of potentially genotoxic activity with the SOS Chromotest. The main characteristics of real PCT wastewater are shown in Table 1 [19].

As seen in Table 1, pollutant concentrations are extremely high in PCT wastewater and consequently, the treatability of this wastewater is very difficult in conventional treatment plants. As the wastewater contains

Table 1 Main characteristics of real paracetamol (PCT) wastewater [19]

Parameter	Value
pH	9.0
Chemical oxygen demand, COD (mg/L)	40,000
5-day biological oxygen demand, BOD$_5$ (mg/L)	19,320
Total organic carbon, TOC (mg/L)	12,000
Paracetamol, PCT (ppm)	107
Para-amino phenol, PAP (ppm)	1818
Aniline (ppm)	2915

different chemicals, the presence of the toxic effect derived from chemical products that could not be removed through conventional biological treatment methods, and also the low BOD/COD ratio show that the wastewater must be pretreated chemically. Badawy et al. [20] indicated that wastewater with a BOD/COD ratio between 0.25 and 0.30 cannot be treated biologically.

Characteristics of wastewater treated by the Fenton process
The aim of the Fenton process, which has been previously studied, is obtaining the highest COD removal efficiency using the optimum chemical dosages. In this study, wastewater treated by the Fenton process was used as the second sample to investigate potential genotoxic activity with the SOS Chromotest. Characteristics of the wastewater treated by the Fenton process are shown in Table 2 [19].

SOS Chromotest
The SOS Chromotest is a colorimetric assay of the enzymatic activities that occur after incubating the test strain of bacteria in the presence of various amounts of experimental samples [21]. The test utilizes a genetically engineered bacterium, *E. coli* PQ37, to detect DNA-damaging agents. In this assay, the β-galactosidase (β-gal) gene (*lacZ*) of the *E. coli* PQ37 tester strain is fused to the bacterial *sfiA* SOS operon. Thus, *lacZ* is concomitantly expressed

Table 2 Characteristics of the wastewater treated by the Fenton process [19]

Parameter	Value
pH	6.0
Chemical oxygen demand, COD (mg/L)	2920
5-day biological oxygen demand, BOD$_5$ (mg/L)	870
Total organic carbon, TOC (mg/L)	880
Paracetamol, PCT (ppm)	1
Para-amino phenol, PAP (ppm)	2
Aniline (ppm)	17

during the bacterial SOS response, and this gene expression can be photometrically determined by the induction of β-gal. The amount of β-gal induction is indicative of the extent of SOS induction and bacterial genotoxicity. Bacterial *alkaline phosphatase* (AP) activity was used to determine the range of bacterial cytotoxicity. The ratio of β-galAP activity was defined as the induction factor (IF), and this ratio was used to indicate the extent of SOS induction for the tested compounds [20]. The test is available as a test kit, which includes all of the necessary materials. No special measuring devices, with the exception of a plate reader, were required to complete this assay. This test can also be used as a qualitative test, based on the use of a color scale. The assay can be completed within 24 h, including the revival of the bacteria. The test detects any primary DNA damage that is caused by genotoxins, and the test can be used for various types of aqueous samples.

In this study, the SOS Chromotest was performed, without metabolic activation, as described by Quillardet and Hofnung [22]. The *E. coli* PQ37 tester strain was kindly provided by Environmental Bio-Detection Products Inc. [21]. Four different dilutions (100, 50, 6.25, and 0.078 % v/v.) for two samples, in triplicate, and the testing began with a 100 mL sample that was equal for each cuvette. The test was performed at 37 °C, and the cuvettes were read after 2 h with a spectrophotometer. Spectrophotometer equipped with 600 nm filter and using 1 cm light-path rectangular cuvettes (for preparation of the bacterial suspension). Growth bacteria suspension was required OD of 0.05 at 600 nm by the spectrophotometer before use in the assay depending upon the degree of growth obtained. The bacteria was grown in 37 °C, incubator to an OD (optical density) of 0.05 to 0.06 in approximately 4 h and the test was run. When this method was used the bacteria were still in log phase growth and the colour development, when exposed to a genotoxin, would have occured within an hour or so. If the OD was is 0.05 colour development would have taken approximately 1.5 h. If the OD was closer to 0.07 the colour development would have occurred within half and hour because of the increased cell density [21, 23].

For the direct assay, the negative control was composed of a 10 % DMSO (dimethyl sulfoxide) solution in sterile, ultrapure water, and the positive control was 4-nitro-quinolineoxide (4NQO).

Determination of genotoxic activity

The SOS Chromotest involves incubating the bacteria with the experimental sample and assessing the β-galactosidase (β-gal) activity (i.e. the level of SOS induction). *Alkaline phosphatase* (AP) activity is also measured and serves as a control for toxicity [15]. AP reduction factors

(RF), β-gal induction factors (IF), and corrected induction factors (CIF = IF/RF) were calculated as described by Legault [24].

$$RF = \frac{(OD_{405})_{mean,t}}{(OD_{405})_{mean,c}} \tag{1}$$

$$IF = \frac{(OD_{620})_{mean,t}}{(OD_{620})_{mean,c}} \tag{2}$$

$$CIF = \frac{RF}{IF} \tag{3}$$

where (OD_{405}) mean and (OD_{620}) mean are the means of the optical density (OD) readings that were taken at 620 nm (β-gal) and 405 nm (AP), and t and c refer to the test and the control dilutions, respectively. Bombardier et al. [25] reported that the RF and IF values account for the background activity of the control. The ratio of IF to RF units yields an estimate of β-gal activity that is corrected for toxicity. The criterion that was used to consider a sample as "positive" in the SOS Chromotest differs between authors [13, 24, 26, 27]. In the present study, significant genotoxic activity was defined as having a corrected induction factor that was equal to or greater than 1.2–1.5, as suggested by most of the previously published studies [13, 24, 28].

All SOS Chromotest analyses were conducted according to the EBPI (Environmental Bio-Detection Products Inc.) protocols [21]. The experimental equipment (i.e. micropipettes, Eppendorf pipettes) were autoclave sterilized at 121 °C and 10.6 bar for 15 min (Nuve OT 032). A minishaker with an orbit of 4.5 mm and a speed range of 0–2500 rpm (IKA Labortechnik, Staufen, Germany) was used to centrifuge the bacteria at a fixed agitation speed of 1500 rpm. The bacteria were grown at a stable temperature of 37 °C in a temperature-controlled incubator (Memmert, Germany). This incubator was also used for the development of the enzymatic activities. The bacteria cultures were grown, and the optical density values (600 nm) were measured using an UV–VIS Spectrophotometer (Shimadzu, UV-1202, UV–VIS) with a special quartz cuvette that allowed for a light path length of 1 cm. ATP and b-gal activities were measured, using a Biotek PowerWave XS Microplate ELISA Reader (BioTek Instruments Inc., Winooski, VT, USA) with data analysis software (Gen 5), at 405 nm (OD_{405}) and 620 nm (OD_{620}), respectively. The pH values of the samples were measured with a pH meter (Jenway 3040 Ion Analyzer) and a pH probe (HI1230, Hanna Instruments, USA). Deionized and sterile ultrapure water was used in the experiments, and the water was supplied from a TKA-GenPure water purification system (Niederelbert, Germany). The physicochemical analyses of the surface water samples were conducted by the procedures described in the Standard Methods [23, 29].

Results and discussion

In the present study, the SOS Chromotest based on *Escherichia coli* PQ37 activities was used for the assessment of genotoxicity of samples of real PCT wastewater before and after the wastewater was treated by the Fenton process. SOS responses were determined as corrected induction factors (CIF) for all samples and presented in Table 3. The tests demonstrate that these PCT samples pre- and post-treatment present a genotoxic effect. Indeed, out of a total of eight samples tested, four were positive (50 %). It is difficult to compare these results with other studies because many parameters can influence the genotoxicity test response (composition of the samples, nature of paracetamol, nature of chemicals used in the Fenton process, nature of the genotoxicity test, etc.) [13].

An appraisal of the genotoxicity of the PCT samples before and after treatment is as follows: (a) for the real PCT wastewater: of a total of four samples tested, one was positive that CIF = 1.24 (25 %), and (b) for the PCT wastewater treated by the Fenton process: of a total of four samples tested, three were positive that CIF = 1.25, 1.50, and 1.72 (75 %). Some of the calculated CIF values were determined to be above the level that is considered to be the 1.2 threshold level. The CIF values of all the PCT wastewater samples (non-treated and treated by Fenton process) were observed within the 1.24–1.72 range. When the CIF for any of the test concentrations reached 1.2, the test substance was scored as significantly genotoxic. However, the SOS Chromotest results clearly indicated that genotoxic effects that were found in the PCT wastewater (non-treated and treated by the Fenton process) samples. Table 4 summarizes the genotoxic activity levels and the corresponding threshold values that were defined in the different studies. In the present study, significant genotoxic activity was defined as having a corrected induction factor that was equal to

Table 4 Genotoxic activity levels and the corresponding threshold values defined in different studies

Genotoxic activity levels	Corrected induction factors (CIF)	References and region
SOS response	>1.2	Legault et al. (1996), Canada
Genotoxic	>1.2	Kocak et al. (2010), Turkey
β-galactosidase activity significantly increases compared with the solvent control	>1.5	Jolibois and Guerbet (2005), France
Genotoxic	>1.5	Mersch-Sundermann et al. (1992), Germany
Genotoxic	>1.5	Margulis et al. (2003), Russia

or greater than 1.2, as suggested by the published studies authors [13, 14, 24, 28, 30].

The performance data revealed that a wide range of CIF values were observed, and the range of CIF values depends on the characteristics of wastewater matrices of PCT and chemical dosage for the Fenton process. High genotoxic activity values are probably due to the presence of several mutagenic and carcinogenic agents, which include persistent components, soluble DNA-damaging products, recalcitrant substances, and other undesirable impurities that are present in the wastewater samples. It is apparent from previous studies that various chemical compounds have been widely used in numerous industrial and environmental applications. However, relatively few genotoxicological investigations are available in the literature. Therefore, additional studies that use genotoxicological data, in addition to the contaminant monitoring data, will be necessary to identify the sources of the toxicants and to ensure that more environmental risk assessments can be verified. Both of these points have been suggested by other researchers [13, 31].

Table 3 Results of the SOS Chromotest on the PCT wastewater pre- and post-treatment by the Fenton process

Sample name	V (%)	CIF	SD
Real PCT wastewater (non-treated)	100	0,64	0,07
	50	**1,24**	0,03
	6,25	0,84	0,05
	0,078	0,92	0,24
PCT wastewater treated by the Fenton proces	100	**1,50**	0,04
	50	**1,72**	0,29
	6,25	**1,25**	0,04
	0,078	0,99	0,15

V, tested concentration; CIF, corrected induction factor; SD, standard deviation
Genotoxic samples are indicated in bold letters

Conclusions

This study showed that the real PCT wastewater before and after treatment are genotoxic. Especially after the Fenton process, genotoxicity generally increased. As a consequence of the different chemical species present in the paracetamol wastewater, the Fenton process was able to increase wastewater genotoxicity; especially after the Fenton process, genotoxicity generally increased. The success of this assay was, at least in part, due to its simplicity and rapidity. The SOS Chromotest responses clearly indicated that there were potential genotoxic impacts, in terms of CIF values, found in the PCT wastewater. Some of the calculated CIF values were determined to be above the 1.2 threshold level. During the PCT process, the CIF variations were much lower than CIF variations that were observed during the Fenton process. These variations possibly depend on the chemical dosing during the Fenton process.

It is noted that the work described here is the first report from an integrated study investigating genotoxicity on PCT (non-treated and treated with Fenton process) wastewater. Although the SOS Chromotest responses indicated that the PCT wastewater was found to have genotoxic effects on the aquatic environment, further investigations will be conducted on other in vitro tests to better characterize the genotoxicity responses. This study can provide useful information to medical and water managers and health authorities in evaluation of water quality strategies for reduction of genotoxic compounds in the PCT wastewater before and after treatment.

Competing interests

The author declares that he/she has no competing interests.

Acknowledgements

The authors would like to thank Research Project Coordinator of Yildiz Technical University for supporting this study financially under the number 2012-05-02-GEP01 and also the factory for providing the wastewater. The author would like to thank our great sorrow for the death of dear professor and one of the members of our project group, Ferruh Erturk, on August 24, 2011. He was a great scientist and at the same time he was a great person with humanity and warm kindness to all of us.

References

1. Ribeiro AVFN, Belisário M, Galazzi RM, Balthazar DC, Pereira MG, Ribeiro JN. Evaluation of two bioadsorbents for removing paracetamol from aqueous media. Electron J Biotechnol 2011. DOI: 10.2225/vol14-issue6-fulltext-8

2. Kulik N, Trapido M, Goi A, Veressina Y, Munter R. Combined chemical treatment of pharmaceutical effluents from madical oinment production. Chemosphere. 2008;70:1525–31.

3. Bautitz IR, Nogueira RFP. Photodegradation of lincomycin and diazepam in sewage treatment plant effluent by photo-Fenton process. Catal Today. 2010;151:94–9.

4. Pérez-Estrada LA, Malato S, Gernjak W, Agüera A, Thurman EM, Ferrer I, et al. Photo-fenton degradation of diclofenac: Identification of main intermediates and degradation pathway. Environ Sci Technol. 2005;39:8300–6.

5. Klavarioti M, Mantzavinos D, Kassinos D. Removal of residual pharmaceuticals from aqueous systems by advanced oxidation processes. Environ Int. 2009;35:402–17.

6. Arslan-Alaton I, Dogruel S. Pre-treatment of penicillin formulation effluent by advanced oxidation processes. J Hazard Mater. 2004;B112:105–13.

7. Tekin H, Bilkay O, Ataberk SS, Balta TH, Ceribasi IH, Sanin FD, et al. Use of Fenton oxidation to improve the biodegradability of a pharmaceutical wastewater. J Hazard Mater. 2006;B136:258–65.

8. Richardson ML, Bowron JM. The fate of pharmaceutical chemicals in the aquatic environment. J Pharm Pharmacol. 1985;37:1–12.

9. Halling-Sorensen BH, Nors Nielsen S, Lanzky PF. Occurrence, fate and effects of pharmaceutical substances in the environment—a review. Chemosphere. 1998;36:357–93.

10. Koivusalo M, Jaakkola JJ, Vartiainen T. Drinking water mutagenicity and gastrointestinal and urinary tract cancers: an ecological study in Finland. Am J Public Health. 1994;84:1223–8.

11. Koivusalo M, Vartiainen T, Hakulinen T. Drinking water mutagenicity and leukemia, lymphomas, and cancers of the liver, pancreas, and soft tissue. Arch Environ Health. 1995;50:269–76.

12. Koivusalo M, Pukkala E, Vartiainen T. Drinking water chlorination and cancer—a historical cohort study in Finland. Cancer Causes Control. 1997;8:192–200.

13. Jolibois B, Guerbet M. Evaluation of industrial, hospital and domestic wastewater genotoxicity with the Salmonella fluctuation test and the SOS chromotest. Mutat Res. 2005;565:151–62.

14. Kocak E, Yetilmezsoy K, Gonullu MT, Petek M. A statistical evaluation of the potential genotoxic activity in the surface waters of the golden horn estuary. Mar Pollut Bull. 2010;60:1708–11.

15. Bombardier M, Bermingham N, Legault R, Fouquet A. Evaluation of an SOS-chromotest-based approach for the isolation and detection of sedimentassociated genotoxins. Chemosphere. 2001;42:931–44.

16. Giuliani F, Koller T, Wurgler FE. Detection of genotoxic activity in native hospital waste water by the umuC test. Mutat Res. 1996;368:49–57.

17. Steger-Hartmann T, Kümmerer K, Hartmann A. Biological degradation of cyclophosphamide and its occurrence in sewage water. Ecotoxicol Environ Saf. 1997;36:174–9.

18. Hartmann A, Golet EM, Gartisier S. Primary DNA damage but not mutagenicity correlates with ciprofloxacin concentrations in German hospital wastewaters. Arch Environ Contam Toxicol. 1999;36:115–9.

19. Dalgic G. Application of advanced oxidation processes for removal of polluting parameters from paracetamol-containing wastewater, MSc Thesis Yildiz Technical University Graduate School of Natural and Applied Sciences. 2013.

20. Badawy MI, Wahaab RA, El-Kalliny AS. Fenton-biological treatment processes for the removal of some pharmaceuticals from industrial wastewater. J Hazard Mater. 2009;167:567–74.

21. Environmental Bio Detection Products Inc. (EBPI). The SOS-Chromotest version 6.3, Instruction for Use, Canada (2008).

22. Quillardet P, Hofnung M. The SOS Chromotest, a colorimetric bacterial assay for genotoxins: procedures. Mutation Research. (1985);147:65–78.

23. Kocak E. Investigation of genotoxic effects on active organism DNA for the contamination of the golden horn, Ph.D. Thesis. Istanbul: Yildiz Technical University, Institute of Natural and Applied Sciences; 2010.

24. Legault R, Blake C, Trottier S. Detecting genotoxic activity in industrial effluents using the SOS chromotest microplate assay. Environ Toxicol Water Qual. 1996;11:151–65.

25. Jolibois B, Guerbet M, Vassal S. Detection of hospital wastewater genotoxicity with the SOS chromotest and Ames fluctuation test. Chemosphere. 2003;51:539–43.

26. Cachot J, Geffard O, Augagneur S, Lacroix S, Le Menach K, Peluhet L, et al. Evidence of genotoxicity related to high PAH content of sediments in the upper part of the Seine Estuary (Normandy, France). Aquat Toxicol. 2006;79:257–67.

27. Gebel T, Koenig A. Impact of dimethyl sulfoxide and examples of combined genotoxicity in the SOS chromotest. Mutat Res. 1999;444:405–11.

28. Mersch-Sundermann V, Kern S, Wintermann F. Genotoxicity of nitrated polycyclic aromatic hydrocarbons and related structures on E.coli PQ37 (SOS chromotest). Environ Mol Mutagen. 1992;18:41–50.

29. APHA. Standard methods for the examination of water and wastewater. Washington DC: American Public Health Association 19th ed; 1995.

30. Margulis AB, Il'inskaya ON, Kolpakov AI, El'-Registan GI. Induction of SOS response by autoregulatory factors of microorganisms. Russ J Genet. 2003;39:993–6.

31. Guzzella L, Di Caterino F, Monarca S, Zani C, Feretti D, Zerbini I, et al. Detection of mutagens in water-distribution systems after disinfection. Mutat Res. 2006;19:72–81.

10

Mercury pollution for marine environment at Farwa Island, Libya

Adel A.S. Banana[1], R. M. S. Radin Mohamed[2] and A. A. S. Al-Gheethi[2,3*]

Abstract

Background: Farwa is an Island in Libya receives petrochemical wastes generated from General Company of Chemical Industries (GCCI) since more than 40 years.

Aim: The present work aimed to determine the concentrations of mercury (Hg^{+2}) in fish, marine plants and sediment collected from Farwa lagoon to evaluate effect of industrial wastewater from GCCI on the marine environment.

Methods: Hundred and twelve samples of fish, pearl oyster, cuttlefish sediments and marine plants were analyzed to determine Hg^{2+} concentration during the period from January to August 2014 by using Atomic Absorption Spectrometer (AAS).

Results: The highest concentration of Hg^{2+} was detected in *Pinctada radiata* (11.67 ± 3.30 µgg^{-1}) followed by *Serranus scriba* (6.37 ± 0.11 µg g^{-1}) and *Epinephelus marginatus* (6.19 ± 0.02 µg g^{-1}). About 75 % of marine plants contained the maximum contaminations during the summer season. In fish samples Hg^{2+} concentrations exceeded the levels provided by international standards.

Conclusions: The fish at Farwa lagoon is heavily contaminated with Hg^{2+} which may represent a source for mercury poisoning for human.

Keywords: Mercury, Fishes, Contamination, Farwa Island

Background

The increasing of industrial activities has led to increase the contamination of environment with several types of pollutants as due to discharge of industrial wastewater into the environment and aquatic system. petrochemicals factors is among various of industrials process which produce heavily contaminated wastewater. In the developed countries the wastewater are treated using advanced technologies such as reverse osmosis, nanotube carbon, adsorption process using different types of adsorbents as well as photo-degradation processes of degradable toxic compounds. Those technologies have high efficiency to remove most toxic substances from wastewater before final disposal into the environment. Others technologies such as multi-walled carbon nanotube/tungsten oxide (MWCNT/WO$_3$) and alumina nano-particles polyamide membrane

still under investigation and they exhibited high efficiency for the removal and degrade various types of pollutants based on the lab scale experiments [1–7].

In the term of heavy metals contamination, the petrochemical industries represent one of the main sources for generation of these toxics into the environment. the adsorption process materials is the most common treatment process to remove heavy metals from wastewater. Recently, some authors focused on improvement this process to be high efficiency. Gupta et al. [8] has combined the magnetic properties of iron oxide with adsorption properties of carbon nanotubes to increase the removal of Cr^{2+} ions.

Heavy metals are groups of elements with high molecular weights that are not degraded when taken into the body; instead, they accumulate in specific body organs and cause illness. Heavy metals have the potential to disrupt the metabolism and biological activities of many proteins because it can oxidize the sulfhydryl groups [9]. Among several of heavy metals, mercury (Hg^{2+}) is the most toxic

---

* Correspondence: adelalghithi@gmail.com
[2]Faculty of Civil & Environment Engineering, UTHM, Parit Raja, Malaysia
[3]High Institute of health sciences, Sana'a, Yemen
Full list of author information is available at the end of the article

element for organisms [10–12]. Hg^{2+} is very toxic pollutant that contaminates fish around the world, therefore fish represent the main source of Hg^{2+} for human [13]. The studies indicated that mercury accumulation in the oceans correlates with the rising tide of mercury pollution. The most serious Hg^{2+} poisoning has been occurred due to consumption of Hg^{2+} contaminated fish and other seafood polluted by industrial wastewater [14]. However, information for mercury contamination of fishes and marine environment in Libya is unavailable; this might due to absence of academic research for more than 40 years. Therefore, the present work aimed to evaluate the concentrations of Hg^{2+} in fish, sea woods and sediments at Farwa Island, Libya that received petrochemical wastes generated from General Company of Chemical Industries (GCCI) for more than 40 years.

Methods
Study area
Farwa Island is located on the Mediterranean in West Zawya, Libya (33° 04′ N, 1° 50′ E to 33° 08′ N and 11° 32′ E) from Abu- Kamash east to the Tunisian border in the west (Fig. 1). It comprises Farwa lagoon that covering an area of 32 km^2 and is the largest lagoon on the Libyan coast. GCCI is located at Abu- Kamash chemical complex. GCCI was opened in 1970s and consist of 3 units that produce 104,000 tonnes/year Ethylene di-chloride, 60, 000 tonnes poly vinyl chloride (PVC), 50,000 tonnes caustic soda and 45,000 tonnes chlorine. In addition to sodium carbonate, sodium hypochlorite and HCl. GCCI has four dumping sites, two of them are located on the west while another two are located on the east.

Collection and analysis of samples
Hundred ninety two samples (in triplicate, 3 sample/month) of fishes, oysters, cuttlefish, magnoliophyta plants and sediments were collected from marine environment around of Farwa Island, Libya during the period from January to August 2014. The marine organisms collected samples included ten types of fishes, only one type of oyster and one type of cuttlefish. These samples were collected using local fishermen. Magnoliophyta plant samples were collected from different location around Farwa Lagoon, Zone I (100 m), Zone II (1000 m) and Zone III (3000 m). These locations represent the distance between the sampling point and the factory and they were selected because its very close to GCCI and the possibility to heavy contamination with Hg^{2+} is high. The samples were transported inside ice box to the laboratory and kept in deep freezer at −20 °C until analysis.

Sample preparation and analysis were carried out according to Bernhard [15]. Liver, muscle, gill, heart, air sac and stomach-intestine were removed before the analysis [16]. Fish samples were homogenized in a blender. Magnoliophyta plants were cut out into small pieces (5 mm in diameter) and then homogenized in a blender. A weight of 10 g of homogenate for each of fish and magnoliophyta plants was digested according to APHA [17]. In briefly; five mL of HNO_3 (65 %) and 5 mL of H_2SO_4 were added into sample placed inside flask (100 mL). The mixture was heated on a hot plate (70–80 °C) for 30 min to the lowest volume (20 mL) before precipitation occurs. The digestion step was continued until light colored, clear solution was observed. The flask walls was washed with distilled water and filtered using Whatman, 125 mm Ø, filter papers (Cat No. 1001 England). The filtrate was

Fig. 1 Google map of study area, A) GCCI B) site of fish samples collection C) site of marine plant samples collection (100 m); D) site of marine plant samples collection (1000 m); E) site of marine plant samples collection (3000 m); F) site of sediment samples collection (100 m, W); G) site of sediment samples collection (500 m, W); H) site of sediment samples collection (1000 m, W); I) site of sediment samples collection (3000 m, W); J) site of sediment samples collection (100 m, E); K) site of sediment samples collection (500 m, E); L) site of sediment samples collection (1000 m, E); site of sediment samples collection (3000 m, E)

transported into a volumetric flask (100 mL) with 10 mL water and mixed thoroughly.

Sediment samples (1 kg) were collected by using grab sampler from eight sites located on the west and east of GCCI. Samples were transported to the laboratory and dried in oven at 50 °C. After that, sediment samples were powdered and passed through 160 μm sieve. The samples packed in paper bags and stored in deep freezer at −20 °C prior to analysis. The mercury was extracted from the samples with 10 mL HNO_3/HCl (1:3 v/v) by using a microwave digestion system as described above.

The Hg^{2+} concentrations in the digested samples were determined by an atomic absorption spectrophotometer (AAS) (Model P.E.A ANALYST 100, HGA-800 and MHS-10, Perkin Elmer, USA).

Table 1 ANOVA Analysis of Hg^{2+} concentrations in different fish samples during the period of study from January to August 2014

Sample		Sum of Squares	df	Mean Square	F	Sig.
Serranus scriba	Between groups	54.369	7	7.767	1257.814	.000
	Within groups	.099	16	.006		
	Total	54.468	23			
Oedalechilus labeo	Between groups	28.478	7	4.068	2697.193	.000
	Within groups	.024	16	.002		
	Total	28.502	23			
Diplodus vulgaris	Between groups	6.588	7	.941	1146.640	.000
	Within groups	.013	16	.001		
	Total	6.602	23			
Dicentrarchus labrax	Between groups	4.697	7	.671	712.616	.000
	Within groups	.015	16	.001		
	Total	4.712	23			
Lithognathus mormyrus	Between groups	70.353	7	10.050	2138.378	.000
	Within groups	.075	16	.005		
	Total	70.428	23			
Epinephelus marginatus	Between groups	99.308	7	14.187	16690.375	.000
	Within groups	.014	16	.001		
	Total	99.321	23			
Sarpa salpa	Between groups	11.577	7	1.654	301.838	.000
	Within groups	.088	16	.005		
	Total	11.664	23			
Sciaena umbra	Between groups	3.448	7	.493	467.260	.000
	Within groups	.017	16	.001		
	Total	3.465	23			
Pagrus pagrus	Between groups	3.373	7	.482	98.591	.000
	Within groups	.078	16	.005		
	Total	3.451	23			
Caranx crysos	Between groups	5.237	7	.748	949.947	.000
	Within groups	.013	16	.001		
	Total	5.249	23			
Pinctada radiata	Between groups	310.444	7	44.349	6809.846	.000
	Within groups	.104	16	.007		
	Total	310.548	23			
Sepia officinalis	Between groups	2.703	7	.386	639.161	.000
	Within groups	.010	16	.001		
	Total	2.713	23			

The concentrations of heavy metals was calculated ($\mu g\ g^{-1}$) using Eqs. (1)

$$MetalConcentration = A \times B/C$$

Where

A = concentrations of metals in digested solution $\mu g\ g^{-1}$
B = final volume of digested solution mL
C = sample size, gram

Data analysis

The data were not normally distributed, therefore, they were log transformed and subjected to parametric statistics. The differences in Hg^{2+} concentrations of samples investigated were tested by ANOVA. The statistical analyses was performed SPSS (version 11.5).

Results and discussion

The present study investigated the mercury contamination of marine environment included fishes, oysters, cuttlefish, magnoliophyta plants and sediments at Farwa Island, Libya that are received industrial wastewater generated from GCCI since 40 years ago. The concentration of mercury at this place has not reported before, thus the current work was conducted to evaluate the effect of petrochemical wastewater on the environment. The results revealed that the Hg^{2+} concentration differed significantly ($p < 0.05$) during the period of study (Table 1). These variables may be due to the climatic conditions of the area, winter season extends from November to March and is generally cold and rainy with unstable winds blowing from different directions which lead to cause dilution of Farwa lagoon, while summer season (May to September) is rather hot and dry [18]. The mean of Hg^{2+} concentrations in fish, oysters and cuttlefish samples collected during the period study are presented in Table 2. It can be noted that the Hg^{2+} concentrations ranged from 3.13 ± 1.5 $\mu g\ g^{-1}$ in *Serranus scriba* to 0.34 ± 0.33 $\mu g\ g^{-1}$ in *Sciaena umbra*. The distributions of Hg^{2+} concentrations for each species in the period from January to August 2014 are depicted in Fig. 2. It shown that the highest concentration of Hg^{2+} was detected in *Pinctada radiata* (11.67 ± 3.30 μgg^{-1}) in August, followed by *Serranus scriba* (6.37 ± 0.11 $\mu g\ g^{-1}$) in July and *Epinephelus marginatus* (6.19 ± 0.02 $\mu g\ g^{-1}$) in February. The *Serranus scriba* have high concentration of Hg^{2+} during the study period from January to July followed by *Epinephelus marginatus*, the average was 2.83 *vs.* 2.18 μg g-1. The lowest Hg^{2+} concentrations were detected in *Pagrus pagrus* (0.001 $\mu g\ g^{-1}$) and *Sciaena umbra* (0.01 $\mu g\ g^{-1}$). Both types contained the lowest average concentrations during the period of study (0.33 and 0.36 $\mu g\ g^{-1}$ respectively). *Lithognathus mormyrus* has the highest Hg (3.59 ± 0.19 $\mu g\ g^{-1}$) among the fish samples collected in April, whereas *Oedalechilus labeo* has the highest Hg (3.59 ± 0.01 $\mu g\ g^{-1}$) among the fish samples collected in May. In June, the highest Hg^{2+} was determined in *Lithognathus mormyrus* (4.97 ± 0.04 $\mu g\ g^{-1}$).

The analysis for association between Hg^{2+} concentrations in fish, oysters as well as cuttlefish samples and months indicated that the concentration of Hg^{2+} in *Serranus scriba*, *Dicentrarchus labrax*, *Sciaena umbra* and *Pinctada radiata* associated significantly ($p < 0.05$) to the seasons with R^2 0.64, 0.24, 0.21 and 0.34 respectively (Table 3). The Hg^{2+} concentrations in magnoliophyta plants are presented in Table 4. It can be noted that the maximum concentration was detected in the samples

Table 2 Hg^{2+} concentrations in Fishes collected from Farwa lagoon, Libya which received petrochemical wastes from General Company of Chemical Industries (GCCI), (±SD represent the standard division from the mean, $n = 24$ for each sample)

Sample No.	Family name	English name	Science name	Hg concentration ($\mu g\ g^{-1}$)
1	Serranidae	Painted comber	Serranus scriba	3.13 ± 1.5
2	Mugilidae	Boxlip Mullet	Oedalechilus labeo	1.4 ± 1.1
3	Sparidae	Common Two-Banded Seabream	Diplodus vulgaris	1.4 ± 0.53
4	Moronidae	European seabass	Dicentrarchus labrax	0.89 ± 0.45
5	Sparidae	Striped sea bream	Lithognathus mormyrus	1.5 ± 0.7
6	Serranidae	Dusky Grouper	Epinephelus marginatus	1.9 ± 2.0
7	Sparidae	Salema	Sarpa salpa	0.99 ± 0.71
8	Sciaenidae	sculpin	Sciaena umbra	0.34 ± 0.33
9	Dentex macrophthalmus	Red porgy	Pagrus pagrus	0.39 ± 0.38
10	Carangidae	blue runner	Caranx crysos	0.78 ± 0.48
11	Oyster	Rayed Pearl Oyster	Pinctada radiata	2.3 ± 3.6
12	Cuttlefish	common cuttlefish	Sepia officinalis	0.63 ± 0.34

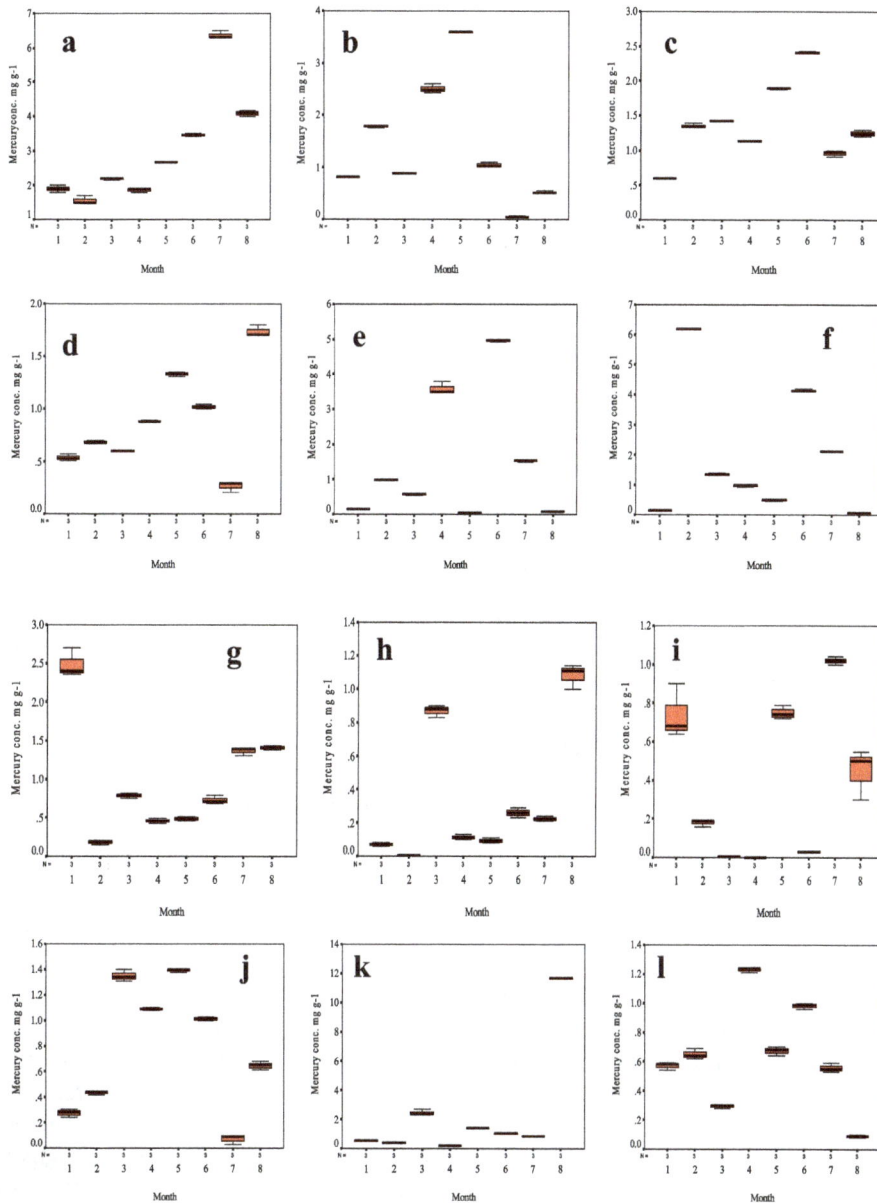

Fig. 2 Seasonal distribution of Hg^{2+} concentrations in fish, oysters and cuttlefish samples collected from Farwa lagoon, Libya; **a**) *Serranus scriba*; **b**) *Oedalechilus labeo*, **c**) *Diplodus vulgaris*; **d**) *Dicentrarchus labrax*; **e**) *Lithognathus mormyrus*; **f**) *Epinephelus marginatus*; **g**) *Sarpa salpa*; **h**) *Sciaena umbra*; **i**) *Pagrus pagrus*; **j**) *Caranx crysos*; **k**) *Pinctada radiate*; **l**) *Sepia officinalis*

collected from place near of GCCI (100 m). The highest concentrations were determined in samples collected during April, where, 2.33 ± 0.60, 1.44 ± 0.42 and 0.96 ± 0.12 µg g^{-1} were determined in samples collected from zone I, II and III respectively. The lowest Hg^{2+} was recorded in samples collected from zone III (0.02 ± 0.00 µg g^{-1}) during January. In comparison with the study conducted by Pergent-Martini [19] which was carried out on the mercury contamination in the *Posidonia oceanica* Collected from mediterranean sea.

It can be noted that the Hg^{2+} concentrations in this study was quite high. There would be due to dispose of wastewater generated from GCCI into the sea without treatment process since 40 years ago.

The present study revealed that the concentrations of Hg^{2+} in all types of fish samples were more than the standards limits recommended by FDA and FAO-WHO [20, 21]. According to U.S. EPA [22], Hg^{2+} should be less than 0.3 µg g^{-1} wet fish muscle tissue for protection of human health. However, Zaza et al. [13] reported that

Table 3 Measures of association between Hg^{2+} concentrations in fish, *oysters as well as cuttlefish* samples and months

Sample*month	R	R Squared	Eta	Eta Squared	Significance (p value)
Serranus scriba	0.80	0.64	0.99	0.99	0.01
Oedalechilus labeo	−0.228	0.05	1.00	0.99	0.14
Diplodus vulgaris	0.327	0.11	0.99	0.99	0.06
Dicentrarchus labrax	0.493	0.24	0.99	0.99	0.01
Lithognathus mormyrus	0.191	0.04	0.99	0.99	0.18
Epinephelus marginatus	−0.173	0.03	1.00	1.00	0.21
Sarpa salpa	−0.068	0.01	0.99	0.99	0.38
Sciaena umbra	0.454	0.21	0.99	0.99	0.01
Pagrus pagrus	0.215	0.05	0.98	0.98	0.16
Caranx crysos	0.006	0.00	0.99	0.99	0.45
Pinctada radiata	0.587	0.34	1.00	1.00	0.01
Sepia officinalis	−0.188	0.04	0.99	0.99	0.19

the minimum level of Hg^{2+} is 0.5 µg g^{-1} for fish species. In the present study, the minimum concentration of Hg^{2+} was 1 µg g^{-1} in *Pagrus pagrus*. Fish consumption is one of the major factors of Hg^{2+} intake for humans [23, 24]. Hg^{2+} is very dangerous for pregnant woman because mercury is most harmful to developing foetuses, infants, and young children.

High Hg^{2+} concentration was detected in sediment samples collected from the West of GCCI than those collected from the East. However, both sites contain high concentration of Hg^{2+}. The Hg^{2+} concentration decreased significantly ($p < 0.05$) as the site distance from GCCI, the maximum Hg^{2+} was noted in sediment samples taken from the west (100 m from GCCI) where 11.14 ± 4.11 µg g^{-1} was recorded in April 2014 (Table 5). Among the sediment samples collected from the east, the samples taken in June contain 4.67 ± 1.62 µg Hg^{2+} g^{-1}. The

Table 4 Hg^{2+} concentrations in magnoliophyta plant samples collected from different distance of GCCI at Farwa lagoon, Libya (±SD represent the standard division from the mean, $n = 3$ for each sample per month)

Sample/month	Hg^{2+} concentrations (µg g^{-1})		
	Zone I (100 m)	Zone II (1000 m)	Zone III (3000 m)
1	0.82 ± 0.20	0.93 ± 0.13	0.02 ± 0.00
2	0.39 ± 0.12	0.50 ± 0.08	0.10 ± 0.09
3	2.10 ± 0.91	0.57 ± 0.16	0.82 ± 0.20
4	2.33 ± 0.60	1.44 ± 0.42	0.96 ± 0.12
5	1.06 ± 0.18	0.79 ± 0.12	0.75 ± 0.19
6	1.00 ± 0.13	1.38 ± 0.92	0.54 ± 0.20
7	0.87 ± 0.30	0.25 ± 0.09	0.64 ± 0.21
8	1.40 ± 0.42	1.07 ± 0.13	0.71 ± 0.15

pollution of environmental area around GCCI represent a serious problem due to that the surrounding areas are used for agricultural purpose such as for Grapes, olives and almonds. More than 1500 people are living around the GCCI.

Farwa Island has high fishery production, but this Island had been exposed for heavy pollution due to GCCI for more than 40 years. Farwa Island is the most important coastal and marine site in western Libya, in terms of its high marine and coastal biodiversity based on several surveys and studies during the last years. However, no information was recorded according to mercury pollution. This region is characterized by an exceptional importance in terms of fish and artisanal fisheries, aquaculture, sea birds, sea grass meadows, land/seascape features and, above all, as one of the few regions in the Mediterranean to experience active tidal movements. In addition to some endangered species which makes it an important area for larva and juvenile protection. In the term of biodiversity, Farwa has many economically important species and certain endangered species are recognized [18].

In the term of toxic pollutants in industrial wastewater and their environmental impact and health effect, the sea water around of GCCI should be treated to remove of Hg^{2+} ions. Variety of biological and physico-chemical methods for wastewater treatment has been developed. Among those technologies, reverse osmosis, activated carbon, advanced oxidation, alumina-coated carbon nanotubes, tire derived carbons, porous carbon, carbon nanotubes and fullerene and CNT/magnesium oxide composite have exhibited high efficiency for removal of heavy metals from different aqueous solution [25–33].

Table 5 Hg^{2+} concentrations in sediment samples collected from the west and east GCCI during the period January to August 2014 (±SD represent the standard division from the mean, $n = 3$ for each sample per month)

Sediments sample/month	Hg^{2+} Concentration (µg g^{-1})								
	West of GCCI				East of GCCI				Control
	<100 m	500 m	1000 m	3000 m	100 m	500 m	1000 m	3000 m	Zwara city (20 km)
1	3.54 ± 1.10	1.38 ± 0.14	1.09 ± 0.04	0.64 ± 0.32	1.96 ± 1.01	0.55 ± 0.06	0.03 ± 0.0	0.01 ± 0.0	0.001 ± 0.0
2	6.74 ± 3.14	4.06 ± 1.81	1.07 ± 0.52	0.32 ± 0.19	1.55 ± 0.71	1.12 ± 0.24	0.75 ± 0.18	0.01 ± 0.0	0.001 ± 0.0
3	4.58 ± 1.43	3.04 ± 0.93	0.06 ± 0.09	0.02 ± 0.01	2.07 ± 0.91	0.75 ± 0.18	0.05 ± 0.01	0.001 ± 0.0	0.001 ± 0.0
4	11.14 ± 4.11	2.64 ± 0.83	1.95 ± 0.31	0.56 ± 0.08	3.65 ± 1.04	1.47 ± 0.91	0.01 ± 0.0	0.001 ± 0.0	0.001 ± 0.0
5	8.50 ± 2.84	5.07 ± 2.81	1.01 ± 0.41	0.75 ± 0.11	3.12 ± 1.11	0.06 ± 0.01	0.003 ± 0.0	0.001 ± 0.0	0.001 ± 0.0
6	2.46 ± 1.31	3.65 ± 1.71	0.95 ± 0.71	0.07 ± 0.22	4.67 ± 1.62	1.5 ± 0.46	0.001 ± 0.0	0.01 ± 0.0	0.003 ± 0.0
7	5.21 ± 2.28	4.06 ± 1.93	0.50 ± 0.01	0.43 ± 0.17	2.13 ± 0.98	0.04 ± 0.0	0.01 ± 0.0	0.002 ± 0.0	0.001 ± 0.0
8	1.56 ± 0.73	3.25 ± 0.88	0.36 ± 0.03	0.05 ± 0.0	3.17 ± 0.51	1.45 ± 0.29	0.003 ± 0.0	0.002 ± 0.0	0.001 ± 0.0

Conclusions

It can be concluded that the heavily contamination of fish, oysters as well as cuttlefish in marine environment around GCCI represent a main source for food poisoning among peoples living in this area. Therefore, the contaminated area should be treated to prevent health risk associated with mercury contamination.

Competing interests

The authors declared that they have no competing interest.

Authors' contributions

AA, RMS and AAS: conceived and planned the experiment. AA collected the samples and carried out the experiment. AAS: formulated the objectives, analysed the data, and drafted the manuscript. RMS has proofread and edited the Manuscript, provided guidance and improved the quality of the final manuscript. All authors read and approved the final manuscript. (AAB, RMSRM and A-GAAS).

Acknowledgments

The authors would like to thank Badu Society for Protection Marine Biology and Wild for allow us to use their laboratories. In addition, to express their appreciations to the Universiti Tun Hussein Onn Malaysia (UTHM), School of Civil and Environmental Engineering for Postdoctoral Fellowship to Adel Al-Gheethi.

Author details

[1]Environment Engineering Department, Subrata College, University of Zawia, Zawia, Libya. [2]Faculty of Civil & Environment Engineering, UTHM, Parit Raja, Malaysia. [3]High Institute of health sciences, Sana'a, Yemen.

References

1. Mittal A, Kaur D, Malviya A, Mittal J, Gupta VK. Adsorption studies on the removal of coloring agent phenol red from wastewater using waste materials as adsorbents. J Coll Interface Sci. 2009;337:345–54.
2. Mittal A, Mittal J, Malviya A, Kaur D, Gupta VK. Decoloration treatment of a hazardous triarylmethane dye, Light Green SF (Yellowish) by waste material adsorbents. J Coll Interface Sci. 2010;342:518–27.
3. Mittal A, Mittal J, Malviya A, Gupta VK. Removal and recovery of Chrysoidine Y from aqueous solutions by waste materials. J Coll Interface Sci. 2010;344:497–507.
4. SaleH TA, Gupta VK. Functionalization of tungsten oxide into MWCNT and its application for sunlight-induced degradation of rhodamine B. J Coll Interface Sci. 2010;362:337–44.
5. Saleh TA, Agarwal S, Gupta VK. Synthesis of MWCNT/MnO$_2$ and their application for simultaneous oxidation of arsenite and sorption of arsenate. Appl Catal B Environ. 2011;106:46–53.
6. Saleh TA, Gupta VK. Synthesis and characterization of alumina nano-particles polyamide membrane with enhanced flux rejection performance. Sep Pur Technol. 2012;89:245–51.
7. Vinod K, Gupa VK, Jain R, Nayak A, Agarwal S, Shrivastava M. Removal of the hazardous dye—Tartrazine by photodegradation on titanium dioxide surface. Mat Sci Eng: C. 2011;31:1062–7.
8. Gupta VK, Nayak A. Cadmium removal and recovery from aqueous solutions by novel adsorbents prepared from orange peel and Fe$_2$O$_3$ nanoparticles. Chem Eng J. 2012;180:81–90.
9. Gupta VK, Agarwal S, Saleh TA. Chromium removal by combining the magnetic properties of iron oxide with adsorption properties of carbon nanotubes. Water Res. 2011;45:2207–12.
10. Epstein E: Health issues related to beneficial use of biosolids. In: 16th Annual Residuals and Biosolids Management Conference of the Water Environment Federation, Texas. 2002, p. 9.
11. Clarkson TW. The toxicology of mercury. Crit Rev Clin Lab Sci. 1997;34:369–403.
12. Magos L. Physiology and toxicology of mercury. Metal Ions Biol Sys. 1997;34:321–70.
13. Langford N, Ferner R. Toxicity of mercury. J Hum Hypertens. 1999;13(10): 651–6.
14. Zaza S, de Balogh K, Palmery M, Pastorelli AA, Stacchini P. Human exposure in Italy to lead, cadmium and mercury through fish and seafood product consumption from Eastern Central Atlantic Fishing Area. J Food Comp Anal 2015, (Accepted). doi: 10.1016/j.jfca.2015.01.007.
15. Harada M. Minamata disease: methylmercury poisoning in Japan caused by environmental pollution. Crit Rev Toxicol. 1995;25:1–24.
16. Bernhard M. Manual of methods in aquatic environment research, part 3: sampling and analyses of biological material. Rome: FAO Fish Tech Paper No. 158, UNEP; 1976.
17. Öztürk M, Özözen G, Minareci O, Minareci E. Determination of heavy metals in fish, water and sediments of Avsar dam lake in Turkey, Iran. J Environ Health Sci Eng. 2009;6(2):73–80.
18. APHA. Standard methods for the examination of water and wastewater. 20th ed. Washington, DC: American Public Health Association; 1998.
19. Haddoud DA, Rawag AA. Marine protected areas along Libyan coast; In Report of the MedSudMed expert consultation on marine protected areas and fisheries management. MedSudMed Technical Documents No. 3 Rome (Italy), June 2007 GCP/RER/010/ITA/MSM-TD-03. 2007.
20. Pergent-Martini C. Posidonia oceanica: a biological indicator of past and present mercury contamination in the mediterranean sea. Marine Environ Res. 1998;45:101–11.
21. Hall RA, Zook EG, Meaburn GM. National Marine Fisheries Service. Survey of trace elements in the fishery resource. U.S. Department of Commerce National Oceanic and Atmospheric Administration National Marine Fisheries Service. 1978.
22. Voegborlo RB, Methnani AME, Abedin MZ. Mercury, cadmium and lead content of canned Tuna fish. Food Chem. 1999;67(4):341–5.

23. EPA US. Water quality criterion for the protection of human health-methylmercury. Office of Science and Technology, Office of Water, USEPA: U.S. Environmental Protection Agency; 2001.

24. Santos ECO, Camara VM, Jesus IM, Brabo ES, Loureiro ECB, Mascarenhas AFS, Fayal KF, Sa Filho GC, Sagica FES, Lima MO, Higuchi H, Silveira IM. A contribution to the establishment of reference values for total mercury levels in hair and fish in Amazonia. Environ Res. 2002;90:6–11.

25. Yasutake A, Matsumoto M, Yamaguchi M, Hachiya N. Current hair mercury levels in Japanese: survey in five districts. Tohoku J Exper Med. 2003;199:161–9.

26. Gupta VK VK, Srivastava SK, Mohan D, Sharma S. Design parameters for fixed bed reactors of activated carbon developed from fertilizer waste for the removal of some heavy metal ions. Waste Manag. 1997;17:517–22.

27. Gupta VK, Agarwal S, Saleh TA. Synthesis and characterization of alumina-coated carbon nanotubes and their application for lead removal. J Hazar Mat. 2011;185:17–23.

28. Saleh TA, Gupta VK. Photo-catalyzed degradation of hazardous dye methyl orange by use of acomposite catalyst consisting of multi-walled carbon nanotubes and titanium dioxide. J Colloid Interface Sci. 2012;371:101–6.

29. Saleh TA, Gupta VK. Processing methods, characteristics and adsorption behavior of tire derived carbons: A review. Adv Colloid Interface Sci. 2014;211:93–101.

30. Gupta VK, Kumar R, Nayak A, Saleh TA, Barakat MA. Adsorptive removal of dyes from aqueous solution onto carbon nanotubes: A review. Adv Colloid Interface Sci. 2013;193–194:24–34.

31. Gupta VK, Ali A, Saleh TA, Nayak A, Agarwal S. Chemical treatment technologies for waste-water recycling—an overview. RSC Adv. 2012;2:6380–8.

32. Gupta VK, Saleh TA. Sorption of pollutants by porous carbon, carbon nanotubes and fullerene- An overview. Environ Sci Poll Res. 2013;20:2828–43.

33. Saleh TA, Gupta VK. Column with CNT/magnesium oxide composite for lead(II) removal from water. Environ Sci Pollut Res Int. 2012;19:1224–8.

11

Optimization of DRASTIC method by artificial neural network, nitrate vulnerability index, and composite DRASTIC models to assess groundwater vulnerability for unconfined aquifer of Shiraz Plain, Iran

Mohammad Ali Baghapour[1], Amir Fadaei Nobandegani[1*], Nasser Talebbeydokhti[2], Somayeh Bagherzadeh[3], Ata Allah Nadiri[4], Maryam Gharekhani[4] and Nima Chitsazan[5]

Abstract

Background: Extensive human activities and unplanned land uses have put groundwater resources of Shiraz plain at a high risk of nitrate pollution, causing several environmental and human health issues. To address these issues, water resources managers utilize groundwater vulnerability assessment and determination of protection. This study aimed to prepare the vulnerability maps of Shiraz aquifer by using Composite DRASTIC index, Nitrate Vulnerability index, and artificial neural network and also to compare their efficiency.

Methods: The parameters of the indexes that were employed in this study are: depth to water table, net recharge, aquifer media, soil media, topography, impact of the vadose zone, hydraulic conductivity, and land use. These parameters were rated, weighted, and integrated using GIS, and then, used to develop the risk maps of Shiraz aquifer.

Results: The results indicated that the southeastern part of the aquifer was at the highest potential risk. Given the distribution of groundwater nitrate concentrations from the wells in the underlying aquifer, the artificial neural network model offered greater accuracy compared to the other two indexes. The study concluded that the artificial neural network model is an effective model to improve the DRASTIC index and provides a confident estimate of the pollution risk.

Conclusions: As intensive agricultural activities are the dominant land use and water table is shallow in the vulnerable zones, optimized irrigation techniques and a lower rate of fertilizers are suggested. The findings of our study could be used as a scientific basis in future for sustainable groundwater management in Shiraz plain.

Keywords: Composite DRASTIC, Nitrate vulnerability, Artificial neural network, Shiraz aquifer

* Correspondence: amirfadaei66@gmail.com
[1]Department of Environmental Health Engineering, School of Health, Shiraz University of Medical Sciences, Shiraz, IR, Iran
Full list of author information is available at the end of the article

Background

Shiraz plain located in southwest of Iran is highly dependent on groundwater for its economic and demographic development. However, urbanization and agricultural activities have caused groundwater contamination by several types of pollutants such as nitrate. Presence of nitrate in the water resources can pose health risks to humans. Therefore, water resources managers are concerned about health and ecological effects of water contaminated with nitrate [1, 2]. Nitrated water can cause blue baby syndrome and certain types of cancer, including cancer of digestive system, stomach, colon, bladder, ovaries, and testicles [3]. Therefore, assessment of groundwater vulnerability to detect the vulnerable areas of aquifers is very important in order to manage groundwater resources.

The concept of aquifer vulnerability was first introduced by Marget. This concept refers to the sensitivity of an aquifer to deterioration due to an external action and is based on the assumption that physical environment may provide some degrees of protection to groundwater against contaminants entering the subsurface zone. Consequently, some land areas are more vulnerable to groundwater contamination than others [4, 5].

There are two main types of vulnerability assessment: intrinsic vulnerability and specific vulnerability. The first term refers to the intrinsic property of groundwater system to human or natural impacts. The most leading model of the intrinsic vulnerability is DRASTIC index. DRASTIC index was introduced by United States Environmental Protection Agency (USEPA) for the first time [6, 7]. It is an abbreviation for seven main parameters in hydro-geological system, which control groundwater contamination. These parameters are Depth to water table (D), net Recharge (R), Aquifer media (A), Soil media (S), Topography (T), Impact of the vadose zone (I), and hydraulic Conductivity (C). In contrast, specific vulnerability is defined as the risk of pollution due to the potential impact of land uses. Based on this definition, it seems that model used in the specific vulnerability are more appropriate for forecasting groundwater vulnerability to nitrate pollution [8]. The Composite DRASTIC is the most widespread method of evaluation of the specific vulnerability. The CD model for the first time was proposed by Secunda et al [9]. In this model, in order to evaluate the potential risk of groundwater nitrate pollution, land use parameter is added to seven main hydrogeological parameters of DRASTIC index, then they were integrated through an additive formulation to estimate the specific vulnerability of a certain area. This model was successfully applied in Sharon region of Israel [9], Azraq basin of Jordan [10] and Hajeb-Jelma aquifer of Tunisia [11]. However, it was claimed that the additive formulation might fail to reflect the protective effect of land uses that do not have any adverse effects on groundwater quality [12]. Thus, a new approach based on a multiplicative model and focused on nitrate pollution has been proposed which is called Nitrate Vulnerability (NV) index. For the first time, Martinez-Bastida [12] suggested to utilize the NV index for assessing the risk of nitrate pollution in Central Spain. He declared using the NV index results in greater accuracy in estimations of specific vulnerability and designation of nitrate vulnerable zones in comparison to the CD index.

Both NV and DC methods rely to expert for assigning weights and rates of the parameters. Recently artificial intelligence (AI) models including artificial neural networks successfully utilized to decrease subjectivity in assessment of groundwater vulnerability [13–15]. ANN is a universal approximator to surrogate complex systems [16]. The most commonly used neural network is the multi-layer perceptron (MLP) with supervised training method that consists of one input layer, hidden layers, and one output layer [17, 18]. To apply ANN to the DRASTIC index, there are seven neurons in the input layer corresponding to the input data (D,R,A,S,T,I, and C), four neurons in the hidden layer, and one neuron in the output layer.

There is an ongoing discussion in the literature regarding the performance of different models in generating the specific vulnerability maps. This study aims to clarify the issue by comparing the results of CD, NV and ANN models for specific vulnerability assessment. The study uses Shiraz unconfined aquifer as a real world example for conducting this comparison.

Methods

The study area

The study area was Shiraz plain located in Fars province, Iran (Fig. 1), which is a part of Maharlu lake catchment. The Shiraz plain area is approximately 300 km^2 and it lies between longitudes 52^0 $29^{'}$ and 52^0 $36^{'}$ E and altitudes 29^0 $33^{'}$ and 29^0 $36^{'}$ N. From the north and northwest, the Shirza Plain, is limited by the Baba Koohi Kaftrak mountain heights and Drak Mountain, respectively. From the south, the Shiraz Plain extends to the Maharlu Lake.

Studies have shown that Shiraz plain is an alluvial aquifer with sequences of sand and clay, where the groundwater exists in the sand layers. Alluvial deposits are a sequence of sand and clay/silt layers with various thicknesses.

In addition, geophysical explorations in the plain have demonstrated that Shiraz plain's water-bearing layer in greater depths, it suffers from inappropriate quality. Previous studies indicated that the Shiraz Plain groundwater system is consist of a surface unconfined aquifer and a deep aquifer. Shallow groundwater goes down to a water table to the depth of 40 m, while deep groundwater ranges from about 40 m of depth to 200 m [19]. In the present study, three different models; i.e., CD and NV and ANN

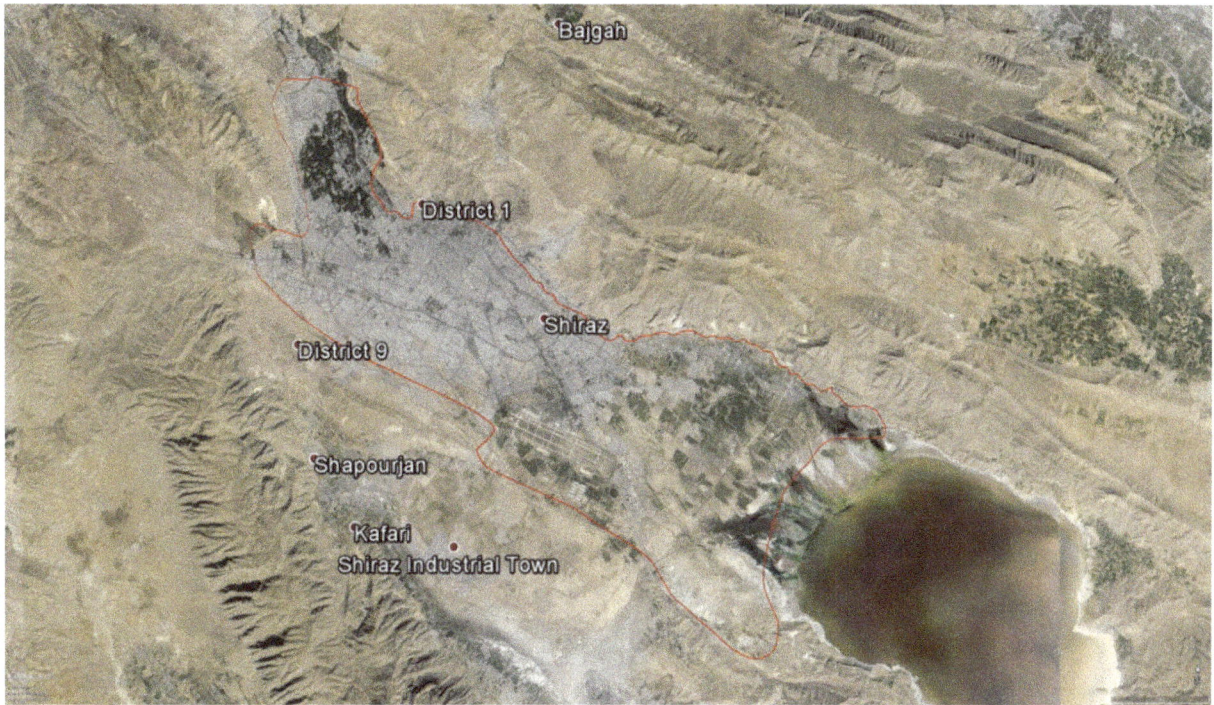

Fig. 1 Satellite image of Shiraz plain

models, were used for assessment of vulnerability and determination of groundwater protection zones of Shiraz plain unconfined (shallow) aquifer.

Two seasonal rivers of Khoshkrud and Chenarrahdar exist in Shiraz area. Chenarrahdar River comes from Rahdar bridge area into Shiraz and goes through the southern angle of the city and after absorbing the surface runoff, it joins Maharlu Lake.

The seasonal Khoshkrud River starts from Golestan and Ghalat mountains heights and after joining Khoshk River, it enters the Shiraz plain. Agricultural runoff and industrial wastewater enter this river at the south eastern part of Shiraz city and it finally joins Maharlu Lake. With respect to geology, the formations that outcrop in the study area, from old to new, are: Tarbor formation (Campanian to Maastrichtian), Pabdeh–Gurpi formation (Paleocene), Sachun formation (Paleocene), Jahrom formation (Eocene), Asmari formation (Oligocene), Razak formation (L. Miocene), Agha Jari formation (U. Miocene to L. Pliocene), Bakhtyari formation (U. Pliocene to L. Pleistocen), and quarternary alluvial deposits. The Asmari limestone formation has the most outcrops in the study area. The Shiraz Plain is located in a semi-arid climate zone with the average height of 1540 (m), annual mean precipitation of 365.3 (mm) and average temperature of 18.04 °C.

The Composite DRASTIC index (additive model)

CD index is an adaption of the DRASTIC index with the addition of a new parameter (L) to define the risk associated with land use (L). The objective of this approach is to evaluate the potential effect of extensive land use on groundwater quality resulting from alteration of the soil matrix and unsaturated zone media over time.

The DRASTIC Index takes into account seven parameters of the geological and hydrological environments, namely depth to water table (D), net Recharge (R), Aquifer media (A), Soil media (S), Topography (T), Impact of the vadose zone (I), and hydraulic Conductivity (C).

According to the effects of parameters on the probable vulnerability, a relative numerical weight from 1 to 5 is given to each parameter, with numbers 1 and 5 representing the least and the most effective, respectively. In addition, these seven parameters are divided into ranges and then receive a number from 1 to 10 according their influence on vulnerability. At the end, after collecting and digitizing the hydro-geological information using GIS, in order to prepare vulnerability maps, the information is overlaid and integrated and the result is a new layer called DRASTIC index (equation 1).

$$\begin{aligned} \text{DRASTIC index} = &\, Dr\ Dw\ + Rr\ Rw + Ar\ Aw \\ &+ Sr\ Sw + Tr\ Tw + Ir\ Iw \\ &+ Cr\ Cw \end{aligned} \tag{1}$$

In this equation: D, R, A, S, T, I, and C are the abbreviations of the seven effective hydro-geological parameters. Besides, subscripts "r" and "w" represent the corresponding ratings and weights in Table 1 [20–22].

Table 1 Ratings and weights given to the DRASTIC parameters [6]

Depth to water table (m)		Topography slope (%)		Hydraulic conductivity (m day -1)	
Range	Rating	Range	Rating	Range	Rating
0.0–1.5	10	0–2	10	0–4.1	1
1.5–4.6	9	0–2	9	4.1–12.2	2
4.6–9.1	7	6–12	5	12.2–28.5	4
9.1–15.2	5	12–18	3	28.5–40.7	6
15.2–22.9	3	>18	1	40.7–81.5	8
22.9–30.5	2				
>30.5	1				
Soil media		Aquifer media		Impact of the vadose zone	
Range	Rating	Range	Rating [a]	Range	Rating [a]
Thin or absent	10	Massive shale	1–3 (2)	Confining layer	1
Gravel	10	Metamorphic/igneous	2–5 (3)	Silt/clay	2–6 (3)
Sand	9	Weathered metamorphic/igneous	3–5 (4)	Shale	2–6 (3)
Peat	8	Glacial till	4–6 (5)	Limestone	2–5 (3)
Shiniking and/or aggregated clay	7	Bedded sandstone, limestone and shale sequence	5–9 (6)	Sandstone	2–7 (6)
Loam	5	Massive sandstone	4–9 (6)	Bedded limestone, sandstone and shale	4–8 (6)
Silty loam	4	Massive limestone	4–9 (8)	Sand and gravel with significant silt and clay	4–8 (6)
Clay loam	3	Sand and gravel	4–9 (8)	Sand and gravel	4–8 (8)
Muck	2	Basalt	2–10 (9)	Basalt	2–10 (9)
Non-shrinking and non-aggregated clay	1	Karst limestone	9–10 (10)	Karst limestone	8–10 (10)
Parameters	Relative weight				
Depth to water table	5				
Impact of the vadose zones	5				
Net recharge	4				
Aquifer media	3				
Hydraulic conductivity	3				
Soil media	2				
Topography slop	1				

[a]Typical rating in parentheses according to Aller et al. [6]

To determine Composite DRASTIC index, an additional parameter is added to DRASTIC index, which is land use. Therefore, the CD index is calculated using the following equation:

$$CD\ index = Dr\ Dw + Rr\ Rw + Ar\ Aw \\ + Sr\ Sw + Tr\ Tw + Ir\ Iw \\ + Cr\ Cw + Lr\ Lw \qquad (2)$$

Where L_r is the rating of the potential risk associated with land use, L_w is the relative weight of the potential risk associated with land use (according to Table 2) and the rest of the parameters are the same as equation 1.

The final outputs are ranged from 28 to 280 and are classified according to Table 3.

Nitrate vulnerability index (multiplicative model)

NV is another adaptation of the DRASTIC index, which was developed to achieve greater accuracy in estimation of the specific vulnerability to nitrate pollution. NV index is based on the real impact of each land use. This model attempts to integrate the risks of groundwater pollution by nitrate considering the land use as a potential source of nitrogen. The model incorporates potential negative and protective impacts uses that do not contribute to significant quantities of nitrate and do not enhance leaching of land uses overtime, such as the protected natural areas. It is based on a multiplicative model, involving the addition of a new parameter called the "potential risk associated with land use" (LU), which is calculated according to the following equation:

Table 2 Ranges and ratings applied to the potential risk associated with land use (L) according to the CD index [9]

Land use category[a]	Lr
Urban areas	8
Irrigated field crops	8
Orchards	6
Uncultivated land	5
Lw = 5	

[a] Main land uses observed in Shiraz Plain

Table 4 Ranges and ratings applied to the potential risk associated with land use (LU) as a source of nitrate pollution for the NV index [12]

Range	LU
Irrigated field crops	1.0
Urban areas	0.8
Uncultivated land and Semi-natural areas	0.3
Forests and natural areas	0.2

$$NV\ index = (Dr\ Dw + Rr\ Rw + Ar\ Aw + Sr\ Sw + Tr\ Tw + Ir\ Iw + Cr\ Cw).LU \tag{3}$$

Where LU refers to the potential risk associated with land use (according to Table 4) and the rest of the parameters are the same as equation 1.

The final outputs are classified according to Table 3.

Preparation of the needed layers

Depth to water table (D)

It is the depth from the ground surface to the water table. To get this layer, the most recent data regarding water level of 31 wells existing in the area of Shiraz plain was used. The interpolation method was applied to change the mentioned point data into raster map of water level. Finally, depth layer was prepared and was classified according to Table 1.

Net Recharge (R)

Is the amount of surface water that infiltrates to the ground and reaches the groundwater level. This study utilizes the Piscopo method [23] to prepare the net recharge layer for the Shiraz Plain based on the Table 5 and equation (4), below:

$$Recharge\ index = Soil\ permeability + Rainfall + Slope\ (\%) \tag{4}$$

In equation (4), the slope (%) was extracted from a Digital Elevation Model (DEM), which was generated from the topographic map of the Shiraz Plain and the Shiraz Plain soil map, log observations and exploration wells were used to quantify the soil permeability.

Table 3 Vulnerability ranges corresponding to the CD index [9] and the NV index [12]

Vulnerability	Ranges (CD index)	Ranges (NV index)
Very low	<100	<70
Low	100–145	70–110
Moderate	145–190	110–150
High	190–235	150–190
Very high	≥235	≥190

Aquifer media (A)

Is an indicator of material characteristics in the saturate zone. In this study, information from 20 well logs from Fars Regional Water Organization (FRWO) was used to prepare the aquifer media layer.

Soil media (S)

Is the top portion of the unsaturated zones that extends to the plants' roots and organic creatures activity areas.

Soil map was prepared by using the soil map of Shiraz plain and log of observation and exploration wells.

Topography (T)

Is an indicator of land slope changes in the area. The abovementioned soil layer was used in preparation of this layer. The Topography Layer was prepared using the same method that was used in provision of net recharge layer and was classified based on the Aller table.

Impact of vadose zone (I)

Vadose zone is a layer in between the aquifer and the soil zone. The vadose zone characteristics show the attenuation behavior of the materials that are located above the groundwater table and below soil. This study used the lithologic data of 20 observation and exploration wells to develop the vadose zone media of the Shiraz Plain. Then, using this information and Table 1, we designed the raster map of Shiraz plain.

Hydraulic conductivity (C)

Is a property of an aquifer that describes the ability of water to move through the aquifer. We used 23 pumping tests data conducted by FRWO to generate the hydraulic conductivity layer. Table 1 was employed to rate the generated hydraulic conductivity layer.

Land use

This layer is imperative since it is required by both CD and NV indices. The 2009 IRS (Indian Remote

Table 5 Ranges and ratings applied to the Net Recharge parameter according to the Piscopo method [23]

Slope		Rainfall		Soil permeability		Recharge value	
Slope %	Factor	Rainfall (mm/year)	Factor	Range	Factor	Range	Rating
<2	4	850<	4	High	5	11–13	10
2–10	3	700–850	3	Moderate to high	4	9–11	8
10–33	2	500–700	2	Moderate	3	7–9	5
>33	1	500>	1	Low	2	5–7	3
				Very low	1	3–5	1

sensing Satellite) data was used to generate this layer and then Table 2 and 3 were applied to rate it for preparing land use (L) map and land use associated risk (Lu), which are required for CD and NV indiced, respectively.

Artificial neural network method (ANN)

In this study, the artificial neural network method (ANN) was used to present a model with higher performance and improve the DRASTIC method. For this purpose, input and output data (vulnerability) of DRASTIC model and the relevant nitrate values were divided into two categories of Train and Test. Vulnerability index values which were the results of drastic model, were corrected by nitrate values and model train was done by these corrected values. To conduct the test phase of the model, drastic parameters in the data of this phase were considered as input and groundwater vulnerability index was assumed as model output and the results were evaluated using nitrate concentration. Artificial neural networks are a mass information processing system that are parallel and have functions like human brain neural network [24]. The following principles are the basis of artificial neural networks: (1) Data processing takes place in individual units called neurons. (2) Signals between neurons are transmitted through communication lines. (3) The weight is assigned to communication line of that line. (4) Each neuron typically has activation functions and convertor to determine output signals from input data of the network [25]. Artificial neural network structure is in traduced by the pattern of connections between neurons, the method of determining the weights of communication and transfer function [26]. Normal structure of an artificial neural network is usually formed by the input layer, middle (hidden) layer and the output layer. The input layer is a transport layer and a mean for supplying the data. The last layer or the output layer includes predicted values by the network and therefore introduces the model output. Middle or hidden layers that are composed by processing neurons, are the place for data processing. The number of hidden layers and neurons in each hidden layer is usually determined by trial and error method. Neurons in adjacent layers in the network are fully linked together. Artificial neural networks are classified in various ways, such as how neurons are connected and data movement in the network [25]. In this study, the Multilayer Perceptron network which is one of the leading networks, was used where the information move input to the output. Neurons in one layer are not connected, but the neurons in one layer are connected to the neurons in the next layer. So the output of a neuron in a layer depends on the signal received from the previous layer, the weight assigned and the type of convertor function. Different steps in a network, are conducted by various mathematical algorithms in which the most important ones are: 1- BP: Back Propagation Algorithm, 2- CG: Conjugate Gradient Algorithm, 3- LM: Levenberg-Marquardt. Among which LM algorithm is the most efficient algorithm [27]. LM algorithm was used in this study.

Nitrate Pollution Map

Groundwater nitrate concentration, as the most common pollutant in the study aquifer, can determine which of the used indexes offers greater precision in predicting the vulnerable areas. It is obvious that the model showing a higher correlation with the nitrate map will be more efficient. 82 groundwater samples were collected from 41 wells in a biannual sampling event, which took place in the months of August and January. The collected samples were analyzed for their nitrate concentration. The nitrate concentration distribution map was prepared using the average nitrate measurement in each point by employing ArcGIS 9.3 for interpolation.

Statistical Analysis

A regression analysis was performed on the measured nitrate concentrations using the SPSS (Statistical Package for Social Science) statistical software (v. 19.0) to compare the three vulnerability indexes and to evaluate the consistency of each index with respect to the spatial distribution of nitrate pollution.

Results and Discussion

Depth to water table (D)

Figure 2 and the DRASTIC model parameters rating table show that the groundwater table in the Shiraz Plain is from few meters to approximately 55 m. The results also indicate that the least effect of groundwater depth on groundwater vulnerability occurs in the northwest of the Shiraz Plain; whereas, the most effect of groundwater depth on vulnerability occurs in the south east part of the Shiraz Plain.

Net recharge (R)

Based on the Pscopo's method the Shiraz Plain was divided in four (4) categories with respect to the net recharge, where the highest net recharge values were observed in southern and southeastern parts of the Shiraz Plain, where, the least net recharge values were observed in the north parts of the Shiraz Plain (Fig. 3).

Aquifer media (A)

Figure 4 shows that majority of the Shiraz Plain is composed of clay and sit, where the coarse deposits can be found around Kaftrak Mountain.

Soil media(S)

Figure 5 shows that an area from northwest to the center of the plain are sandy loam, where, moving to the southeastern parts, the thickness of the soil layer decreases.

Topography (T)

Figure 6 indicates that the Shiraz Plain slope varies from 0 to 2 % with the worst possible rating.

Impact of vadose zone (I)

The results show that clay deposits exist in the north west toward southeast of the Shiraz Plain, where coarser media is seen in north and northeastern parts of the Shiraz Plain (see Fig. 7). The results also indicate that 87 % of the Shiraz Plain aquifer received rating values of one (1) to two (2).

Hydraulic conductivity (C)

The hydraulic conductivity of the Shiraz Plain aquifer was found to be varied from 0.34 m/day to 37 m/day, where majority of the aquifer have the hydraulic conductivity of around 12 m/day (Fig. 8) and high conductivity parts are in the east of the aquifer.

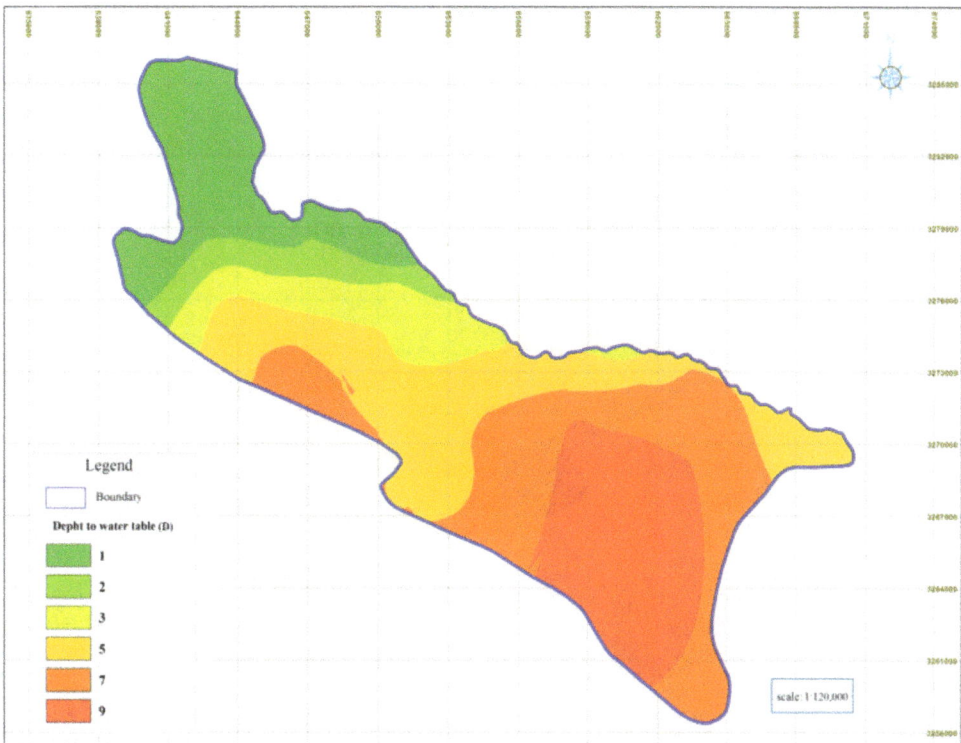

Fig. 2 Rated maps of Depth to water table

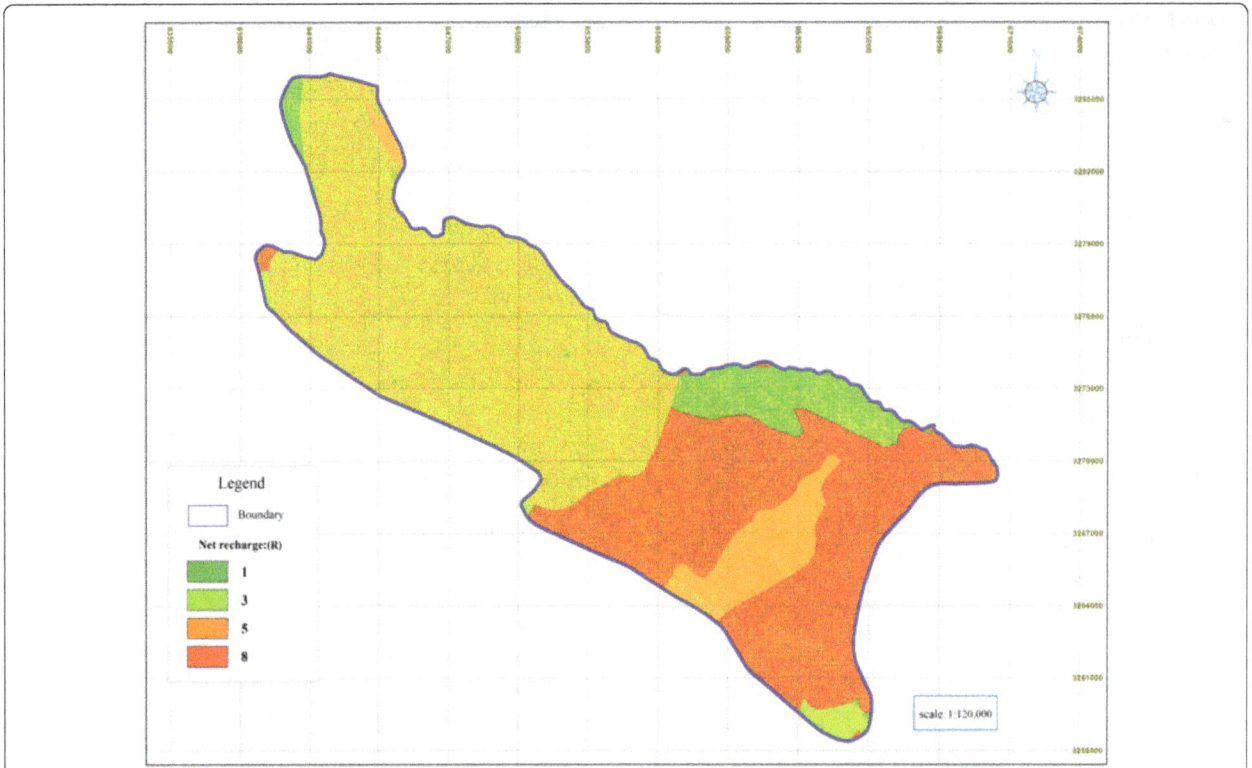

Fig. 3 Rated maps of net recharge

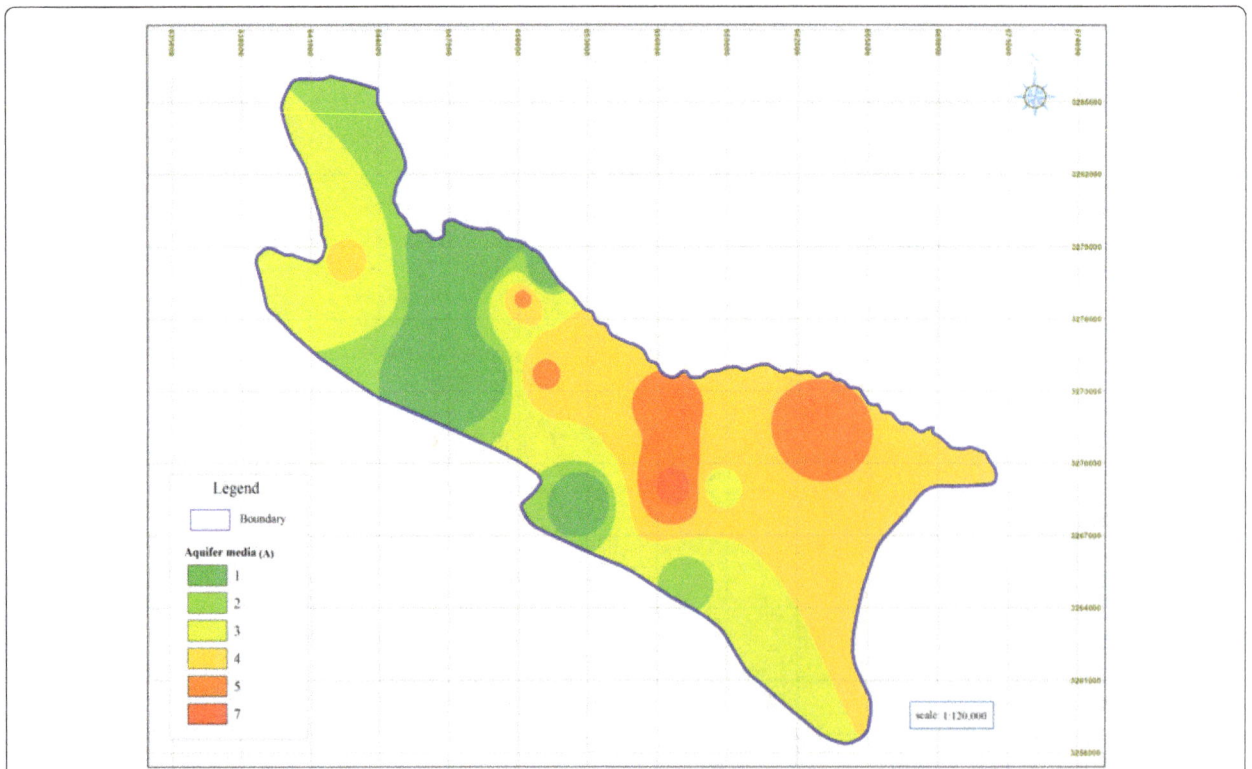

Fig. 4 Rated maps of aquifer media

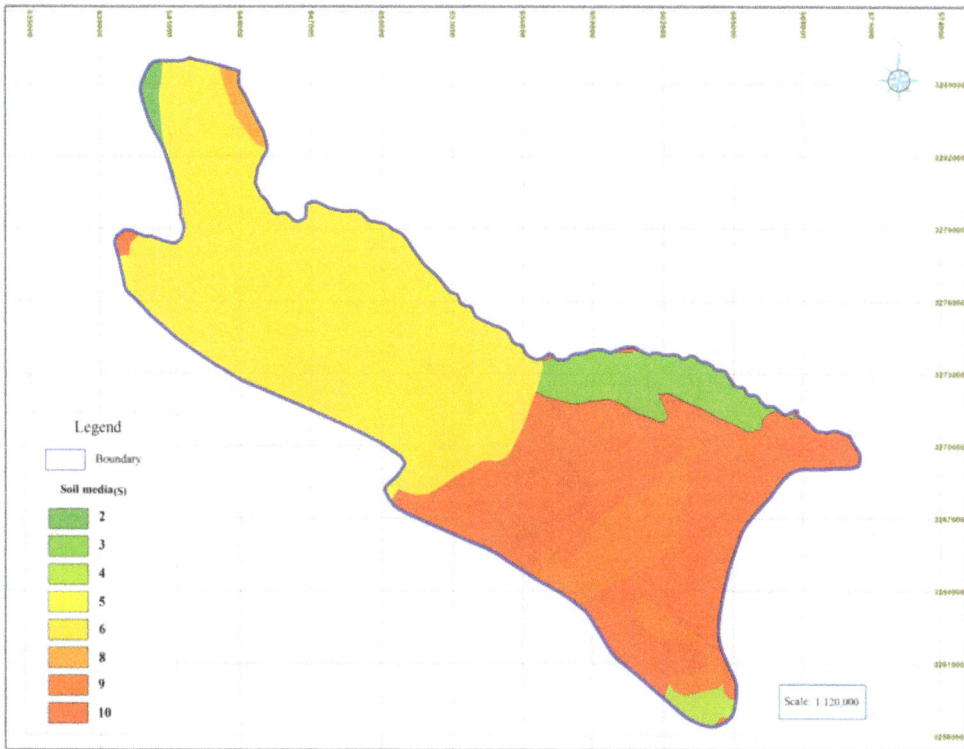

Fig. 5 Rated maps of soil media

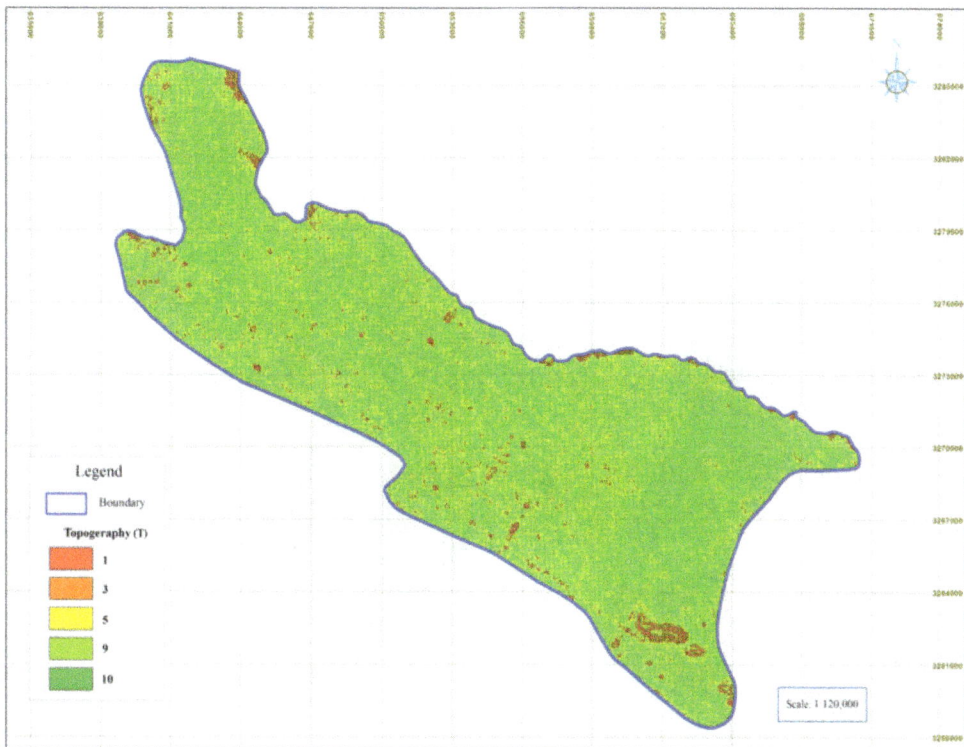

Fig. 6 Rated maps of Topography

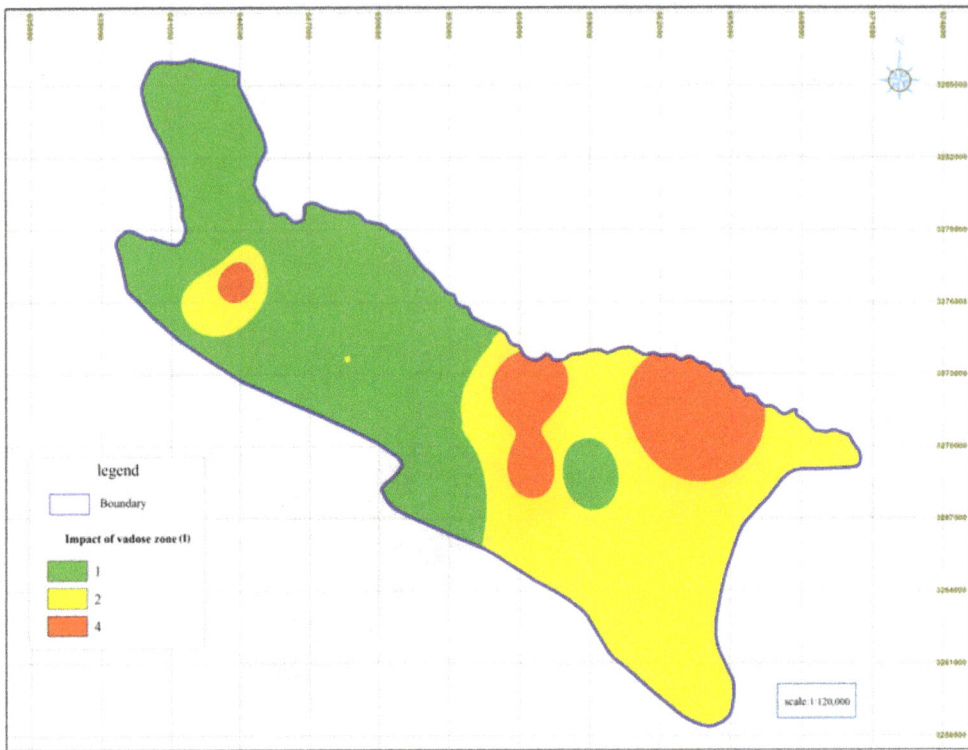

Fig. 7 Rated maps of Impact of vadose zone

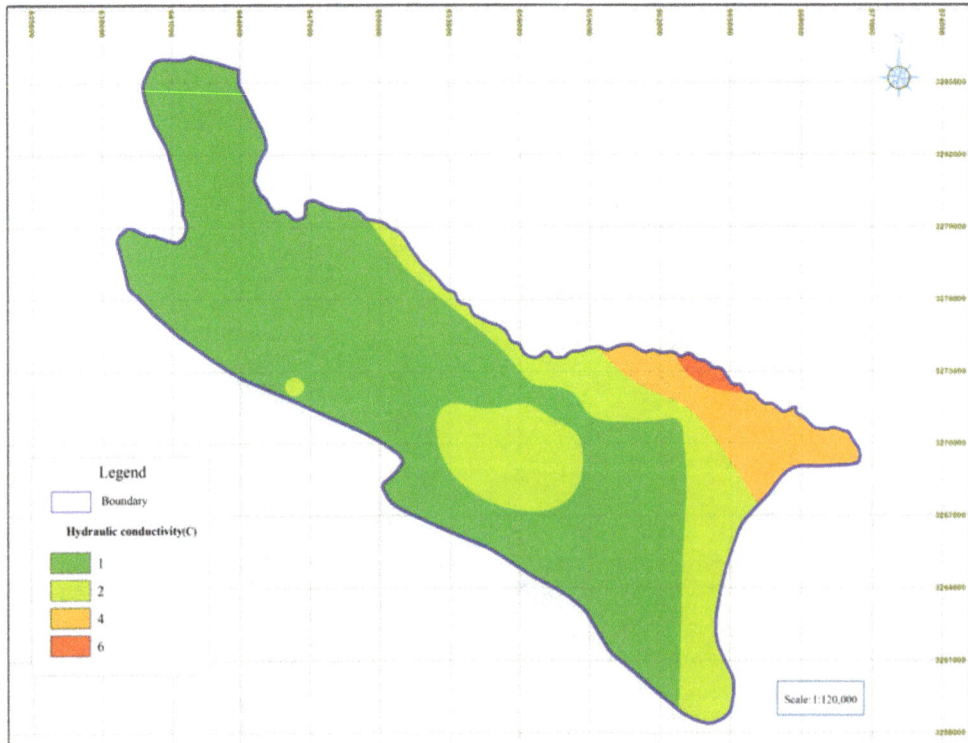

Fig. 8 Rated maps of Hydraulic conductivity

Land Uses (L) and Potential risk associated with land use (LU)
The results in Figs. 9 and 10 show that portions of uncultivated land, urban areas, agricultural lands, and orchards are 45.67, 38.29, 11.32, and 4.72 % respectively.

Specific vulnerability of groundwater to nitrate pollution according to the CD index
The final CD index was calculated using equation 2 through multiplying the rated layers by their weights and integrating them in GIS (Geographic Information System). Then, zoning of Shiraz plain's vulnerability map was done (Fig. 11). Accordingly, CD index for Shiraz aquifer varied from 53 (very low) to 185 (medium) and was divided into three classes as follows: 19 % of the area; i.e., northern and northwestern parts, had very low risk of pollution, 25 % of the area, mainly southeastern areas, showed moderate vulnerability to nitrate pollution, and other parts of the aquifer (56 %) were at low risk.

Specific vulnerability of groundwater to nitrate pollution according to the NV index
The results for specific vulnerability to nitrate pollution according to the NV index have been presented in Fig. 12. The map demonstrated that the NV index of Shiraz plain ranged from 6.4 to 185, and was divided into three classes: very low (<70), low (70–110), and

medium (110–145). On this basis, 6.45 % of the study area located at central and southeastern parts of the plain had moderate vulnerability, whereas the remaining 81.9 % were of very low and low vulnerability. There were no areas in the high and very high vulnerability classes.

Artificial neural network method (ANN)
In order to predict the vulnerability of groundwater by artificial neural networks method, perception perceptron three-layer network with LM algorithm was used. In this method, 7 input parameters including drastic parameters were used as the input layer. The number of neurons in the middle and output layers is 9 and 1, respectively. LM algorithm was used to train the network; the details of training and the calculation process is provided by the American Society of Civil Engineers [25]. The convertor transfer function is in the second layer of the tangential sigmoid (tansig) and the third linear layer (purlin). The number of courses epoch was 100 and the coefficient of determination and RMSE (Root Mean Square Error) values were 0.90 and 7.55, respectively. After training, the model was conducted for test phase step and vulnerability coefficient predicted by nitrate values was obtained. In Fig. 13, vulnerability map is provided by artificial neural network. As seen in the figure, majority of the region has low vulnerability and

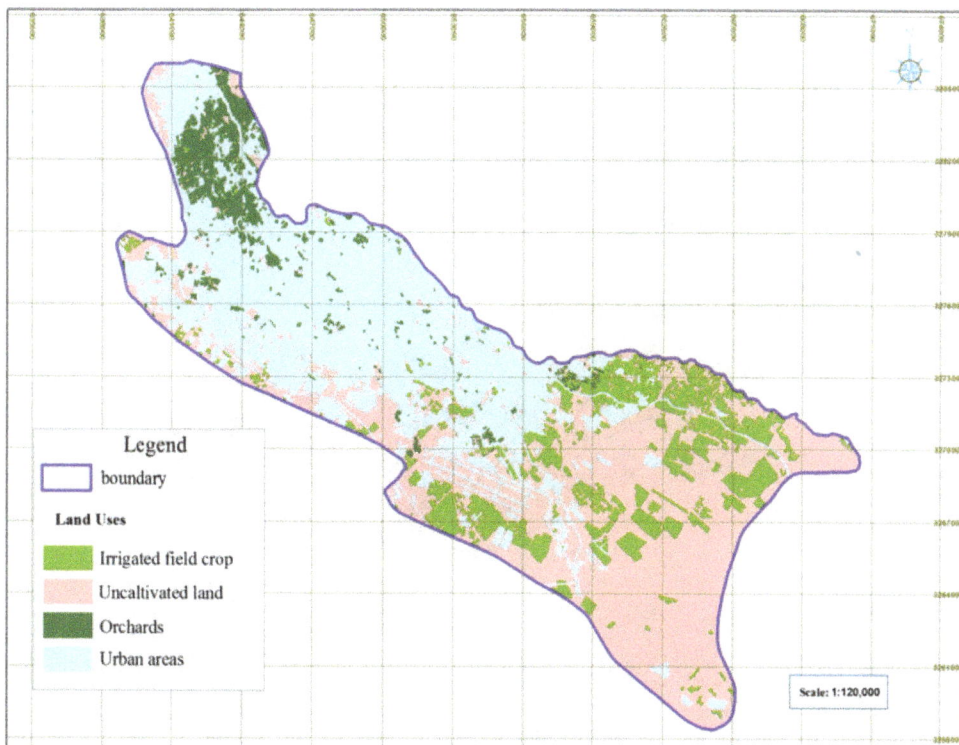

Fig. 9 The major land uses classes in the study area

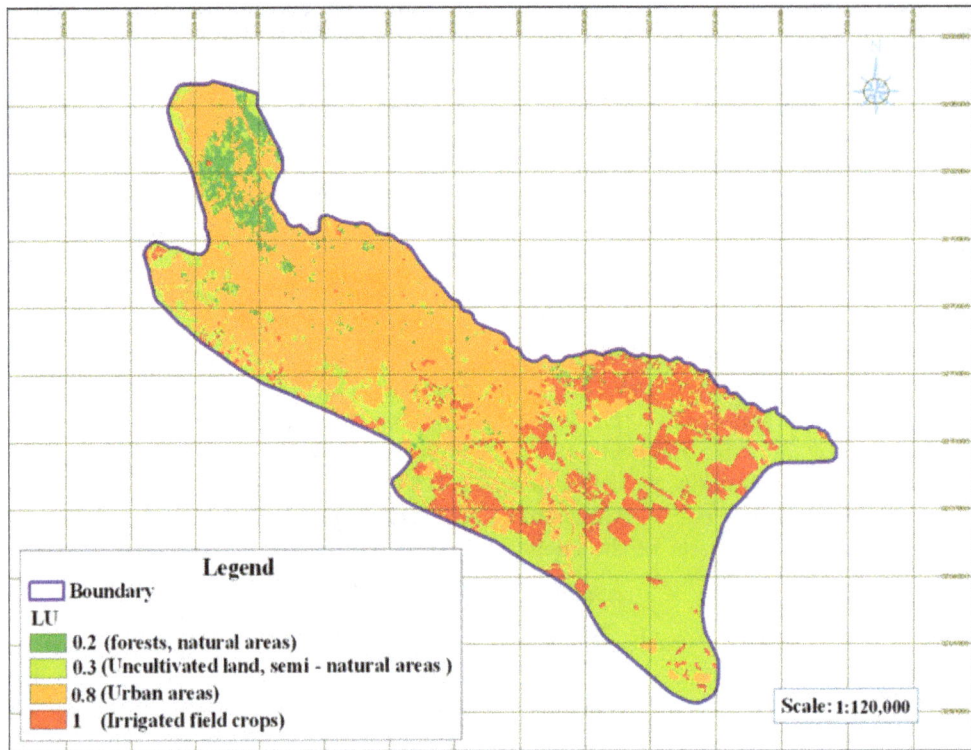

Fig. 10 Rated map of potential risk associated with land use (LU)

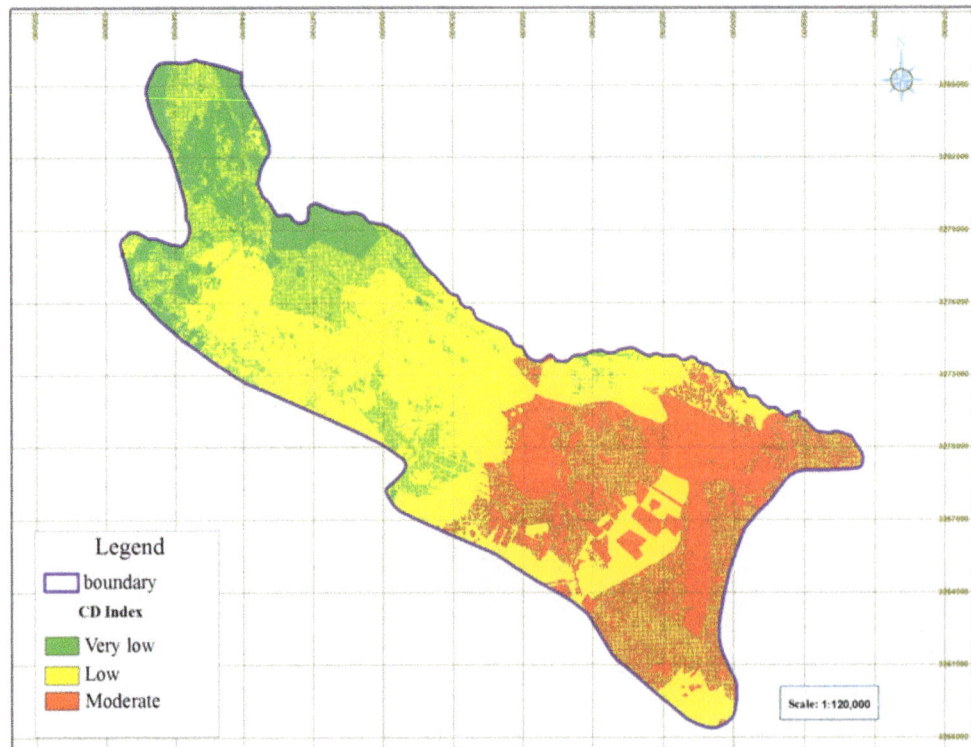

Fig. 11 Map of specific vulnerability to nitrate pollution according to the CD index

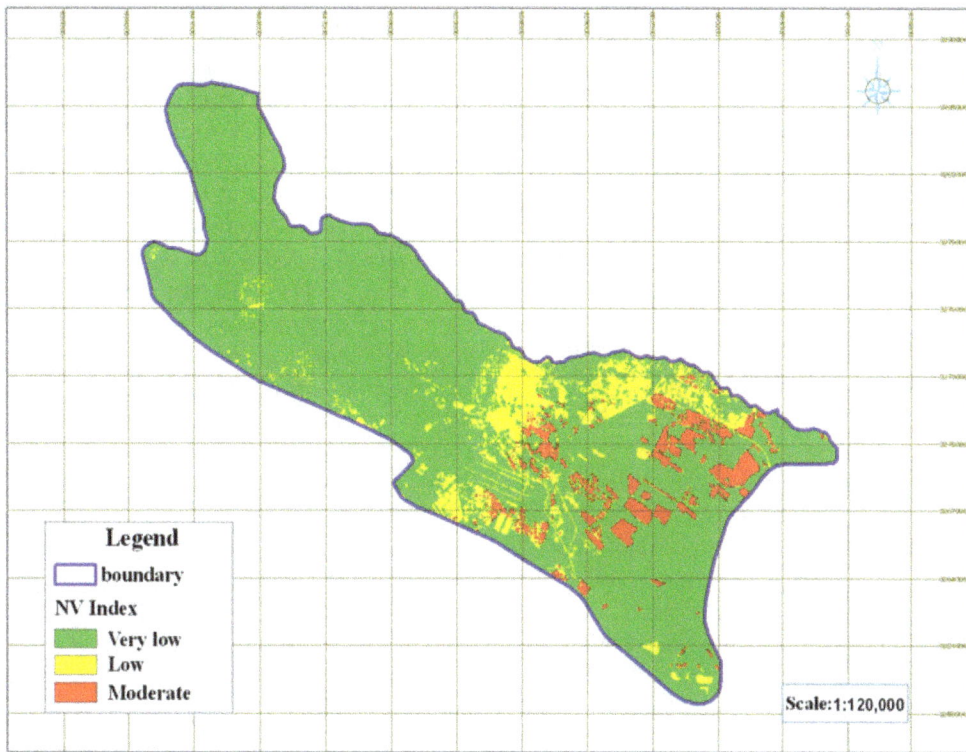

Fig. 12 Map of specific vulnerability to nitrate pollution according to the NV index

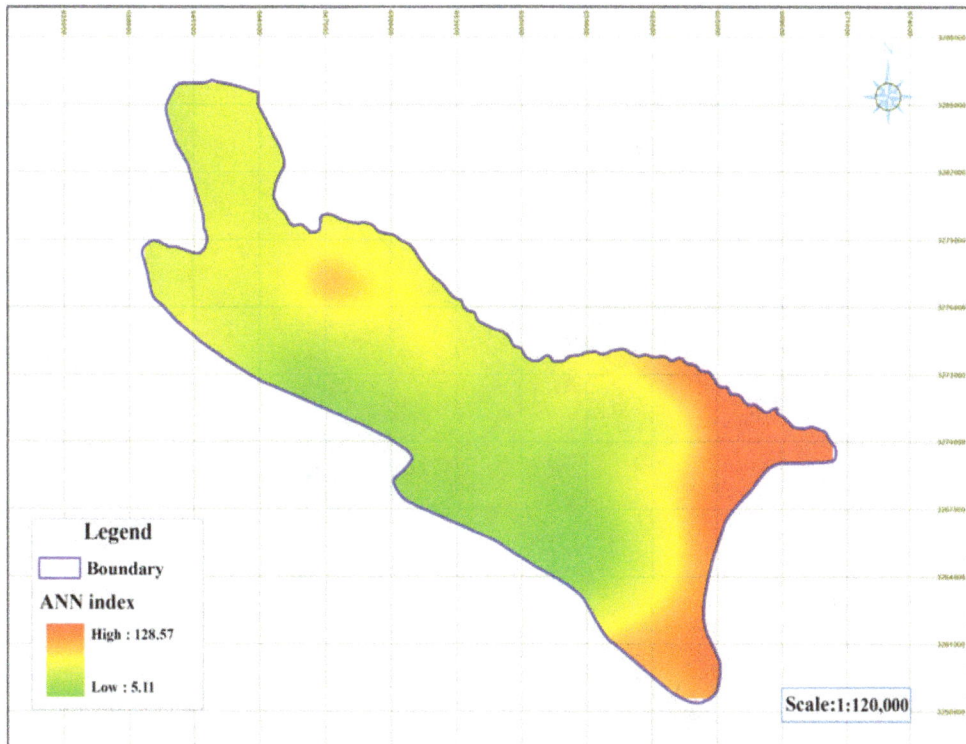

Fig. 13 The vulnerability map using ANN method

the regions adjacent to Maharloo Lake in the South East part of the region and the northern part of the plain have more vulnerability. Comparing this map with the map of nitrate, it can be seen that there is a good conformity between nitrate distribution in the region and vulnerability. Nitrate is increased in the south part of the region and areas adjacent to Maharloo Lake, as well as some parts in the North; and according to ANN map, vulnerability in these areas is higher than other estimated areas.

Comparison of CD, NV and ANN models

The groundwater nitrate pollution map for Shiraz aquifer is presented in Fig. 14. Based on this results the southeast part of the study area has these highest nitrate concentration and the western part of the study area has the lowest nitrate concentration. This confirms the results of NV, CD and ANN models, since the areas with higher concentration of nitrate have higher risk in comparison to the areas with lower risk. The vulnerability maps based on these three indexes showed similar results, identifying the southeastern part of the aquifer as the vulnerable zone. However, the percentage of the areas in the moderate class in the NV index was lower compared to that in the CD index. The results of regression analysis indicated a significant quadratic non-linear

relationship between groundwater nitrate concentration in the study area and NV and CD and ANN model values ($P < 0.01$). Moreover, the ANN models showed better r-squared value and greater accuracy compared to the NV and CD indexes regarding nitrate distribution (Table 6).

The results showed that the ANN model was the most effective model to improve the DRASTIC index, compared with CD and NV indexes. This can be justified by three reasons: (1) ANN as a nonlinear model is able to better explain nonlinear behavior of aquifer that is a complex system, (2) In contrast to NV and CD methods, supervised training method adopted corrected vulnerability to train the ANN model, (3) The ANN model reduces subjectivity of model by using LM optimization method to reach to suitable DRASTIC model. In addition according to our findings, the NV index allowed improved accuracy in determination of the vulnerable areas and showed better correlation coefficient compared to the CD index. This result was consistent with the observations of Martinez-Bastida J et al. [12] who proposed the NV model for the first time and concluded that their new type of multiplicative model offered greater accuracy. With respect to their viewpoint, this is due to the fact that the NV index incorporates both the negative impacts of some land uses over time and also the protective effects that others may have

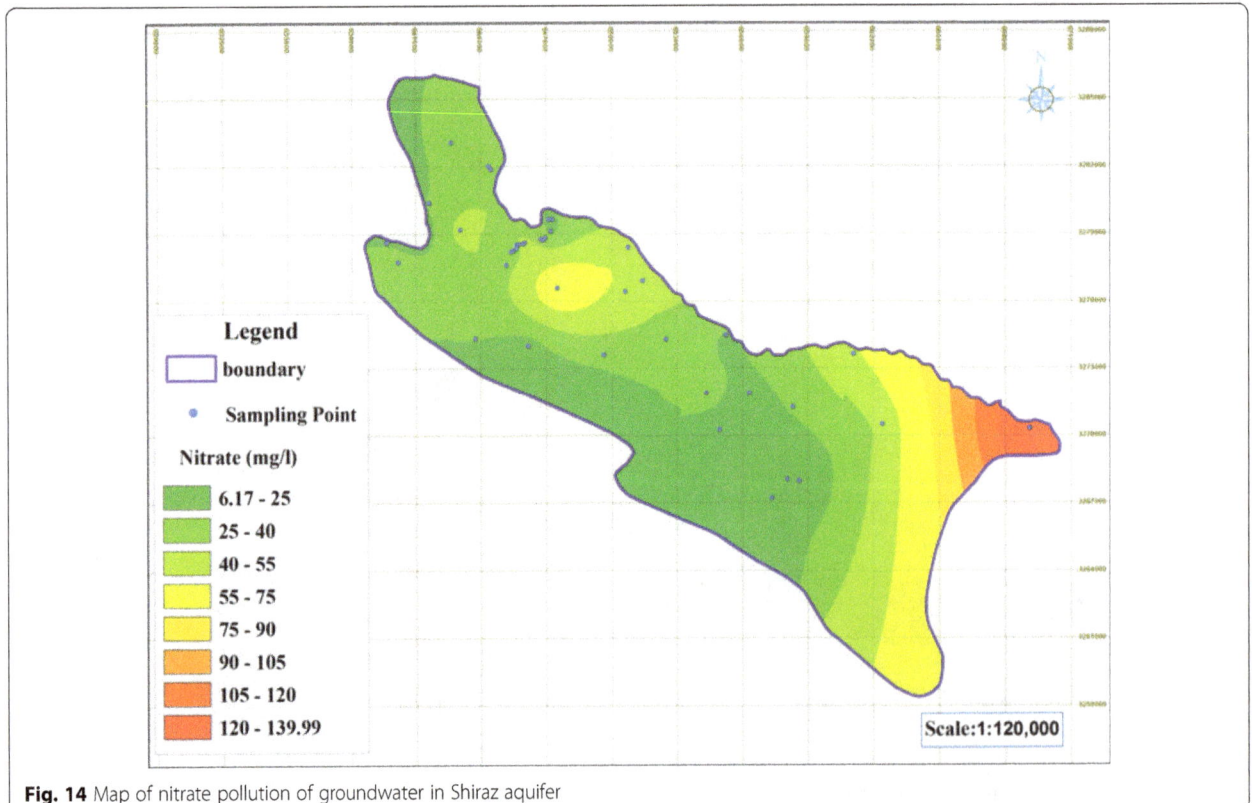

Fig. 14 Map of nitrate pollution of groundwater in Shiraz aquifer

Table 6 Regression analysis results using the mean nitrate concentration of groundwater in the Shiraz aquifer

Model	R Square
CD	0.292
NV	0.303
ANN	0.80

upon the aquifer media (uses that do not contribute to significant quantities of nitrate and do not enhance leaching, such as the protected natural areas. Nevertheless, it seems that the real reason for higher accurate of the NV index is that in this index, maximum rating (1) is assigned to irrigated field crops and lower rating (0.8) is allocated to urban areas. In the CD index, on the other hand, a same rating (8) is assigned to both agricultural and urban areas, but construction of sewerage systems in the recent years has caused leakage of nitrate from urban uses to decrease, so that in some regions, including Shiraz, this type of land use is not considered as a main nitrate source any more. As a result, it is not reasonable to assign the same rating to land uses with different effects. This seems to be the main reason why the NV index was more accurate in the present study.

As mentioned above, most parts of Shiraz aquifer had very low and low vulnerability. Although depth of water table in these parts was shallow, they contained fine-grained sediments which led to decrease of surface recharge and the possibility of increase in occurrence of attenuation process, including chemical degradation, absorption, and dispersion [28, 29]. This can be noticed from the rated maps of the aquifer media, vadose zone, and hydraulic conductivity (Figs. 4, 7 and 8). On the other hand, NV, CD, and ANN models determined the southeastern parts of the study area as the most vulnerable areas. This can be explained by the shallow groundwater depth of aquifer, low thickness of the saturated area, lack of soil layer, the general slope of the Shiraz Plain, which is toward this area, and the high volume of agricultural activities in the area. These facts imply that the traditional irrigation methods, such as flood irrigation, result in low N-use efficiency with high risks to the southeastern part of the study aquifer. The impact of irrigation on enhanced nitrate leaching has also been reported by some other researchers [30–33]. In one study, it was found that before irrigation, NO_3 concentration was less than 20 mg/L, while irrigation accompanied by fertilizer application caused NO_3 concentration in the upper layers of the aquifer to reach 65 mg/L [34]. It is necessary to say that nitrate leaching can be reduced through: (i) the use of crops with high N-use efficiency during the cropping

period, (ii) a fertilization strategy aimed at synchronizing fertilizer application to meet crops demand through split applications or the use of slow release nitrate-N fertilizers, (iii) an effective irrigation and water management system to minimize water losses and increase water use efficiency by the crops, (iv) an integrated approach using a good water and fertilizer management system and crops with high N-use efficiency, and (v) avoiding intensive agriculture in vulnerable areas where nitrate leaching potential is high.

Another main reason for the higher nitrate concentrations in the southeastern parts of the plain was elevated groundwater level. In order to overcome this problem, three drainage lines are being constructed in southern and southeastern parts of Shiraz plain during the recent years. However, the efficiency of the drainage system in this area decreases due to the fine grain soil in the area near Maharlu Lake and at the end of the drainage.

Conclusion

In this study, we assessed the vulnerability of Shiraz unconfined aquifer using CD (based on additive formulation), NV (based on multiplicative formulation), and ANN (based on artificial intelligence) models. The results confirmed the usefulness of these models for evaluating the risk of nitrate pollution in Shiraz plain. However, the map of specific vulnerability based on the ANN model proved to be more accurate with respect to the real nitrate distribution in groundwater within the study area. Based on the findings, more than two-thirds of Shiraz aquifer had very low and low vulnerability. Three models confirmed that areas with high volume of agricultural activities and shallow groundwater depth were the most vulnerable zones for nitrate contamination. Based on these results, optimized irrigation techniques and lower consumption of fertilizers are suggested for the vulnerable zones. The results of our study could serve as a scientific basis in future for sustainable land use planning and groundwater management in Shiraz plain.

Abbreviations

AI, Artificial Intelligence; ANN, Artificial Neural Networks; BP, Back Propagation Algorithm; CD index, Composite DRASTIC index; CG, Conjugate Gradient Algorithm; DEM, Digital Elevation Model; FRWO, Fars Regional Water Organization; GIS, Geographic Information System; IRS, Indian Remote sensing Satellite; LM, Levenberg-Marquardt; MLP, Multi-Layer Perceptron; NV index, Nitrate vulnerability index; RMSE, Root Mean Square Error; SPSS, Statistical Package for Social Science; USEPA, United States Environmental Protection Agency.

Acknowledgements

This article was extracted from Amir Fadaei's M.Sc. thesis approved by Shiraz University of Medical Sciences and was financially supported by the Research Vice-chancellor of the University. The authors would like to appreciate Dr. Behroz, manager of Basic Studies of Water Resources office of Fars Regional Water Co., and Mrs. Hayati, head of library in this Organization, for their sincere cooperation in this project. They are also grateful for Ms. A. Keivanshekouh at the Research Improvement Center of Shiraz University of Medical Sciences for improving the scientific writing of the manuscript.

Funding
This research has been supported financially by Shiraz University of Medical Sciences.

Authors' contributions
MAB devised the main design of the study and supervised it. AF carried out all practical works of the study and wrote the article. NTB participated in the design and helped to draft the manuscript. SB helped to perform GIS works. All authors read and approved the final manuscript.

Competing Interests
The authors appreciate the financial support by Shiraz University of Medical Sciences.

Author details
[1]Department of Environmental Health Engineering, School of Health, Shiraz University of Medical Sciences, Shiraz, IR, Iran. [2]Department of Civil Engineering, College of Engineering, Shiraz University, Shiraz, IR, Iran. [3]Department of Hydrogeology, Ab Ati Pazhooh Consulting Engineers, Shiraz, IR, Iran. [4]Department of Earth Science, Faculty of Science, University of Tabriz, Tabriz, East Azarbaijan, IR, Iran. [5]Research Engineer at EnTech Engineering, PC11 broadway 21st floor, New York, NY 10004, USA.

References
1. Tilahun K, Merkel BJ. Assessment of groundwater vulnerability to pollution in Dire Dawa, Ethiopia using DRASTIC. Environ Earth Sci. 2010;59:1485–96.
2. Sinan M, Razack M. An extension to the DRASTIC model to assess groundwater vulnerability to pollution: application to the Haouz aquifer of Marrakech (Morocco). Environ Geol. 2009;57:349–63.
3. Jamaludin N, Sham SM, Smail SNS. Health risk assessment of nitrate exposure in well water of residents in intensive agriculture area. AJAS. 2013;10:442–8.
4. Rahman A. A GIS based DRASTIC model for assessing groundwater vulnerability in shallow aquifer in Aligarh, India. J Appl Geography. 2008;28(1):32–53.
5. Manos B, Papathanasiou J, Bournaris T. A multicriteria model for planning agricultural regions within a context of groundwater rational management. J Environ Manage. 2010;91:1593–600.
6. Sener E, Sener S, Davraz A. Assessment of aquifer vulnerability based on GIS and DRASTIC methods: a case study of the Senirkent-Uluborlu Basin (Isparta, Turkey). H ydrogeol J. 2009;17:2023–35.
7. Babiker IS, Mohamed MAA, Hiyama T, Kato K. A GIS-based DRASTIC model for assessing aquifer vulnerability in Kakamigahara Heights, Gifu Prefecture, central Japan. Sci Total Environ. 2005;345:127–40.
8. Stigter TY, Ribeiro L, Carvalho Dill AMM. Evaluation of an intrinsic and a specific vulnerability assessment method in comparison with groundwater salinisation and nitrate contamination levels in two agricultural regions in the south of Portugal. Hydrogeol J. 2006;14:79–99.
9. Secunda S, Collin ML, Melloul AJ. Groundwater vulnerability assessment using a composite model combining DRASTIC with extensive agricultural land use in Israel's Sharon region. J Environ Manage. 1998;54:39–57.
10. Al-Adamat RAN, Foster IDL, Baban SMJ. Groundwater vulnerability and risk mapping for the Basaltic aquifer of the Azraq basin of Jordan using GIS, Remote sensing and DRASTIC. J Appl Geography. 2003;23:303–24.
11. Saidi S, Bouri S, Ben Dhia H. Groundwater vulnerability and risk mapping of the Hajeb-jelma aquifer (Central Tunisia) using a GIS-based DRASTIC model. Environ Earth Sci. 2010;59:1579–88.
12. Martínez-Bastida JJ, Arauzo M, Valladolid M. Intrinsic and specific vulnerability of groundwater in central Spain: the risk of nitrate pollution. Hydrogeol J. 2010;18:681–98.
13. Dixon B. Applicability of neuro-fuzzy techniques in predicting ground-water vulnerability: a GIS-based sensitivity analysis. J Hydrolo. 2005;309:17–38.
14. Dixon B. Groundwater vulnerability mapping: a GIS and fuzzy rule based integrated tool. J Appl Geography. 2005;25:327–47.
15. Dixon B. A case study using support vector machines, neural networks and logistic regression in a GIS to identify wells contaminated with nitrate-N. Hydrogeol J. 2009;17:1507–20.
16. Maier HR, Jain A, Dandy GC, Sudheer KP. Methods used for the development of neural networks for the prediction of water resource variables in river systems: current status and future directions. Environ Model Softw. 2010;25:891–909.
17. Nourani V, Mogaddam AA, Nadiri AA. An ANN-based model for spatiotemporal groundwater level forecasting. Hydrol Processes. 2008;22:5054–66.
18. Chitsazan N, Nadiri AA, Tsai FTC. Prediction and structural uncertainty analyses of artificial neural networks using hierarchical Bayesian model averaging. J Hydrology. 2015;528:52–62.
19. Baghapour MA, Talebbeydokhti N, Tabatabee H, Nobandegani AF. Assessment of Groundwater Nitrate Pollution and Determination of Groundwater Protection Zones Using DRASTIC and Composite DRASTIC (CD) Models: The Case of Shiraz Unconfined Aquifer. J Health Sci Surveillance Sys. 2014;2:54–65.
20. Shirazi SM, Imran HM, Akib S, Yusop Z, Harun ZB. Groundwater vulnerability assessment in the Melaka State of Malaysia using DRASTIC and GIS techniques. Environ Earth Sci. 2013;70:2293–304.
21. Yin L, Zhang E, Wang X, Wenninger J, Dong J, Guo L, Huang J. A GIS-based DRASTIC model for assessing groundwater vulnerability in the Ordos Plateau, China. Environ Earth Sci. 2013;69:171–85.
22. Jamrah A, Al-Futaisi A, Rajmohan N, Al-Yaroubi S. Assessment of groundwater vulnerability in the coastal region of Oman using DRASTIC index method in GIS environment. Environ Monit Assess. 2008;147:125–38.
23. Chitsazan M, Akhtari Y. A GIS-based DRASTIC model for assessing aquifer vulnerability in Kherran Plain, Khuzestan, Iran. Water Resour Manage. 2009;23:1137–55.
24. Hopfield JJ. Neural networks and physical systems with emergent collective computational abilities. Proc Natl Acad Sci U S A. 1982;79:2554–8.
25. Fijani E, Nadiri AA, Mogaddam AA, Mogaddam AA, Tsai FTC, Dixon B. Optimization of DRASTIC method by supervised committee machine artificial intelligence to assess groundwater vulnerability for Maragheh–Bonab plain aquifer, Iran. J Hydrolo. 2013;503:89–100.
26. Nadiri A, Hassan MM, Asadi S. Supervised Intelligence Committee Machine to Evaluate Field Performance of Photocatalytic Asphalt Pavement for Ambient Air Purification. Transportation Research Record: TRB. 2015;2528:96–105.
27. ASCE Task Committee on Application of Artificial Neural Networks in Hydrology. Artificial neural network in hydrology, part I and II. J Hydraul ENG-ASCE. 2000;5:115–37.
28. Brahim FB, Khanfir H, Bouri S. Groundwater vulnerability and risk mapping of the Northern Sfax Aquifer, Tunisia. Arab J Sci Eng. 2012;37:1405–21.
29. Prasad RK, Singh VS, Krishnamacharyulu SKG, Banerjee P. Application of drastic model and GIS: for assessing vulnerability in hard rock granitic aquifer. Environ Monit Assess. 2011;176:143–55.
30. Mahvi AH, Nouri J, Babaei AA, Nabizadeh R. Agricultural activities impact on groundwater nitrate pollution. Int J Environ Sci Tech. 2005;2:41–7.
31. Dahan O, Babad A, Lazarovitch N, Russak EE. Nitrate leaching from intensive organic farms to groundwater. Hydrol Earth Syst Sci. 2013;18:333–41.
32. Kraft GJ, Stites W. Nitrate impacts on groundwater from irrigated-vegetable systems in a humid north-central US sand plain. Agric Ecosyst Environ. 2003;100:63–74.
33. Esmaeili A, Moore F, Keshavarzi B. Nitrate contamination in irrigation groundwater, Isfahan, Iran. Environ Earth Sci. 2014;72:2511–22.
34. Shen Y, Lei H, Yang D, Kanae SH. Effects of agricultural activities on nitrate contamination of groundwater in a Yellow River irrigated region. Water Quality: Current Trends and Expected Climate Change Impacts. IAHS Publ. 2011;348:73–80.

12

Bisphenol A removal from aqueous solutions using novel UV/persulfate/H$_2$O$_2$/Cu system: optimization and modelling with central composite design and response surface methodology

S. Ahmad Mokhtari[1,3], Mehdi Farzadkia[1,2], Ali Esrafili[1,2], Roshanak Rezaei Kalantari[1,2], Ahmad Jonidi Jafari[1,2], Majid Kermani[1,2] and Mitra Gholami[1,2*]

Abstract

Background: Bisphenol A is a high production volume chemical widely used in manufacturing polycarbonate plastics and epoxy resins used in many industries. Due to its adverse effects on human health as an endocrine disruptor and many other effects on the various organs of the human body as well as aquatic organisms, it should be removed from the aquatic environments. This study aimed to mineralisation of BPA from aquatic environments by application of novel UV/SPS/H$_2$O$_2$/Cu system and optimization and modelling of its removal using central composite design (CCD) from response surface methodology (RSM).

Methods: CCD from RSM was used for modeling and optimization of operation parameters on the BPA degradation using UV/SPS/HP/Cu system. Effective operation parameters were initial persulfate, H$_2$O$_2$, Cu^{2+} and BPA concentration along with pH and reaction time, all in three levels were investigated. For analysis of obtained data ANOVA test was used.

Results: The results showed that a quadratic model is suitable to fit the experimental data ($p < 0.0001$). Analysis of response surface plots showed a considerable impact of all six selected variables which BPA and Cu^{2+} initial concentrations have been the highest and the least impact on the process, respectively. *F*-value of model was 54.74 that indicate significance of the model. The optimum values of the operation parameters were determined. The maximum removal of BPA was achieved 99.99 % in optimal conditions and in that condition TOC removal was about 70 %. Finally, validation and accuracy of the model were also evaluated by graphical residual analysis and the influential diagnostics plots. The higher relevance between actual and predicted values demonstrated the validation and applicability of the obtained equation as the model.

Conclusions: According to the results, UV/SPS/HP/Cu system is an effective process in degradation and mineralisation of BPA and CCD methodology is a convenient and reliable statistical tool for optimizing BPA removal from aqueous solutions.

Keywords: BPA, UV/SPS/HP/Cu, Aqueous solutions, CCD

* Correspondence: gholamim@iums.ac.ir; gholamimitra32@gmail.com
[1]Research Center for Environmental Health Technology, School of Public Health, Iran University of Medical Sciences, Tehran, Iran
[2]Department of Environmental Health Engineering, School of Public Health, Iran University of Medical Sciences, Tehran, Iran
Full list of author information is available at the end of the article

Background

During the last decade, the so-called emerging contaminants (CECs or ECs) have assigned the majority of environmental researches. This is due to concerns about the potential effects of these chemicals on human health and the environment [1–4]. A group of these pollutants is endocrine disrupting chemicals (EDCs) and potential EDCs are mostly man-made, found in various materials such as pesticides, metals, additives or contaminants in food, and personal care products. BPA (CAS No: 80-05-7), one of the most important EDCs, has a solid state with white colour which often used as an intermediate in the production of epoxy, polycarbonate, polysulfone and resins of special polyester. In recent centuries because of BPA presence in variety of products such as packaging of beverage and food, adhesives, construction materials, retardants, electronic equipment, and paper coatings [5, 6]. It has shown that BPA has hormone-like properties at higher concentrations, so causes an increase in concerns about its entrance to the environment as well as its suitability in different applications such as consumption goods and packaging for some food products [3].

Several studies have been shown BPA's side and adverse health effects on human and animals. Human contact with BPA can occurs often through ingestion of contaminated food and water, dust from inhaling gases and particulates in the air, and also via the skin [7, 8]. BPA is a nonsteroidal xenoestrogen which shows about 10–4 times higher than estradiol activation [9]. Another work indicates that BPA may be has an equivalent effect of estradiol in triggering some responses of receptors [10] and probably it act as antagonist for androgen receptor [11, 12]. Other important health issues related to BPA that reported in the literature are high rates in the incidence of diabetes, breast and prostate cancers, and decreased sperm count, sexual problems, premature puberty, obesity, nerve problems and etc. [13]. According to the EPA and FDA, intake limits for BPA in human health assessment was determined 0.05 mg/kg.d and 0.005 – 0.5 mg/kg.d respectively; also European Union has determined 1.5 μg/L as PNEC[1] for aquatic organisms. Based on current information all toys have plastic in their structure must not contain BPA.

BPA can enter wastewater treatment plants through both discharge directly to the sewage system and surface runoff, and in the next steps it has been entered in influent and effluent also in residual sludge in wastewater treatment plants (WWTPs) [14]. In fact, WWTPs are considered to be the main secondary sources of BPA pollution due to their imperfect and defective performance of treatment process [15–17].

It has been demonstrated that several processes including ultrafiltration, RO, NF, photo-catalysis and membrane filtration, hollow fibre microfiltration membrane, single walled carbon nanotubes–ultrafiltration, coagulation and adsorption, laccase catalyzation and atc are capable of removing BPA from aqueous environments with different removal efficiencies [18–27].

Newly, another oxidation method was introduced based on PS that has high redox potential (E_0 = 2.01 V), and it is applied less frequently compared to ozone (E_0 = 2.07 V); The dissociation of PS in the aquatic environment is a major stage for yielding sulphate radicals (SO_4^-, E_0 = 2.60 V), for strong oxidation, and these radicals also react with water for generating hydroxyl radicals that both of generated radicals can react with organic matter and degrade it to the products [15, 28–32]:

$$S_2O_8^{-2} + \text{initiator} \rightarrow SO_4^{-\bullet} + \left(SO_4^{-\bullet} \text{or } SO_4^{-2}\right) \quad (1)$$

$$SO_4^{-\bullet} + H_2O \rightarrow H^+ + SO_4^{2-} + HO^{\bullet} \quad (2)$$

Oxidation of organic matters with PS has some advantages than other oxidizing agents due of its convenient storage and transportation, being in solid state at the normal temperatures, high ability of water solubility, relatively lower costs, and high stability; another advantage of AOPs (as well as sulphate base radicals) is degradation and mineralization of recalcitrant and/or toxic organic pollutants and lack of leaving residues and sludge [33–36].

There are few methods for activation of PS ion, initiation, and then sulphate radical production, which are included conventional methods (*heating, catalysis with transition metal, and UV irradiation*) and new methods (*photochemical, iron and chelated iron ion, zero-valent iron, minerals, activated carbon, microwave, and integrated* activation) [33].

In this study, it was used an integrated method for degradation and mineralization of BPA which consist of UV, SPS (Sodium persulfate), H_2O_2, and Cu cation. This study was conducted with the propose of evaluating BPA removal efficiency by UV/SPS/HP/Cu system and survey the effect of factors, including BPA initial concentration, time, pH, PS, peroxide and Cu ion concentrations and finally optimization of BPA removal with their factors by CCD.

Generally, treatment processes of pollution are optimised using "*one factor at the time*", such as most of other processes in industries with variating of its variables. Moreover, this method assumes that various treatment parameters do not interact and that the response variable is only function of the single varied parameter. In addition, the response obtained from pollution treatment procedure, for instance finding from the interactive effects of the various variables. When a mix of multiple independent variables and interactions of them affect given responses, response surface methodology (RSM) is a proper and efficient tool in optimization of the response [37]. In RSM an experimental design is used like CCD for a fitting model by minimum squares technique [38, 39]. In the next stage, adequacy of the obtained model is demonstrated by using the diagnostic tests presented by analysis of variance (*ANOVA*). The plots of response surface can be applied to survey the surfaces and location of the optimized point. In most process of industrials, RSM is normally applied for evaluation of the results and efficacy of the operations [40, 41].

In present study, a CCD in the form of full factorial was applied for developing mathematic relations, in terms percentage of removal, providing quantitative aassessment of the UV/SPS/HP/Cu system used to treat BPA.

Methods

Chemicals and samples

The model contaminant, BPA ≥ 99 %, was purchased from Aldrich (SKU-Pack Size). Analytical grade n-hexane and acetone, methanol were all provide by Merck (Darmstdat, Germany). Other chemicals such as SPS, HP, copper sulphate (Cu_2SO_4), sulphuric acid (H_2SO_4), Sodium hydroxide (NaOH), monosodium phosphate and etc. also purchased from Merck. Deionized water was supplied by a home-made deionizer. Stock standard solutions of BPA (10,000 mg/L) were prepared in ethanol and were then diluted with water (as working solutions) and stored at −18 ± 2 °C. The calibration curve was constructed of six calibration standard solutions (1–50 mg/L; CS) were prepared by adding the diluted working solution into the deionized water samples. The validation tests were carried out using two quality controls (QC, 2 and 5 mg/L). Both CS and QC samples were stored at −18 ± 2 °C until analysis. Taken samples from system at the specified times, immediately was injected to the HPLC; If necessary, to stop the reaction, methanol was added to the samples.

Instrumentation

The photochemical-degradation studies were carried out in a batch reactor system. The reactor used in this study, the tubular steel, was made from stainless steel, which was available for the high reflection of the radiation, resistant to chemical and cylindrical shape, was designed and built with a diameter about 8 cm and a length of 90 cm. Irradiation was achieved by a low-pressure mercury lamp (Philips, TUV 30 W/GC T8) with UV radiation 12 W located axially in the centre of reactor and was placed inside the quartz tube. The light length of the lamp and quartz sleeve diameter of 30 mm was same with reactor that emits approximately 90 % of its radiation at 253.7 nm. The chemicals used in each run along with aqueous solution with a volume of 2 l was added in the reactor, which was placed between the reactor walls and UV lamp system. To make a uniform solution a shaker was used with 100 rpm.

The samples containing BPA, were prepared from the stock solution, and their residual concentrations in the samples were analysed by HPLC (Cecil CE 4100) using a Hypersil C18 column (250 mm × 4.6 mm i.d, with 5 μm particle size) with a UV detector (Cecil CE 4200) at 230 nm. The Mobile phase consisted of a mixture of methanol and deionized water with a ratio of 0.8:0.2 (v:v) at a flow rate of 1 mL/min. 20 μL of sample was injected using a Hamilton syringe to the injector. The detection limit of method for BPA was 1 μg/L.

Other equipment used in this study were TOC analyser (analytikjena, multi N/C 3100), GC-MS (Agilent Technologies 7890A, USA), pH meter (HQ11D, Portable), distillation apparatus (GFL, Germany), spectrophotometer (DR 6000 UV/VIS, HACH, USA), laboratory precision scale (BEL Engineering, Italy) and variety of laboratory glass containers and accessories.

Experimental design and data analysis

A CCD in the form of full factorial design was used, in which six independent variables were converted to dimensionless ones (x_1, x_2, x_3, x_4, x_5, x_6), with the coded values at 3 levels: −1, 0, +1 (Table 1). Preliminary optimal variable levels were obtained using multi-simplex method. The arrangement of CCD in mentioned 86 runs, was in such a way that allows the development of the appropriate empirical equations (second order polynomial multiple regression equations):

$$y = \beta_0 + \beta_1 x_1 + \cdots + \beta_6 x_6 + \beta_{11} x_1^2 + \cdots + \beta_{66} x_6^2 \\ + \beta_{12} x_1 x_2 + \cdots + \beta_{56} x_5 x_6 \tag{3}$$

And therefore, the predicted response (y, or BPA removal percentage) was correlated to the set of regression coefficients (β): the intercept (β_0), linear (β_1, β_2, β_3), interaction (β_{12}, β_{13}, β_{23}) and quadratic coefficients (β_{11}, β_{22}, β_{33}).

The "Design expert" (version 7), SPSS (version 18), and excel 2013 softwares were used for regression and graphical analyses of the obtained data.

Table 1 Independent variables and their levels for the central composite design used in this study

Variable	Unite	Symbol	Coded variable levels		
			−1	0	1
C persulfate	*mg/L*	x_1	25	65	105
C peroxide	*mg/L*	x_2	5	10	15
pH	-	x_3	3	7	11
Cu⁺	*mg/L*	x_4	5	17.5	30
Time	*min*	x_5	5	122.5	240
C BPA	*mg/L*	x_6	5	22.5	40

Toxicity test

To evaluate the acute toxicity of effluent containing BPA and its intermediate products before and after treatment and degradation by UV/SPS/HP/Cu system, bioassay test with Daphnia Magna, as an indicator, was done as detailed in standard methods for the examination of water and wastewater. Exposure times 12, 24, 48, 72, and 96 h was considered, and the number of immobilized and dead infants was recorded and then LC_{50} values were calculated by usage of PROBIT regression in the SPSS software (version 23) and the toxic unit (TU) of effluent was determined by $TU = 100/LC_{50}$. Also for identify intermediate products and their possible role in creation of toxicity, GC-MS analysis was carried out.

Results and discussion

Selecting the ratio for reagents

The overall efficiency of the UV/SPS/HP/Cu system was determined with multi-simplex method as a results of our pre-test experiments (Results not shown). Based of obtained results from this work and similar work, it was found PS ion, hydrogen peroxide, Cu^{2+} ion, pH, reaction time and initial BPA concentration all are key parameters and contribute to the proper performance of system and its removal efficiency depends on them and the their relationships in terms of formation and utilization of sulphate and hydroxyl radicals [15, 42]. In the present study, with the aim of reducing the use of chemicals and their related cost saving, and take advantage of potential synergistic property, chemicals and other applied parameters such as UV irradiation applied simultaneously. Initially different levels of chemicals and parameters were selected and applied alone in the wider ranges for each parameter (UV lamp: 30 Wat as fixed; SPS: 0.1–10 as molar ratio; HP (Hydrogen peroxide): 0.1–5 as molar ratio; Cu^{2+}: 0.1–2; BPA: 5–40 mg/L); in this stage it was shown that use of the each variable alone, have the less efficiency in removal and degradation of BPA and for more effectiveness PS must be active with activation factors such as UV, Cu^{2+} and hydrogen peroxide [43–45]. Excess amounts of hydrogen peroxide and PS is not recommended since unutilized hydrogen peroxide contributes to COD (1 mg/L of H_2O_2 contributes 0.25–0.27 mg/L to the COD concentration), and the remaining of PS can be exhibit as contaminant in the effluent [46, 47]. Also, PS and H_2O_2 in large quantity act as radical scavenger and lead to duplicate of reagent consumption for activation of PS and therefore probably reducing removal efficiency of system [48, 49]. Moreover, used ratios and concentrations applied in the present study were in the range reported by others when the AOPs based sulphate and hydroxyl radicals have been applied to the removal and degradation of BPA in the water and wastewater [15, 44, 50]. According to results, some levels and ratios of variables was excluded and the following levels were chosen: SPS in ranges 0.5–2, HP in ranges 0.5–2, Cu^{2+} in ranges 0.1–1 as molar ratio to BPA, and finally BPA in range 5–40 mg/L.

CCD and fitted regression models as related to the BPA removal

In this work, the relationship between one criteria of the pollutant removal (named BPA removal) and six controllable factors (namely SPS, HP and Cu^{2+} initial concentrations, pH, reaction time, and BPA initial concentration) was studied. CCD with 86 runs allows us the development of mathematical equation where response variable y (BPA removal) is assessed as a function of SPS concentration (x_1), HP concentration (x_2), Cu^{2+} concentration (x_3), pH (x_4), reaction time (x_5) and BPA initial concentration (x_6) and calculated as the sum of a constant number, six first-order effects (terms in x_1, x_2, ... and x_6), thirteen interaction effects (terms in x_1x_2, x_1x_3, ... and x_5x_6) and six second-order effects (x_{12}, x_{22}, ... and x_{62}) according to the Eq. (4).

$$\begin{aligned} Y\,(_{BPA\ Removal}) &= 84.75 + 2.18\,x_1 + 1.74\,x_2 + 2.64\,x_3 \\ &\quad + 0.9\,x_4 + 4.36\,x_5 + 4.52\,x_6 - 0.12\,x_1x_2 \\ &\quad - 0.04\,x_1x_3 - 0.083\,x_1x_4 - 0.20\,x_1x_5 \\ &\quad + 0.13\,x_1x_6 - 0.042\,x_2x_3 + 0.082\,x_2x_5 \\ &\quad + 0.074\,x_2x_6 + 0.23\,x_3x_5 + 0.41\,x_3x_6 \\ &\quad + 0.45\,x_4x_5 + 0.13\,x_4x_6 + 0.039\,x_5x_6 \\ &\quad - 4.30\,x_1{}^2 - 2.5\,x_2{}^2 + 7.10\,x_3{}^2 \\ &\quad - 3.85\,x_4{}^2 - 3.5\,x_5{}^2 + 0.3\,x_6{}^2 \end{aligned} \tag{4}$$

By removing the model terms are not significant statistically, obtained model is reduced to the following form that can be improved it.

$$\begin{aligned} Y\,(_{BPA\ Removal}) &= 84.76 + 2.18\,x_1 + 1.74\,x_2 + 2.64x_3 + 0.9x_4 \\ &\quad + 4.36\,x_5 + 4.52\,x_6 + 0.45\,x_4x_5 + 4.30\,x_1{}^2 \\ &\quad - 2.5\,x_2{}^2 + 7.10\,x_3{}^2 - 3.85\,x_4{}^2 - 3.5\,x_5{}^2 \end{aligned} \tag{5}$$

Then in the next stage, to evaluation of fitness for achieved model, the obtained results were analyzed using ANOVA (analysis of variance). The model for BPA removal (Y) was considerable significant by using F-test with confidence level 5 % (0.05 > F < Prob). Obtained F-value of 54.74 for model indicate its significance. There is just 0.01 % possibility that a "Model F-Value" could cause by the error. Equation 4 as the fitted regression model (in terms of coded values for the variables) were applied to quantitative analysis of the impact of SPS, HP, and Cu^{2+} concentration, pH, time and initial BPA concentration on the description of UV/SPS/HP/Cu system for the BPA removal.

The statistical parameters of the ANOVA for the model of BPA removal are provided in Table 2. Since R^2 always reduces whenever a variable is dropped from a regression model, in statistical modelling the adjusted R^2 which takes the number of regressor variables into account, is usually selected [51]. The R^2 coefficient gives the proportion of the total variation in the response variable explained or accounted for by the predictors (x's) included in the model [52]. In this work, the adjusted and predicted R^2 were calculated to 0.9447 and 0.9215, respectively. Adequate precision index for the model was obtained 32.885; this index measures "signal to noise ratio" in the model, which statistically ratio that higher than four is desirable [53] and obtained ratio for the model is much higher than this (32.85). The R^2 factor in this work certified a suitable match of the quadratic model to the experimental data.

As it has been provided in Table 2, the C.V. for model has been calculated to be %2.19. The C.V. as the ratio of standard error of estimate for the mean value of the observed response (as a percentage) is a measure of reproducibility of the model and as a general rule can be considered reasonably reproducible if its CV is not greater than 10 % [37, 53]. The model fitted for BPA removal had a relatively good CV. This model had high R^2 value and showed no lack of fit (Lack of fit F-value = 0.49). Diagnostic plots were applied for further evaluation of model such as Normal probability plot of the studentized residuals (to check for normality of residuals), Studentized residuals versus predicted values (to check for constant error), Externally Studentized Residuals (to look for outliers, i.e., influential values), and Box-Cox plot (for power transformations); and the assumptions of normality, independence and randomness of the residuals were satisfied. Using Box-Cox plot, Lambda value was obtained equal 1.0, with range Low C.I. = −2.24 and High C.I. = 1.15. Thus the predicted model can be used to navigate the space defined by the CCD.

The relative contribution of each factor to dependent variable (y: BPA removal) was directly measured by the respective coefficient in the fitted model. A positive sign for the coefficients (β) in the fitted model for Y indicated that the level of pollutant removal increased with increased levels of factors from x_1 to x_6. A review of these coefficients shows that related coefficient with x_6 factor (BPA initial concentration) has the highest value and therefor the highest effect and obtained model has the most sensitivity for x_6 with the equivalent 4.52; And other factors, including x_5, x_3, x_1, x_2, and x_4 have been allocated the next ranks in sensitivity of the model for them. On the other hand, these coefficient's ranks and arrangement of them in the model indicate that the removal of BPA as a model pollutant is affected by these factors by the mentioned order. Moreover, the existence of a negative sign for the regression coefficient of some factors, indicates that the ability of the system decreases with that factor's value in the removing of the BPA. Of course, this issue is true just about some coefficients related to interactions and quadratic form of factors. In order to better understanding of the obtained results, the predicted model is provided in Figs. 2, 3, 4, 5 and 6 as the three-dimensional response surface plots. However, to validate the model, 15 separate experiments with different condition of applied parameters was done. The predicted values for BPA removal in this study are given in Table 3, which includes the measured data for these response variables. Comparing experimental values with the values predicted by the mathematical model indicate 92.56 % correlation between them, that which represents

Table 2 Statistical parameters obtained from the ANOVA for the model

R^2 (R squared)	0.9622
R^2 adjusted	0.9447
Predicted R-Squared	0.9215
Prob > F	<0.0001
Std. Dev.	1.74
C.V. %	2.19
PRESS	365.41
Adequate precision	32.885

Table 3 Experimental and theoretically predicted values for BPA removal based on obtained model

Experiment no.	Actual value	Predicted value
1	91.59	88.85
2	71.47	73.30
3	75.19	76.50
4	89.21	87.10
5	83.22	81.60
6	72.88	75.95
7	82.35	87.60
8	81.50	78.90
9	90.91	88.65
10	81.14	82.95
11	81.20	85.15
12	85.26	84.10
13	77.83	76.95
14	92.27	97.55
15	85.17	90.74

adequacy and suitability of model in predicting of the responses.

Optimization of BPA removal and response surface plotting
The use of three-dimensional plots of the regression model is highly recommended for the graphical interpretation of the interactions [54–56]. Variables that giving quadratic and interaction terms with the largest absolute coefficients in the fitted model, were chosen for the axes of the response surface plots to account for the curvature of the surfaces. SPS and HP concentration were selected for the RSM plots of BPA removal, while other factors ($x_3 - x_6$) were kept at the central levels (Fig. 1), and also in the next figures (Figs. 2, 3, 4 and 5) SPS concentration and pH, SPS and Cu^{2+} concentrations, SPS concentration and reaction time, and SPS and initial BPA concentrations were selected respectively as variables while four other factors were constant as cantered levels at the same time. Resulted interaction indicates that the effect created by variation of the SPS concentration, as a regressor variable depends on the value of HP as the another regressor variable (Fig. 2), and similarly it depends on the levels of the other variable as regressor ($x_3 - x_6$) and also all of the factors have interactions with each other based on the model. In the obtained model, x_6 is the principle regressor variable

influencing the responses (Highest coefficients, β_1s) whereas this variable had an interaction with x_1 equal to + 0.13 ($\beta 16$); and according to Figs. 1, 2, 3, 4 and 5, SPS concentration (x_1) has interactions with x_2, x_3, x_4 and x_5 equal to −0.12, −0.04, −0.083 and −0.2 respectively. The plots show a relative high degree of curvy of 3-dimensional surfaces (Upwards or downwards depend on interaction type and degree).

The removal of BPA increased when SPS concentration raised to its centred level (65 mg/L, obtained as the ratio to BPA), and increasing of it towards +1 level (105 mg/L) results in remaining constant and slightly reduced in BPA removal; simultaneously with increasing HP concentration, BPA removal starts to increase and about in centre level (10 mg/L) it is reached to maximum level, and then starts to decrease with increasing HP concentration to the high level (15 mg/L) (Fig. 1). In this system, in addition of UV irradiation, that had a constant intensity (30 W UV lamp), other chemicals as variables have been used as compilation and integrated. The reason of efficiency reducing can be attributed to being constant of BPA mole number as reactant that will react with increased PS concentration and therefore sulphate radicals for degradation; this means that possibility of contact between radicals and BPA will not increase so much and increasing of SPS concentration will be result in slightly increase in reaction rate

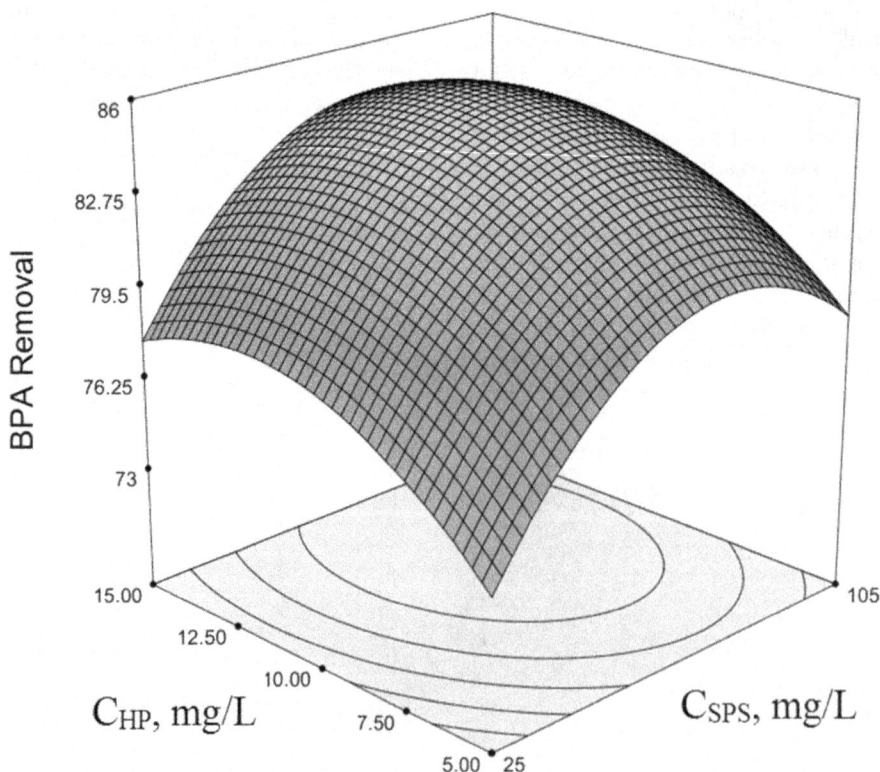

Fig. 1 Second-order response surface plot in the BPA removal for the UV/SPS/HP/Cu system. Dependence of y on the SPS and HP concentrations. (pH = 7, Cu^{2+} = 17.5 mg/L, Time = 122.5 min, BPA = 22.5 mg/L)

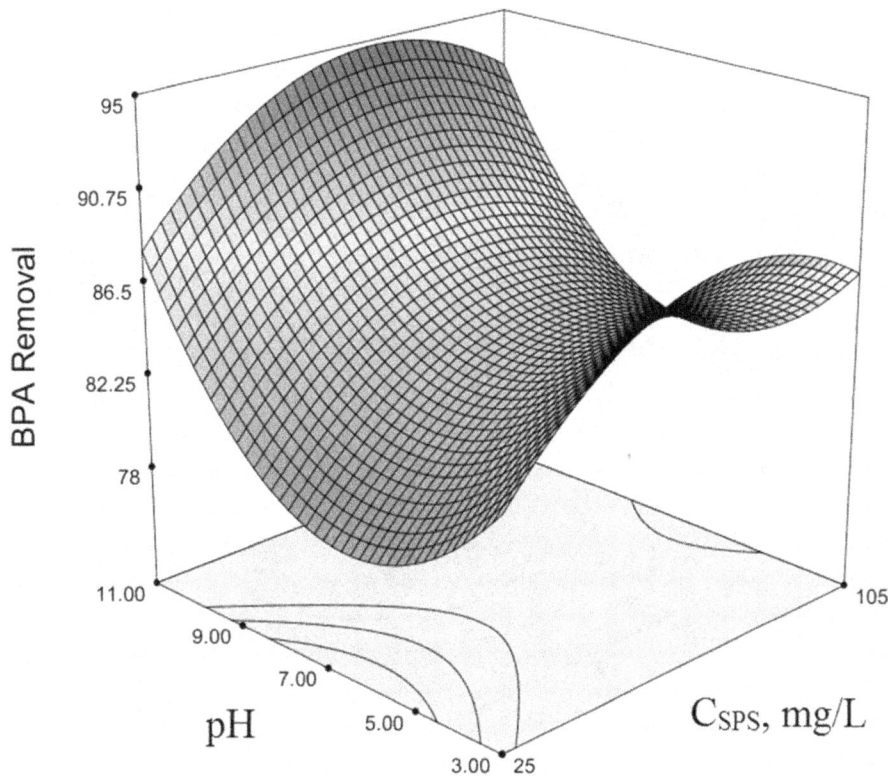

Fig. 2 Second-order response surface plot in the BPA removal for the UV/SPS/HP/Cu system. Dependence of y on the SPS and pH concentrations. (HP = 10 mg/L, Cu^{2+} = 17.5 mg/L, Time = 122.5 min, BPA = 22.5 mg/L)

in the first minutes of process, but by continuing of reaction and reduction in BPA concentration in the reactor, reaction rate again reduce greatly because of reduced numbers of BPA moles; results of Yang (2009), Olmez-Hanci (2013) and Liang (2013) confirmed also obtained results in this work.

However, the presence of excess HP moles as SPS activator and also responsible agent for $HO^•$ radical, can be play as a radical scavenger for $SO_4^{-•}$ and again reconvert it to SO_4^{2-} ion that cause reduced in reaction rate and then BPA removal. An excessively high initial PS concentration can lead to the generation of a higher quantity of $SO_4^{-•}$ that may inhibit BPA oxidation according to the following equation [57–59]:

$$SO_4^{-•} + SO_4^{-•} \rightarrow S_2O_8^{2-} \tag{6}$$

$$SO_4^{-•} + HO^• \rightarrow SO_4^{2-} + OH^- \tag{7}$$

Reviews of Fig. 2 illustrates that BPA oxidation with this system completely depend on pH; and SPS concentration and pH has interaction with each other, which their interaction is reflected in model with coefficient equal 0.04. In addition to the SPS behaviour described in the previous section, pH also plays an important role in the process, and therefore in the neutral pH, the system has the minimum efficiency and at the acidic and basic

pH is more efficient, so that pH =11 had been the highest efficiency.

It has already been demonstrated that in acidic condition $SO_4^{-•}$ are the dominant radical and in alkaline conditions can induce the mechanism of $SO_4^{-•}$ interconversion to $HO^•$ in the PS activation system and $SO_4^{-•}$ can undergo reactions with hydroxide ions in accordance with Eq. (8) to generate hydroxyl radicals, in addition, $SO_4^{-•}$ may react with water at all pHs to produce $HO^•$, according to the following equations. However, Norman et al. reported that the reaction rate constant of Eq. (9) is low in comparison to those of $SO_4^{-•}$ reactions with organic compounds [57, 60–62]:

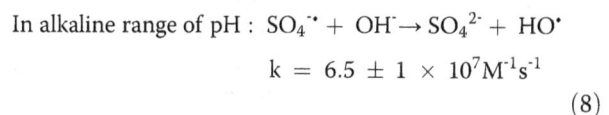

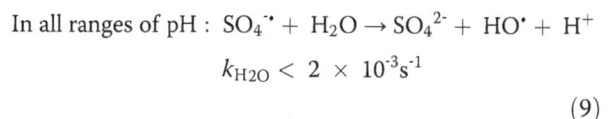

In alkaline range of pH : $SO_4^{-•} + OH^- \rightarrow SO_4^{2-} + HO^•$

$$k = 6.5 \pm 1 \times 10^7 M^{-1}s^{-1} \tag{8}$$

In all ranges of pH : $SO_4^{-•} + H_2O \rightarrow SO_4^{2-} + HO^• + H^+$

$$k_{H2O} < 2 \times 10^{-3}s^{-1} \tag{9}$$

It should be mentioned the hydroxyl radical has a higher redox potential and a less-selective reactivity to contaminant degradation. Therefore, the formation of $HO^•$ potentially increases the degradation rate of the BPA and intermediate compounds in alkaline pH; as well

Fig. 3 Second-order response surface plot in the BPA removal for the UV/SPS/HP/Cu system. Dependence of y on the SPS and Cu^{2+} concentrations. (HP = 10 mg/L, pH = 7, Time = 122.5 min, BPA = 22.5 mg/L)

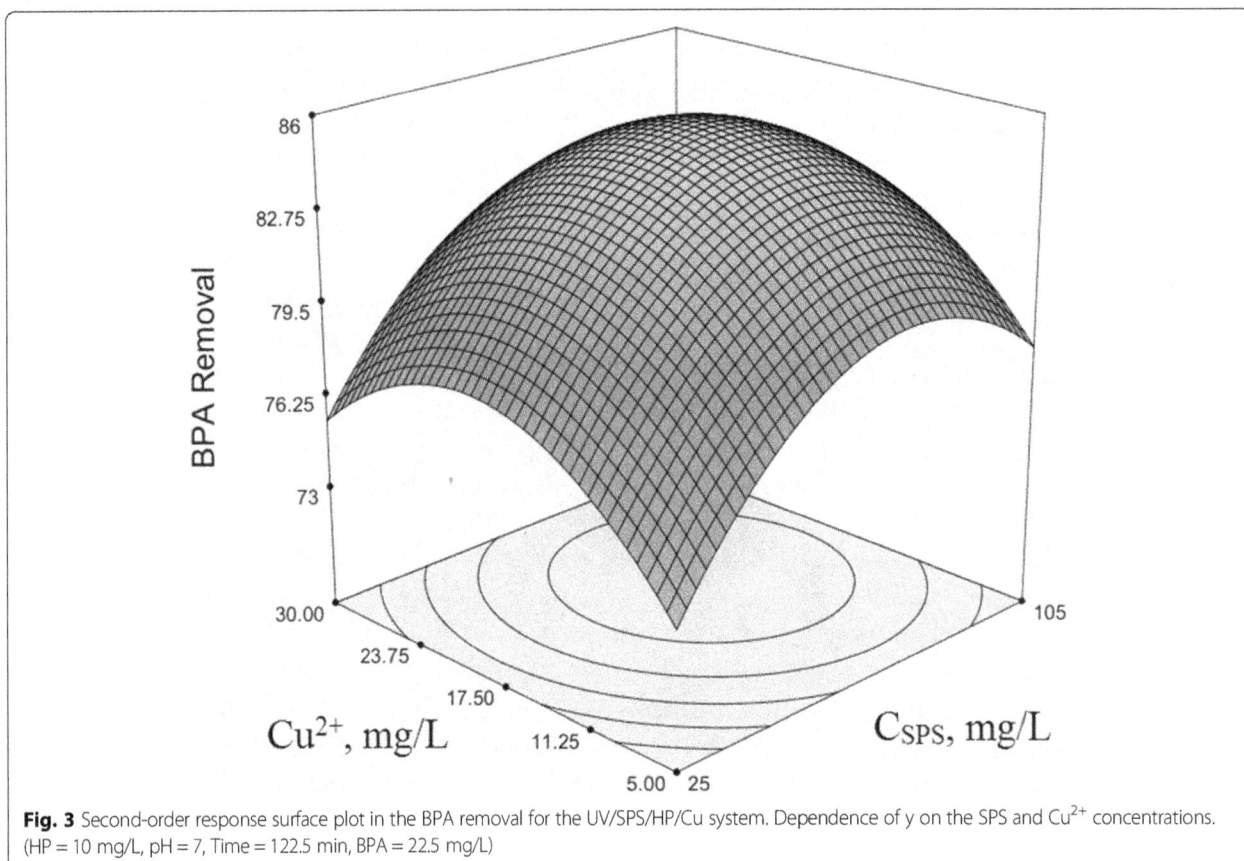

as $SO_4^{-\bullet}$ and $HO^{\bullet}$ react with organic compounds mainly by three main mechanisms: hydrogen abstraction, hydrogen addition, and electron transfer [61], and generally $SO_4^{-\bullet}$ is more likely to participate in electron transfer reactions than $HO^{\bullet}$, which is more likely to participate in hydrogen abstraction or addition reactions [63, 64]; hence reactivity of $HO^{\bullet}$ is less selective than $SO_4^{-\bullet}$. Furthermore, Liang and et al., using a chemical probe method on a thermally activated persulfate oxidation system, identified dominant radical species in different pH conditions: at basic pH hydroxyl radical is dominant, in the neutral pH both $SO_4^{-\bullet}$ and $HO^{\bullet}$ are present, and in the acidic pH dominant specie is $SO_4^{-\bullet}$ [65]. The results of this work about pH effect were confirmed by the results of Ya-Ting et al., on the ultraviolet activated persulfate oxidation of phenol, and results of Sung-Hwan et al., on the oxidation of BPA with UV/HP and UV/SPS [44, 61].

Copper cation (Cu^{2+}) interaction with PS has been investigated in Fig. 3 in accordance of obtained model. As seen in the model the BPA removal increased as Cu^{2+} concentration increased from −1 level to its central level (17.5 mg/L, obtained as the molar ratio to BPA), and then starts to decrease when Cu^{2+} is increasing towards +1 level (30 mg/L); simultaneously PS has almost the similar behaviour in the previous steps.

Usually to enhance the oxidation of contaminants, PS oxidation is conducted under different methods of activation, and transition metals, especially divalent metals (Cu^{2+}, Fe^{2+}, and …) which are commonly present and are important in soil and groundwater systems, also act as electron donors to catalyse the decomposition of SPS through a one-electron transfer reaction analogous to the Fenton initiation reaction, and according to the following equation activation of SPS result in sulphate radical that applied for degradation of BPA. However, excessive Cu^{2+} can significantly scavenge $SO_4^{-\bullet}$ and inhibit radical oxidation of target pollutant, BPA, via Eq. (11) [45, 66, 67].

$$S_2O_8^{2-} + Cu^{2+} \rightarrow SO_4^{-\bullet} + SO_4^{2-} + Cu^{3+} \qquad (10)$$

$$Cu^{2+} + SO_4^{-\bullet} \rightarrow Cu^{3+} + SO_4^{2-} \qquad (11)$$

Results of Krzysztof Kuśmierek et al., about oxidative degradation of 2-chloophenol by PS showed that synergistic activation of PS by heat and Cu^{2+} was more effective than heat and alkali activation and excess Cu^{2+} resulted in a relative decrease in removal efficiency [68]; Results of this work compatible with the present work. And in another work Liu, C. S., et al. studied oxidative degradation of propachlor by ferrous and copper ion activated persulfate. Results showed the activation of PS by Fe^{2+} ions resulted

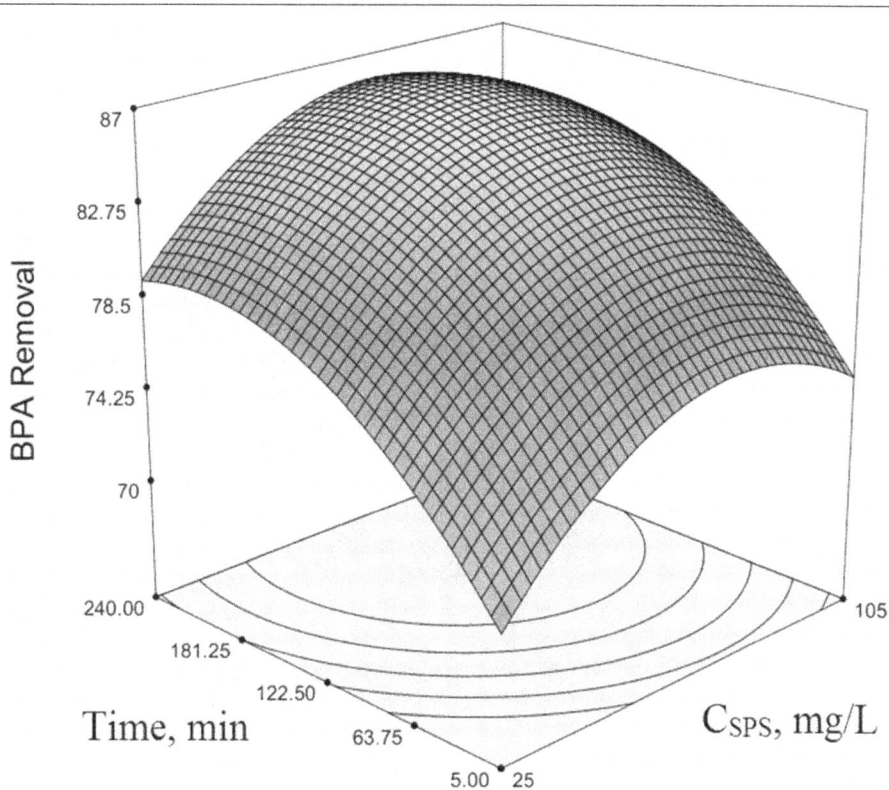

Fig. 4 Second-order response surface plot in the BPA removal for the UV/SPS/HP/Cu system. Dependence of y on the SPS and time. (HP = 10 mg/L, pH = 7, Cu^{2+} = 17.5 mg/L, BPA = 22.5 mg/L)

in rapid degradation of propachlor in the early stage, but was accompanied by a dramatic decrease in efficiency due to the rapid depletion of Fe^{2+} by the sulphate radicals generated, but the Cu^{2+} activated PS had a longer lasting degradation effect and a proportionally greater degradation enhancement at elevated Cu^{2+} concentrations [45].

Figure 4 indicates reaction time (min) interaction with SPS concentration (mg/L) on the BPA degradation rate in the range of 240 min based on the obtained model. Other effective factors were kept constant as cantered levels. As obviously seen in the graph, with the progress of reaction and increasing of time, BPA degradation and its removal was enhanced noticeably, while SPS concentration ranged from −1 to +1 levels as coded values, and in all of SPS levels, BPA has been increased with time reaction. So that at BPA concentrations of 22.5 mg/L, with increasing SPS concentration from −1 to +1 levels, the BPA removal rate first increased and then remained constant and finally declined slightly. In addition, gradient of the graph was up and then slightly was decreased. This is due to consumption and reduce of reactant mole numbers over time, and thereby reducing the probability of collisions between them [58]. The other studies have been similar results on the effect of reaction time [47, 67, 69–72].

Figure 5 shows the second-order response surface plot for the BPA removal by UV/SPS/HP/Cu system that indicates the interaction influence of BPA initial concentration (mg/L) and SPS concentration (mg/L) on the BPA degradation rate at the centred level of other factors. Figure 5 demonstrate that, with increasing SPS and BPA initial concentrations, the BPA degradation rate has considerably improved. Although PS has continued its quadratic effect, BPA initial concentration almost has a linear effect, which means that any increase in its level could lead to increase in degradation rate and removal efficiency. These results suggest that with increasing of BPA concentration under applied process with given conditions, reactants including sulphate and hydroxyl radicals could have more opportunities for encounter with the target pollutant, BPA, that can be lead to increasing the degradation rate and removal efficiency of BPA. In addition, using the increased level of PS can lead to more radicals, and synchronization of this phenomenon along with increased the initial level of BPA, can cause to more faster degradation of BPA [73, 74]. The obtained results of this work are confirmed with similar study's results by application of the PS and other AOPs processes for BPA removal from various fields of environment [57, 61, 71, 75, 76].

For optimization of BPA removal by changing selected factors and with choosing criteria goal as "in range" for all six variables as affecting factors and "maximize" as goal for BPA removal using model, the optimum values of

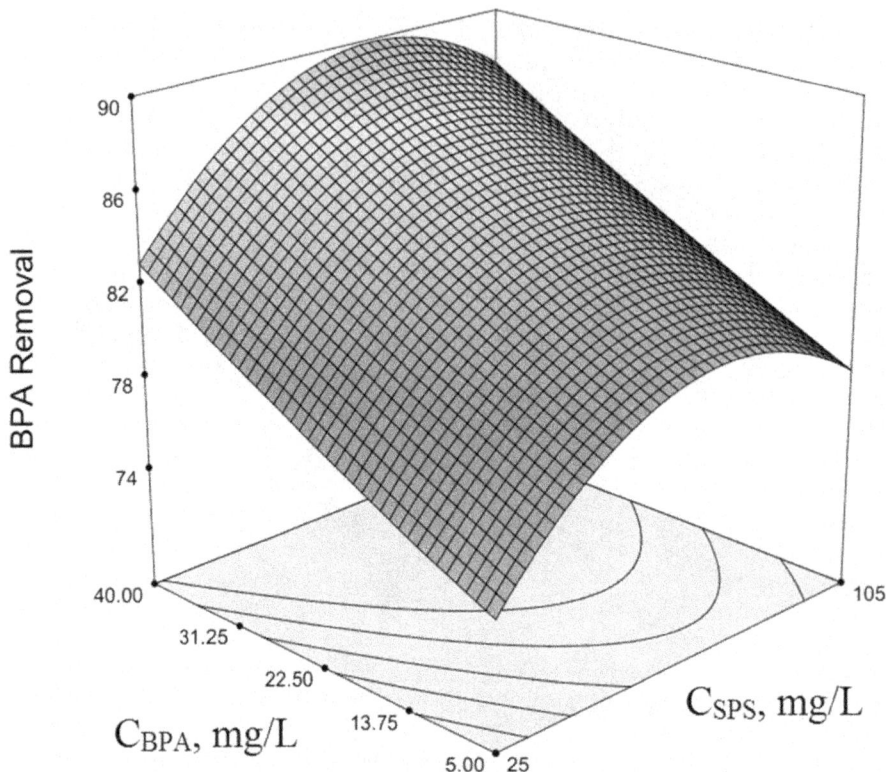

Fig. 5 Second-order response surface plot in the BPA removal for the UV/SPS/HP/Cu system. Dependence of y on the SPS and initial BPA concentrations. (HP = 10 mg/L, pH = 7, Cu^{2+} = 17.5 mg/L, Time = 122.5 min)

operation parameters were obtained as: initial SPS concentration (69.57 mg/L), initial HP concentration (9.85 mg/L), initial pH (10.91), initial Cu^{2+} concentration (20.76 mg/L), reaction time (140.38 min), and BPA initial concentration (39.33 mg/L), respectively, leading to 99.99 % of BPA removal.

According to the kinetic study carried out in optimal condition, BPA degradation using this system follow the first-order kinetic model and reaction rate was a function of time and initial concentration BPA. Reaction rate constant (K) was calculated in range 0.0073–0.0433 min^{-1} for initial BPA concentration in range 5–40 mg/L. Also, study of system as separate process (for example: UV alone) showed that using system as combined form, have synergistic effect than separate processess (Table 4) [77, 78]. Quenching studies with methanol (MA) and tert-butyl alcohol (TBA) were performed to identify the primary radical species formed in the system, showed that K reduced from 0.049 min^{-1} to 0.0116 min^{-1} and 0.0176 min^{-1} for MA and TBA, respectively. These results demonstrate that both $SO_4^{-\bullet}$ and $HO^{\bullet}$ can degrade BPA, but $SO_4^{-\bullet}$ plays the dominant role. Results of Yang and Hanci's study on removal of Azo dye acid Orange 7 (AO7) and BPA matches with results of this work [47, 57].

Finally, to evaluation of system performance in mineralisation of BPA in optimal condition, TOC analysis was performed. Results showed that process was able to

Table 4 Synergistic effect of the UV/SPS/HP/Cu in different arrangements (Optimal Condition)

No.	Process	K (min^{-1})					Synergistic effect	
		UV	SPS	HP	Cu	Combined	S	S'
1	Distinct process	×	×	-	-	-	-	-
2	UV/SPS	×	-	×	-	0.0073	1.7	41.9
3	UV/HP	×	-	-	×	0.006	1.6	37.5
4	UV/Cu	-	×	×	-	0.0034	1.3	22.7
5	SPS/HP	×	-	-	×	0.005	1.7	41.1
6	SPS/Cu	-	×	-	×	0.004	1.6	36.9
7	HP/Cu	×	×	×	-	0.0036	1.9	47.2
8	UV/SPS/HP	×	×	-	×	0.018	3.3	70
9	UV/SPS/Cu	×	-	×	×	0.011	2.4	58.5
10	UV/HP/Cu	-	×	×	×	0.0096	2.5	60
11	SPS/HP/Cu	×	×	×	×	0.0075	2.4	58.5
12	UV/SPS/HP/Cu	×	×	-	-	0.043	7.8	87.14

$*S = \frac{k(A+B)}{k(A)+k(B)}$, $S' = \frac{k(A+B)-[k(A)+k(B)]}{k(A)+k(B)} \times 100$ [78, 79]

remove up to 70 % of TOC in 180 min of reaction time. This means that mineralisation of BPA was done.

Toxicity study and GC-MS analysis

Results of bioassay for start point, and for times 15, 30, 60 and 180 min after beginning of reaction time is presented in Fig. 6. As shown in mentioned figure toxicity of BPA was increased during the reaction and then after 30 min starts to reduce and finally at the end of process reached to a minimum (24 h. toxic was increase from 2.35 TU to 8.34 TU in 30 min and then reduced to 1.58 TU at the end of reaction). This can be attributed to the production of intermediate products (aromatic and aliphatic as shown in Fig. 7) during the reaction, which probably are more toxic than BPA [66, 79]. Dehghani and et al. also was obtained similar results with ultrasonic and hydrogen peroxide processes and reach to < 2 TU [80]. Results of Tugba Olmez-Hanci's study by hot PS process with bioluminescence inhibitor are compatible with present study and has the same trend in toxicity along the process [57].

Several mechanisms have been proposed for better understanding the BPA degradation pathways by HO$^\bullet$-based AOPs [81, 82]. However, so far, there are very few reports related to the BPA oxidation pathway by SO$_4^{-\bullet}$-based AOPs process, especially UV/SPS/HP/Cu process. It must be considered that the degradation of organic pollutants is often done by the formation of some intermediates that can potentially impose toxic effect in the environment.

So, oxidation products formed during the UV/SPS/HP/Cu treatment of BPA were analysed by GC-MS after 0, 15, 30, 60 and 180 min treatment. Based on the obtained mass spectra, the evolution of several oxidation products was evident. Some of identified products include BPA catechol, 3-Hydroxy-4-methyl-benzoic acid, 2,3-dimethyl benzoic acid, 4-hydroxyacetophenone, 4-isopropenylphenol, Benzaldehyde, Succinic acid, p-hydroquinone, Phenol, Ethylene glycol monoformate, and numbers of unknown compounds that some of them heavier and some of them had lighter molar weight than BPA. The finding of similar studies using various AOPs processes, confirm presence of some mentioned compounds and some other similar compounds in treated effluents due to degradation of BPA [42, 49, 57, 82, 83]. Continuing of process results in more reduction in these compound's concentration and therefor complete mineralization of BPA after 180 min (Results are not shown). These trend of BPA degradation and production of intermediates during the process explains changes in effluent toxicity during the treatment; and it has been postulated that the intermediates formed during the initial stages of BPA oxidation are more toxic than BPA itself and according to Zazo et al., the subsequent decrease in the toxicity corresponds to the disappearance of aromatic intermediates giving rise to organic acids of relatively low toxicity [57, 84]. Considering that SO$_4^{-\bullet}$ and HO$^\bullet$ react with organic compounds by hydrogen abstraction, hydrogen addition, and electron transfer mechanisms [61], and as discussed in the last sections, generally SO$_4^{-\bullet}$ is more likely to participate in electron transfer reactions along with HO$^\bullet$ that tends

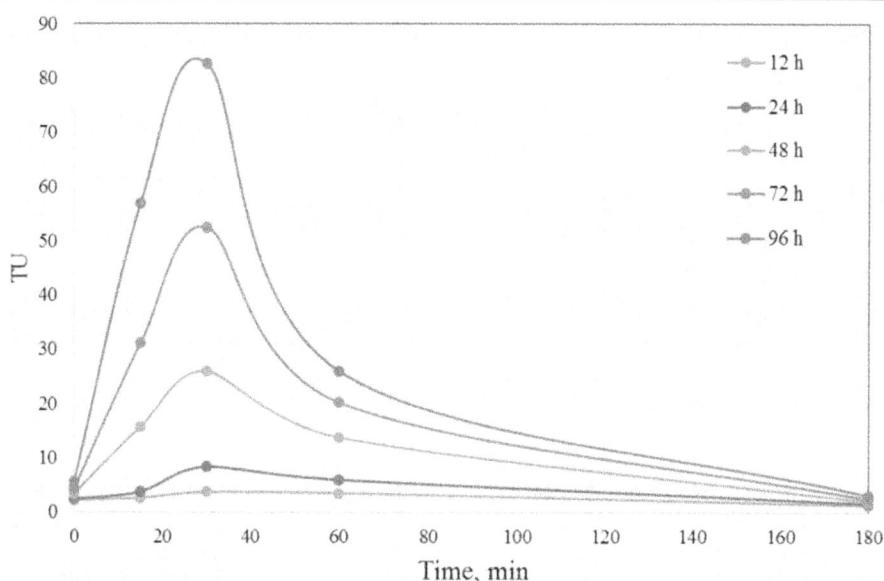

Fig. 6 Toxicity of BPA and its intermediate products during the reaction, based as acute toxicity unite (TUa) in the optimum conditions. [BPA]$_i$ = 0.219 mmol

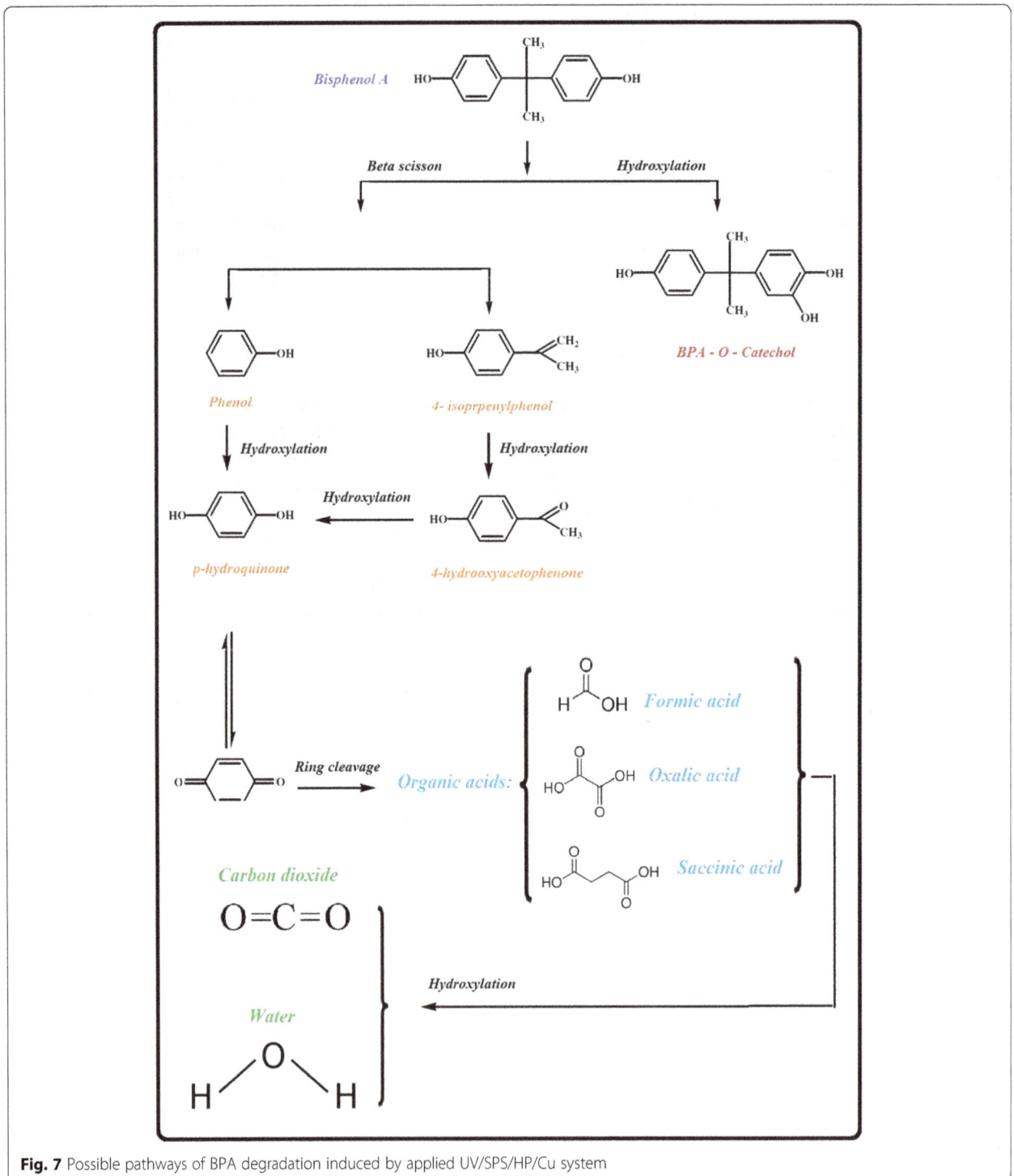

Fig. 7 Possible pathways of BPA degradation induced by applied UV/SPS/HP/Cu system

to two other mechanism, and according to the produced intermediates mentioned above, possible degradation pathway as simplified can be assumed as Fig. 7.

Finally, to evaluate the remaining parameters in the effluents including SPS, HP and Cu^{2+} in optimum conditions and after completing the process, a sample from effluent were analysed based on standard methods. Aaccording to the results, remaining concentrations of PS, HP and Cu^{2+} was respectively 32.5 mg/L, 4.9 mg/L and 8.75 mg/L. Considering that probably after treatment, effluents enter to the receiving waters, and for continuing process and degradation and mineralization of intermediates produced in the process, the mentioned remaining values will be useful, and will also meet effluent standards.

Evaluation of system performance on real effluent

Due to presence of BPA in real effluents with different conditions that distinguish it from synthetic wastewater, obtained optimal conditions were applied for treating a sample from effluent of plastic industry. Initial specifications of mentioned sample were presented in Table 5. Results showed that BPA reduced to 42 µg/L (about 77 % removal) and TOC was removed about 51 %. These explain that UV/SPS/HP/Cu system is proper process in BPA removal and it can be used for treating of the variety effluents contain BPA, but reason of low-performance of system for real conditions can be attributed to the presence of other factors as inhibitors in the content of wastewater. So, if this system applied for real conditions, it is recommended that further experiments shall be done for achieve optimum condition of removal and mineralization of BPA.

Conclusion

The purpose of this study was to investigate degradation and mineralisation of BPA in the aqueous solutions as well as optimization BPA removal from aqueous solution using UV/SPS/HP/Cu process. Based on our knowledge, this work is carried out for the first time using a statistical experimental design for optimization of BPA removal by combined UV/SPS/HP/Cu activated persulfate process. A six-variable, three-level CCD in combination with response surface modelling and quadratic polynomial were used to determine of the effects of initial SPS concentration (25–105 mg/L), initial HP concentration (5–15 mg/L), initial pH (3–11), initial Cu^{2+} concentration (5–30 mg/L), reaction time (5–240 min), and BPA initial concentration (10–40 mg/L) on the degradation of BPA as the response variable. Graphical response surface and contour plot were applied for the optimization of reaction condition.

The results of analysis of the response surfaces revealed that the six independent variables and their interactions have been significant effects on BPA degradation rate. The BPA initial concentration showed the highest effect, followed by reaction time, pH, SPS concentration, HP concentration and Cu^{2+} concentration.

With selecting criteria goal as "in range" for all six variables and "maximize" as goal for BPA removal, the optimum values of operation parameters were obtained as: initial SPS concentration (69.57 mg/L), initial HP concentration (9.85 mg/L), initial pH (10.91), initial Cu^{2+} concentration (20.76 mg/L), reaction time (140.38 min), and BPA initial concentration (39.33 mg/L), respectively, leading to 99.99 % of BPA removal. As well as, the ANOVA results indicated relatively high coefficient of the determination values ($R^2 = 0.9622$ and Adj-$R^2 = 0.9447$). The assessment of the model validation and adequacy also indicated a satisfactory goodness-of-fit obtained between the actual and predictive values by the model.

TOC analysis and bioassay with D. Magna revealed that BPA has been mineralised, toxicity of effluent reduced, and capable to discharge to the environment. Finally, the results of this study showed that UV/SPS/HP/Cu system is a new and efficient method in removal and mineralisation of BPA from aqueous solutions; moreover, CCD from the response surface methodology was an efficient statistical technique for the optimization of BPA removal from aqueous solutions using UV/SPS/HP/Cu process.

Endnotes
[1]Predicted no effect concentration

Abbreviations
µg/L: microgram per litre; BPA: Bisphenol A; C.V.: Coefficient of variance; CCD: Central composite design; CS: Calibration standard; EPA: Environmental Protection Agency; FDA: Food and Drug Administration; HO˙: Hydroxyl Radical; HPLC: High-performance liquid chromatography; mg/L: milligram per litre; NF: Nano filtration; PNEC: Predicted no effect concentration; PS: Persulfate; QC: Quality control; RO: Reverse osmosis; RSM: Response surface methodology; SO_4^-: Sulfate Radical; SPS: Sodium persulfate; TOC: Total organic carbon; UV: Ultraviolet; WWTPs: Wastewater treatment plants

Acknowledgements
The authors thank the sponsor of the project and the staffs of water and wastewater chemistry lab as well as advanced analytics lab of Health faculty.

Funding
This paper was part of a Ph.D thesis of the first author and has been funded by the Vice Chancellor for Research & Technology of Iran University of Medical Science under Project No. 93-4-27 24627.

Authors' contributions
S.A.M was the main investigator, run the system and drafted the manuscript. MG and MF supervised the study. AE, AJ, RR and MK were advisors of the study. All of authors contributed in all steps of work and read and approved the final manuscript.

Table 5 The physicochemical characteristics of used real effluent (Plastic industry, polycarbonate)

Parameter	Concentration, mg/L
pH	6.5
Cu	3.5
Hardness	385
Chloride	340
TOC	1760
Bicharbonate	280
BPA	145 µg/L

Competing interests

The authors declare that they have no competing interests.

Author details

[1]Research Center for Environmental Health Technology, School of Public Health, Iran University of Medical Sciences, Tehran, Iran. [2]Department of Environmental Health Engineering, School of Public Health, Iran University of Medical Sciences, Tehran, Iran. [3]Department of Environmental Health Engineering, School of Public Health, Ardabil University of Medical Sciences, Ardabil, Iran.

References

1. Agency, U.S.E.P. and O.o. Water. Treating Contaminants of Emerging Concern - A Literature Review Database. U.S. EPA; 2010.
2. Martha JM, WellsMartha JM, Wells AM, Katherine Y, Bell Marie-Laure Pellegrin, Lorien J. Fono, Emerging Pollutants. Water Environ Res. 2009; 81(10):2211–2254.
3. USEPA, U. Environmental Protection Agency, Bisphenol A action plan. 2010. CASRN 80-05-7.
4. Dalvand A, Gholami M, Joneidi A, Mahmoodi NM. Investigation of electrochemical coagulation process efficiency for removal of reactive red 198 from colored wastewater. J Color Sci Tech. 2009;3:97–105.
5. Zhang Y, Causserand C, Aimar P, Cravedi JP. Removal of bisphenol A by a nanofiltration membrane in view of drinking water production. Water Res. 2006;40(20):3793–9.
6. Flint S, Markle T, Thompson S, Wallace E. Bisphenol A exposure, effects, and policy: A wildlife perspective. J Environ Manag. 2012;104:19–34.
7. Rochester JR. Bisphenol A and human health: A review of the literature. Reprod Toxicol. 2013;42:132–55.
8. Erler C, Novak J. Bisphenol a exposure: human risk and health policy. J Pediatr Nurs. 2010;25(5):400–7.
9. Sun Y, Huang H, Sun Y, Wang C, Shi X, Hu H, Kameya T, Fujie K. Occurrence of estrogenic endocrine disrupting chemicals concern in sewage plant effluent. Front Environ Sci Eng. 2014;8(1):18–26.
10. Stahlhut RW, Welshons WV, Swan SH. Bisphenol A data in NHANES suggest longer than expected half-life, substantial nonfood exposure, or both. Environmental Health Perspectives, 117.5. Biomed Sci Publ (MU). 2009:784–9.
11. Bjerregaard-Olesen C, Ghisari M, Kjeldsen LS, Wielsøe M, Bonefeld-Jørgensen EC. Estrone sulfate and dehydroepiandrosterone sulfate: Transactivation of the estrogen and androgen receptor. Steroids. 2016;105:50–8.
12. Rachoń D. Endocrine disrupting chemicals (EDCs) and female cancer: Informing the patients. Rev Endocr Metab Disord. 2016:1–6.
13. Cooke PS, Simon L, Denslow ND. Chapter 37 - Endocrine Disruptors. In: Haschek WM, Rousseaux CG, Wallig MA, editors. Haschek and Rousseaux's Handbook of Toxicologic Pathology (Third Edition). Boston: Academic; 2013. p. 1123–54.
14. Mohapatra DP, Brar SK, Tyagi RD, Surampalli RY. Physico-chemical pre-treatment and biotransformation of wastewater and wastewater Sludge–Fate of bisphenol A. Chemosphere. 2010;78(8):923–41.
15. Huang Y-F, Huang Y-H. Identification of produced powerful radicals involved in the mineralization of bisphenol A using a novel UV-Na2S2O8/H2O2-Fe(II, III) two-stage oxidation process. J Hazard Mater. 2009;162(2–3):1211–6.
16. Badi MY, Azari A, Esrafili A, Ahmadi E, Gholami M. Performance evaluation of magnetized multiwall carbon nanotubes by iron oxide nanoparticles in removing fluoride from aqueous solution. J Mazandaran Univ Med Sci. 2015; 25(124):128–42.
17. Gholami M, Nassehinia HR, Jonidi-Jafari A, Nasseri S, Esrafili A. Comparison of Benzene & Toluene removal from synthetic polluted air with use of Nano photocatalytic TiO2/ ZNO process. J Environ Health Sci Eng. 2014;12(1).
18. Bing-zhi D, Lin W, Nai-yun G. The removal of bisphenol A by ultrafiltration. Desalination. 2008;221(1–3):312–7.
19. Fuerhacker M. Bisphenol A emission factors from industrial sources and elimination rates in a sewage treatment plant. Water Sci Tech. 2003;47(10):117–22.
20. Lee H-B, Peart TE, Svoboda ML. Determination of endocrine-disrupting phenols, acidic pharmaceuticals, and personal-care products in sewage by solid-phase extraction and gas chromatography–mass spectrometry. J Chromatogr A. 2005;1094(1):122–9.
21. Kimura K, Amy G, Drewes J, Watanabe Y. Adsorption of hydrophobic compounds onto NF/RO membranes: an artifact leading to overestimation of rejection. J Membr Sci. 2003;221(1):89–101.
22. Chin SS, Lim TM, Chiang K, Fane AG. Hybrid low-pressure submerged membrane photoreactor for the removal of bisphenol A. Desalination. 2007;202(1–3):253–61.
23. Bing-zhi D, Hua-qiang C, Lin W, Sheng-ji X, Nai-yun G. The removal of bisphenol A by hollow fiber microfiltration membrane. Desalination. 2010;250(2):693–7.
24. Heo J, Flora JRV, Her N, Park Y-G, Cho J, Son A, Yoon Y. Removal of bisphenol A and 17β-estradiol in single walled carbon nanotubes–ultrafiltration (SWNTs–UF) membrane systems. Sep Purif Technol. 2012;90:39–52.
25. Joseph L, Boateng LK, Flora JRV, Park Y-G, Son A, Badawy M, Yoon Y. Removal of bisphenol A and 17α-ethinyl estradiol by combined coagulation and adsorption using carbon nanomaterials and powdered activated carbon. Sep Purif Technol. 2013;107:37–47.
26. Asadgol Z, Forootanfar H, Rezaei S, Mahvi AH, Faramarzi MA. Removal of phenol and bisphenol-A catalyzed by laccase in aqueous solution. J Environ Health Sci Eng. 2014;12(1):93.
27. Gholami M, Shirzad-Siboni M, Farzadkia M, Yang JK. Synthesis, characterization, and application of ZnO/TiO2 nanocomposite for photocatalysis of a herbicide (Bentazon). Desalin Water Treat. 2016;57(29):13632–44.
28. Liang C, Lee IL, Hsu IY, Liang CP, Lin YL. Persulfate oxidation of trichloroethylene with and without iron activation in porous media. Chemosphere. 2008;70(3):426–35.
29. Maleki A, Mahvi AH, Ebrahimi R, Zandsalimi Y. Study of photochemical and sonochemical processes efficiency for degradation of dyes in aqueous solution. Korean J Chem Eng. 2010;27(6):1805–10.
30. Mahvi A. Application of ultrasonic technology for water and wastewater treatment. Iran J Public Health. 2009;38(2):1–17.
31. Shakerkhatibi M, Monajemi P, Jafarzadeh MT, Mokhtari SA, Farshchian MR. Feasibility study on EO/EG wastewater treatment using pilot scale SBR. Int J Environ Res. 2012;7(1):195–204.
32. Badi MY, Esrafili A, Kalantary RR, Azari A, Ahmadi E, Gholami M. Removal of diethyl phthalate from aqueous solution using persulfate-based (UV / Na2S2O8 / Fe2+) advanced oxidation process. J Mazandaran Univ Med Sci. 2016;25(132):122–35.
33. Xiang-Rong Xu SL, Qing H, Jin-Ling L, Yi-Yi Y, Hua-Bin L. Activation of Persulfate and Its Environmental Application. Int J Environ Bioenergy. 2012;1(1):22.
34. Dehghani S, Jafari AJ, Farzadkia M, Gholami M. Sulfonamide antibiotic reduction in aquatic environment by application of fenton oxidation process. J Environ Health Sci Eng. 2013;10(1).
35. Bahmani P, Kalantary RR, Esrafili A, Gholami M, Jafari AJ. Evaluation of Fenton oxidation process coupled with biological treatment for the removal of reactive black 5 from aqueous solution. J Environ Health Sci Eng. 2013;11(1).
36. Davoudi M, Gholami M, Naseri S, Mahvi AH, Farzadkia M, Esrafili A, Alidadi H. Application of electrochemical reactor divided by cellulosic membrane for optimized simultaneous removal of phenols, chromium, and ammonia from tannery effluents. Toxicol Environ Chem. 2014;96(9):1310–32.
37. Ahmadi M, Vahabzadeh F, Bonakdarpour B, Mofarrah E, Mehranian M. Application of the central composite design and response surface methodology to the advanced treatment of olive oil processing wastewater using Fenton's peroxidation. J Hazard Mater. 2005;123(1):187–95.
38. Roosta M, Ghaedi M, Daneshfar A, Sahraei R. Experimental design based response surface methodology optimization of ultrasonic assisted adsorption of safaranin O by tin sulfide nanoparticle loaded on activated carbon. Spectrochim Acta A Mol Biomol Spectrosc. 2014;122:223–31.
39. Jafari A, Mahvi AH, Godini H, Rezaee R, Hosseini SS. Process optimization for fluoride removal from water by Moringa oleifera seed extract. Fluoride. 2014;47(2):152–60.
40. Torrades F, García-Montaño J. Using central composite experimental design to optimize the degradation of real dye wastewater by Fenton and photo-Fenton reactions. Dyes Pigments. 2014;100:184–9.
41. Roosta M, Ghaedi M, Daneshfar A, Darafarin S, Sahraei R, Purkait MK. Simultaneous ultrasound-assisted removal of sunset yellow and erythrosine by ZnS: Ni nanoparticles loaded on activated carbon: Optimization by central composite design. Ultrason Sonochem. 2014;21(4):1441–50.
42. Zhang X, Ding Y, Tang H, Han X, Zhu L, Wang N. Degradation of bisphenol A by hydrogen peroxide activated with CuFeO2 microparticles as a heterogeneous Fenton-like catalyst: Efficiency, stability and mechanism. Chem Eng J. 2014;236:251–62.
43. De Heredia JB, Torregrosa J, Dominguez JR, Peres JA. Kinetic model for phenolic compound oxidation by Fenton's reagent. Chemosphere. 2001;45(1):85–90.
44. Yoon S-H, Jeong S, Lee S. Oxidation of bisphenol A by UV/S2O: Comparison with UV/H2O2. Environ Technol. 2012;33(1):123–8.

45. Liu CS, Shih K, Sun CX, Wang F. Oxidative degradation of propachlor by ferrous and copper ion activated persulfate. Sci Total Environ. 2012;416:507–12.

46. Talinli I, Anderson G. Interference of hydrogen peroxide on the standard COD test. Water Res. 1992;26(1):107–10.

47. Yang S, Wang P, Yang X, Wei G, Zhang W, Shan L. A novel advanced oxidation process to degrade organic pollutants in wastewater: Microwave-activated persulfate oxidation. J Environ Sci. 2009;21(9):1175–80.

48. Wang C, Zhu L, Wei M, Chen P, Shan G. Photolytic reaction mechanism and impacts of coexisting substances on photodegradation of bisphenol A by Bi 2 WO 6 in water. Water Res. 2012;46(3):845–53.

49. Sharma J, Mishra IM, Dionysiou DD, Kumar V. Oxidative removal of Bisphenol A by UV-C/peroxymonosulfate (PMS): Kinetics, influence of co-existing chemicals and degradation pathway. Chem Eng J. 2015;276:193–204.

50. Kohzadi O, Reshadi P, Jamshidi N. Comparing the performance of advanced photochemical oxidation technologies for the removal of phenol and bisphenol A from aqueous solutions, in National Conference of wastewater and solid waste management in the oil and energy industries. 2012.

51. Parajo JC, Alonso J, Lage MA, Vazquez D. Empirical modeling of eucalyptus wood processing. Bioprocess Eng. 1992;8(3–4):129–36.

52. Kirk RE. Experimental design. Wiley Online Library; 1982.

53. Beg QK, Sahai V, Gupta R. Statistical media optimization and alkaline protease production from Bacillus mojavensis in a bioreactor. Process Biochem. 2003;39(2):203–9.

54. Mason RL, Gunst RF, Hess JL. Statistical design and analysis of experiments: with applications to engineering and science, vol 474. John Wiley & Sons; 2003.

55. Montgomery DC, Runger GC, Hubele NF. Engineering statistics. John Wiley & Sons; 2009.

56. Vining GG, Kowalski S. Statistical methods for engineers. Cengage Learning Inc.; 2010.

57. Olmez-Hanci T, Arslan-Alaton I, Genc B. Bisphenol A treatment by the hot persulfate process: Oxidation products and acute toxicity. J Hazard Mater. 2013;263, Part 2(0):283–90.

58. Zhang Y-Q, Huang S-B, Hussain I. Oxidative degradation of p-chloroaniline by copper oxidate activated persulfate. Chem Eng J. 2013;218:384–91.

59. Mahvi AH, Maleki A, Alimohamadi M, Ghasri A. Photo-oxidation of phenol in aqueous solution: toxicity of intermediates. Korean J Chem Eng. 2007;24(1):79–82.

60. Norman R, Storey P, West P. Electron spin resonance studies. Part XXV. Reactions of the sulphate radical anion with organic compounds. J Chem Soc B: Phys Organ. 1970;1087–1095.

61. Lin Y-T, Liang C, Chen J-H. Feasibility study of ultraviolet activated persulfate oxidation of phenol. Chemosphere. 2011;82(8):1168–72.

62. Mahvi AH, Maleki A, Rezaee R, Safari M. Reduction of humic substances in water by application of ultrasound waves and ultraviolet irradiation. 2010.

63. Peyton GR. The free-radical chemistry of persulfate-based total organic carbon analyzers. Mar Chem. 1993;41(1):91–103.

64. Rezaee R, Maleki A, Jafari A, Mazloomi S, Zandsalimi Y, Mahvi AH. Application of response surface methodology for optimization of natural organic matter degradation by UV/H 2 O 2 advanced oxidation process. J Environ Health Sci Eng. 2014;12(1):1.

65. Liang C, Su H-W. Identification of sulfate and hydroxyl radicals in thermally activated persulfate. Ind Eng Chem Res. 2009;48(11):5558–62.

66. Huang Y-F, Huang Y-H. Behavioral evidence of the dominant radicals and intermediates involved in Bisphenol A degradation using an efficient Co2 +/PMS oxidation process. J Hazard Mater. 2009;167(1–3):418–26.

67. Lin H, Wu J, Zhang H. Degradation of bisphenol A in aqueous solution by a novel electro/Fe3+/peroxydisulfate process. Sep Purif Technol. 2013;117:18–23.

68. Krzysztof Kuśmierek AŚ, Lidia D. Oxidative degradation of 2-chlorophenol by persulfate. J Ecol Eng. 2015;16(3):115–23.

69. Lee Y-C, Lo S-L, Kuo J, Lin Y-L. Persulfate oxidation of perfluorooctanoic acid under the temperatures of 20–40 °C. Chem Eng J. 2012;198–199:27–32.

70. Yen C-H, Chen K-F, Kao C-M, Liang S-H, Chen T-Y. Application of persulfate to remediate petroleum hydrocarbon-contaminated soil: Feasibility and comparison with common oxidants. J Hazard Mater. 2011;186(2–3):2097–102.

71. Jiang X, Wu Y, Wang P, Li H, Dong W. Degradation of bisphenol A in aqueous solution by persulfate activated with ferrous ion. Environ Sci Pollut Res. 2013;20(7):4947–53.

72. Mahvi AH, Ebrahimi SJ, Mesdaghinia A, Gharibi H, Sowlat MH. Performance evaluation of a continuous bipolar electrocoagulation/electrooxidation–electroflotation (ECEO–EF) reactor designed for simultaneous removal of ammonia and phosphate from wastewater effluent. J Hazard Mater. 2011; 192(3):1267–74.

73. Oh S-Y, Kim H-W, Park J-M, Park H-S, Yoon C. Oxidation of polyvinyl alcohol by persulfate activated with heat, Fe 2+, and zero-valent iron. J Hazard Mater. 2009;168(1):346–51.

74. Xu X-R, Li X-Z. Degradation of azo dye Orange G in aqueous solutions by persulfate with ferrous ion. Sep Purif Technol. 2010;72(1):105–11.

75. Saien J, Osali M, and Soleymani A. UV/persulfate and UV/hydrogen peroxide processes for the treatment of salicylic acid: effect of operating parameters, kinetic, and energy consumption. Desalin Water Treat. 2014;56:1–9.

76. Asgari G, Seidmohammadi A, Chavoshani A. Pentachlorophenol removal from aqueous solutions by microwave/persulfate and microwave/H2O2: a comparative kinetic study. J Environ Health Sci Eng. 2014;12(1):94.

77. Tran N, Drogui P, Brar S. Sonochemical techniques to degrade pharmaceutical organic pollutants. Environ Chem Lett. 2015;13(3):251–68.

78. Kavitha S, Palanisamy P. Photocatalytic and sonophotocatalytic degradation of reactive red 120 using dye sensitized TiO2 under visible light. Int J Civil Environ Eng. 2011;3(1):1–6.

79. Rubalcaba A, Suárez-Ojeda ME, Stüber F, Fortuny A, Bengoa C, Metcalfe I, Font J, Carrera J, Fabregat A. Biodegradability enhancement of phenolic compounds by hydrogen peroxide promoted catalytic wet air oxidation. Catal Today. 2007;124(3):191–7.

80. Dehghani MH, Norozi Z, Nikfar E, Rastkari N. Evaluation of Bisphenol A solution toxicity before and after ultrasonic and hydrogen peroxide processes using Daphnia Magna bioassay. J Kermanshah Univ Med Sci. 2013;17(6):336–42.

81. Guo C, Ge M, Liu L, Gao G, Feng Y, Wang Y. Directed synthesis of mesoporous TiO2 microspheres: catalysts and their photocatalysis for bisphenol A degradation. Environ Sci Technol. 2009;44(1):419–25.

82. Poerschmann J, Trommler U, Górecki T. Aromatic intermediate formation during oxidative degradation of Bisphenol A by homogeneous sub-stoichiometric Fenton reaction. Chemosphere. 2010;79(10):975–86.

83. Deborde M, Rabouan S, Mazellier P, Duguet JP, Legube B. Oxidation of bisphenol A by ozone in aqueous solution. Water Res. 2008;42(16):4299–308.

84. Zazo JA, Casas JA, Molina CB, Quintanilla A, Rodriguez JJ. Evolution of ecotoxicity upon Fenton's oxidation of phenol in water. Environ Sci Technol. 2007;41(20):7164–70.

13

Assessment of tetracycline contamination in surface and groundwater resources proximal to animal farming houses in Tehran, Iran

Allahbakhsh Javid[1], Alireza Mesdaghinia[2*], Simin Nasseri[2,3], Amir Hossein Mahvi[2,4], Mahmood Alimohammadi[2] and Hamed Gharibi[1]

Abstract

Background: Antibiotics have been increasingly used for veterinary and medical purposes. The overuse of these compounds for these purposes can pollute the environment, water resources in particular. Tetracycline, among other forms of antibiotics, is one of the most applied antibiotic in aquaculture and veterinary medicine. The present study aimed to tack the traces of tetracycline in the effluents of municipal and hospital wastewater treatment plants, surface and groundwater resources and finally the drinking water provided from these water resources.

Methods: The samples were taken from Fasha-Foyeh Dam, wells located at Varamin Plain, and Yaftabad; and also, wastewater samples were collected from the wastewater treatment plant effluents of Emam Khomeini Hospital and a municipal wastewater treatment plant which its effluent is being released to the surface water of the area covered in this work. 24 samples were collected in total during July 2012 to December 2012. The prepared samples were analyzed using high-performance liquid chromatography.

Results: Based on the results, mean tetracycline levels in surface and ground water at nearby of animal farms was found to vary from 5.4 to 8.1 ng L^{-1}. Furthermore, the maximum TC concentration of 9.3 ng L^{-1} was found to be at Yaft-Abad sampling station. Although tetracycline traces could not be detected in any investigated Hospital WWTP effluents, it was tracked in MWWTP effluent samples, in the concentration range of 280 to 540 ng l^{-1}.

Conclusion: The results showed that the concentration of TC in water resource near the animal farms is higher than the other sampling stations. This is related to the usage of antibiotic for animals. In fact, it caused the contamination of water resources and could contribute to radical changes in the ecology of these regions.

Keywords: Tetracycline, Groundwater resources, Surface water resources, Drinking water, Animal farms

Background

Antibiotics have been increasingly used for veterinary and medical purposes. The increase in the use of these compounds affects both the environment and human health; in other words, the active forms of the antibiotics are being excreted from the body via urine and/or feces into the environment. Considering this, the overuse of these compounds can pollute the water resources [1]. It

should be noted that there are various pathways in which these compounds enter into both surface and groundwater resources, including run-off, leakage from lagoons, leaching of manure applied to fields, and leaching from animal housing areas [Watanabe et al.]. The presence of antibiotics, TC in particular, in water and soil can cause some allergies and toxicity, since these compounds are still active [2]. For instance, excreted antibiotics in the environment affect almost all the bacterial species forcing them to develop a resistance toward these compounds [3]. Based on the reports of previously conducted studies, the residues of various

* Correspondence: mesdaghinia@sina.tums.ac.ir
[2]Department of Environmental Health Engineering, School of public Health, Tehran University of Medical Sciences, Tehran, Iran
Full list of author information is available at the end of the article

forms of antibiotics have been detected in the samples taken from surface and groundwater resources and also drinking water [4, 5]. Furthermore, the antibiotics have also been found in the samples taken from the effluents of both municipal and hospital wastewater treatment plant [6, 7].

Tetracycline (TC), among other forms of antibiotics, is one of the most applied antibiotics in aquaculture and veterinary medicine [8]. It should be noted that this antibiotic has been applied in livestock and poultry productions more than the aquaculture medicine. In addition, tetracycline is being discharged into the environment, water resources in particular, through wastewater effluent of drug manufacturing companies, disposal of non-consumable compounds and expired drugs containing tetracycline, and also from animal and agricultural wastes [9, 10]. TC has been classified among the antibiotics frequently detected in sewage, domestic wastewaters, surface and groundwater resources, drinking water, and sludge [11]. Considering the increase in the usage of TC and also the inefficiency of most conventional wastewater treatment processes in removing this antibiotic, the surface and ground water resources are now at more risk of being polluted with TC. Furthermore, it should be noted that there is not any regulation for routine sampling and analyzing TC level in the water resources. Previously conducted studies showed that one of the most widely used antibiotic in animals is tetracycline [8].

This study aimed to tack the traces of tetracycline in a specific route; in other words, tetracycline was strived to detect in the samples taken from the effluents of municipal and hospital wastewater treatment plants, surface and groundwater resources and finally the drinking water provided from these water resources. In this work, TC was selected due to its wide usage and persistence in the environment. Figure 1 shows the sampling locations. It should be noted that these locations are related to each other. In fact, the dam represents the surface water resource receiving the effluents of selected municipal and hospital wastewater treatment plants. The wells, in addition, are located in the vicinity of the main animal farming houses in Tehran, Iran and also the dam. Furthermore, drinking water prepared from using the water resources was also studied.

Methods
Sampling stations
In the present work, four samples over 6 months from July 2012 to December 2012 were taken; in total, 24 samples were collected and analyzed in this study. Water samples were taken from tap water, Fasha-Foyeh Dam and also specific wells located at Varamin Plain and Yaftabad. In addition, wastewater samples were collected from the effluents of Emam Khomeini Hospital wastewater treatment plant and a municipal wastewater treatment plant. Figure 1 shows Geographical position of places which the samples were taken.

In this study, two locations were selected to take sample from, namely, Varamin Plain and Yaftabad (i. e. wells nearby animal farming houses) to determine whether TC leached to groundwater resources. In addition, Fasha-Foyeh Dam was selected as surface water resource nearby animal farming houses to assess whether TC has been transported by run-off into the most nearby surface water resource; In other words, the sampling took place at nearby water resources of animal farms which their wastewater effluents are being discharged into the resources. Furthermore, the effluents of municipal and hospital wastewater plants which are being released into the same water resources were considered and analyzed in order to detect TC. The wastewater samples were also taken directly from the effluents of WWTPs to assess the amount of TC released from human sources and compare it with water samples mentioned above. Another spot was also selected (i. e. tap water) to determine whether the drinking water is polluted.

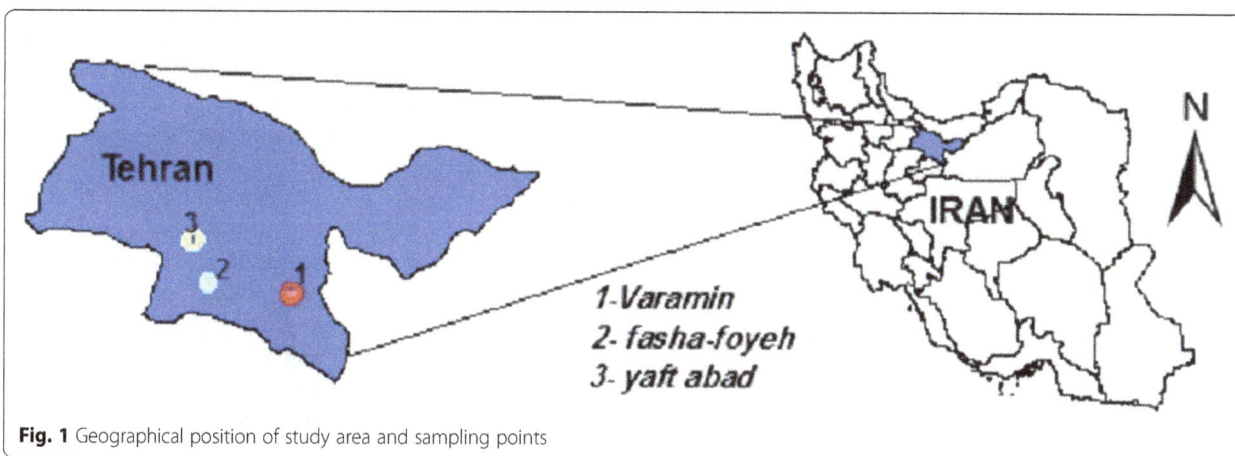

1-Varamin
2- fasha-foyeh
3- yaft abad

Fig. 1 Geographical position of study area and sampling points

Preparing and analyzing the samples

The method of analyzing the samples is based on a method applied and proposed in a study by A Pena et al., [3]. At first, samples were allowed to settle at 4 °C for 12–24 h in the dark. Careful handling prevented re-suspension of settled particles, and the supernatant of the samples was used for analysis instead of a filtered aliquot [12]. Since antibiotics are present in surface and ground waters and other resources at trace levels, pre-conditioning by suitable sorbent is a necessary step in sample preparation. The average recoveries 80 ± 3 were obtained using strata C18-E cartridge. Detection frequency of tetracycline in water sources and WWTP effluent is shown in Table 1.

According to a study conducted by Shalaby et al., the strata C18-E cartridge was applied to buffer extract the residue; after sample loading, the cartridge was washed by 10 ml of 5 % methanol mixed in water, and tetracycline was eluted with a mixture of 10 ml of methanol and 0.01 M oxalic acid [13].

The samples were taken by syringe and filtered through 0.45 μm membrane at pre-selected time intervals; then, the samples were measured using high-performance liquid chromatography (HPLC). HPLC consisted of a Knauer LPG pump, an EZ-chrom HPLC system manager program and a UV detector (k-2500). The UV–detector was set at the maximum absorption wavelength of 365 nm. Aliquots of 100 μL were injected manually using a model SGE injection valve (SGE. Australia). MZ-analysentechnik ODS-3 C18 (4.6 mm × 250 mm) packed with 5 μm spherical particles was used for separation. An Acetonitrile (A) aqueous oxalic acid 0.01 M (B) mixture was used as mobile phase at 30^0C temperature with a constant flow rate of 1.0 mLmin^{-1}. The used mobile phase are shown in Table 1 [3].

Results and discussion

This work is the first survey of the contamination of surface and ground water resources with TC in Tehran, Iran. The results showed that the mean concentration of TC in water resources, both surface and groundwater, varies from 5.4 to 8.1 ng L^{-1}among which Yaft-abad station with 9.3 ng L^{-1} level of TC was found to have the maximum TC concentration; The results are shown in Table 2. The concentration of TC in the samples taken

Table 1 HPLC gradient program used for determination of TC

Time (min)	Mobile phase A (%)	Mobile phase B (%)
0	10	90
7	40	60
7.5	50	50
12.95	10	90
13		End

Table 2 Mean concentrations of tetracycline in water resource and wastewater treatment plant effluent

Location	TC concentration (ngL^{-1})	
Yaft-abad – (well)	Mean	8.1
	Max	9.3
	Min	6.9
	$n = 4$	
Fasha-foyeh- (dam)	Mean	6.4
	Max	8.7
	Min	5.7
	$n = 4$	
Varamin- (well)	Mean	5.4
	Max	7.2
	Min	4.4
	$n = 4$	
Tap water	ND	
	$n = 4$	
HWWTP effluent	ND	
	$n = 4$	
MWWTP effluent	Mean	540
	Max	630
	Min	280
	$n = 4$	

ND non-detectable

from surface water resource (i. e. Fasha-foyeh- (dam)), was found in the range of 5.7 to 8.7 ng/L. Detected levels of this antibiotic in surface water resources could be explained by the fact that tetracycline of veterinary wastes finds its way into the surface and ground water resources. Run-off is one of the main ways in which antibiotics leach into particularly surface water resources [14, 15]. Considering the location of animal farming houses in vicinity of Fasha-foyeh- (dam), this could be a probable cause of finding trace amount of TC in the taken samples from this location. There is also another source of polluting the surface water resources, as taken into account in this study, and it is the release of wastewater effluents into the resources [15]. The concentration of TC in MWWTP effluent samples ranged from 280 to 540 ng l^{-1}. Considering the fact that most of the antibiotics have high affinity to soil compounds [16, 17], tetracycline in particular, the main source of TC in the surface water resource studied in this work could be the effluent of MWWTP. This result is in line with the results of previously conducted studies on the presence of antibiotics in the water resources [1, 13]. In addition, the concentration of TC was negligible in the samples taken from the effluents of Hospital wastewater treatment plant. The main reason is that TC is no longer being used among hospitalized patients. In a study conducted in the

United States of America, the concentration of tetracycline in the effluent of MWWTP was reported to be between 170 and 850 ng L^{-1} [18]. In a study conducted by Miao et al., 150 to 970 ngL^{-1} concentration of TC was reported in Canada [19]. Comparing with these findings, the concentrations of TC detected in the samples taken from WWTPs effluents in this study are at lower levels.

As mentioned above, the highest concentration of TC was found in the samples taken from ground water resources. Although TC has high affinity to soil compounds, the presence of this antibiotic in the samples even in this range is of high importance. It should be noted that the overuse of these drugs causes them to be found in the water resources due mainly to the saturation of soil capacity; in fact, this phenomenon is known as terracumulation [20, 21]. When the soil becomes saturated with TC, the infiltration of water originated from rain into groundwater resources can carry this antibiotic contributing to the pollution of these resources [22]. Considering the fact that TC is still an active antibiotic when it is bounded to soil particles, the presence of it in groundwater resource is indeed a major problem.

In addition, four samples were taken from tap water to determine whether any concentration of TC is in drinking water. The water treated and distributed among the people of the region we took samples from came from a water treatment plant which uses the water resources (i. e. surface and groundwater) mentioned above. As shown in Table 2, the concentration of TC in these samples was negligible.

Conclusion

This work was aimed to survey the contamination of surface and ground water resources with TC in Tehran, Iran. Based on the results, the water sources studied in this work are contaminated with TC and the presence of other types of antibiotics is also probable. The release of TC from the animal farming areas can be implied to be the main source of groundwater resources pollution with this antibiotic. In addition, the release wastewater effluent from WWTP can be considered as a potential source for the contamination of surface water resource with TC. In fact, the presence of this antibiotic in the environment can cause the mutation of the bacterial species and make them resistant to the antibiotics; and also, negatively affect those who use the water from this resource. In this regard, regular monitoring of the presence of antibiotics mostly used in these areas in order to prevent further damage to the environment. In conclusion, so we recommend more monitoring of the existence of antibiotics residues in water resource near the animal farms.

Abbreviations
TC: tetracycline; HPLC: high-performance liquid chromatography; WWTP: wastewater treatment plant; MWWTP: municipal wastewater treatment plant.

Competing interests
The authors declare that they have no competing interests.

Authors' contributions
AJ was the main investigator, collected the data, SN and AM supervised the study. AHM, and MA were advisors of the study. HGH performed the statistical analysis, and drafted the manuscript. All authors read and approved the final manuscript.

Acknowledgments
This work is part of a PhD dissertation and supported by Tehran University of Medical Sciences. The authors wish to thank the staff of laboratory of School of Public Health. The second part has been published as "Performance of photocatalytic oxidation of tetracycline in aqueous solution by TiO2 nanofibers".

Author details
[1]School of Public Health, Shahroud University of Medical Sciences, Shahroud, Iran. [2]Department of Environmental Health Engineering, School of public Health, Tehran University of Medical Sciences, Tehran, Iran. [3]Center for Water Quality Research, Institute for Environmental Research, Tehran University of Medical Sciences, Tehran, Iran. [4]Center for Solid Waste Research, Institute for Environmental Research, Tehran University of Medical Sciences, Tehran, Iran.

References
1. Xu WH, Zhang G, Zou SC, Li XD, Liu YC. Determination of selected antibiotics in the Victoria Harbour and the Pearl River, South China using ighperformance liquid chromatographyeelectrospray ionization tandem mass spectrometry. Environ Pollut. 2007;145:672–9.
2. Kummerer K. The presence of pharmaceuticals in the environment due to human use - present knowledge and future challenges. J Environ Manage. 2009;90:2354–66.
3. Pena A, Albert-Garcia JR, Silva LJG, Lino CM, Calatayud JM. Photo-induced fluorescence of magnesium derivatives of tetracycline antibiotics in wastewater samples. J Hazard Mater. 2010;179:409–14.
4. Yiruhan, Wang QJ, Mo CH, Li YW, Gao P, Tai YP, et al. Determination of four fluoroquinolone antibiotics in tap water in Guangzhou and Macao. Environ Pollut. 2010;158(7):2350–8.
5. Chee-Sanford J, Aminov R, Krapac I, Garrigues-Jeanjean N, Mackie R. Occurrence and diversity of tetracycline resistance genes in lagoons and groundwater underlying two swine production facilities. Appl Environ Microbiol. 2001;67:1494–1502.
6. Brown KD, Kulis J, Thomson B, Chapman TH, Mawhinney DB. Occurrence of antibiotics in hospital, residential, and dairy effluent, municipal wastewater, and the Rio Grande in New Mexico. Sci Total Environ. 2006;366(2-3):772–83.
7. Homem V, Santos L. Degradation and removal methods of antibiotics from aqueous matrices–a review. J Environ Manage. 2011;92(10):2304–47.
8. Sarmah AK, Meyer MT, Boxall ABA. A global perspective on the use, sales, exposure pathways, occurrence, fate and effects of veterinary antibiotics (VAs) in the environment. Chemosphere. 2006;65:725–59.
9. Boxall AB, Kolpin DW, Halling-Sørensen B, Tolls J. Are veterinary medicines causing environmental risks? J Environ Sci Technol. 2003;37:286A–94.
10. Mompelat S, LeBot B, Thomas O. Occurrence and fate of pharmaceutical products and by-products, from resource to drinking water. Environ Int. 2009;35:803–14.
11. Penghua W, Pow-Seng Y, Lim TT. C–N–S tridoped TiO2 for photocatalytic degradation of tetracycline under visible-light irradiation. Appl Catal Gen. 2011;399:252–61.
12. Nödler K, Licha T, Bester K, Sauter M. Development of a multi-residue analytical method, based on liquid chromatography–tandem mass spectrometry, for the simultaneous determination of 46 micro-contaminants in aqueous samples. J Chromatogr A. 2010;1217:6511–21.

13. Shalaby AR, Salama NA, Abou-Raya SH, Emam WH, Mehaya FM. Validation of HPLC method for determination of tetracycline residues in chicken meat and liver. Food Chem. 2011;124:1660–6.

14. Alder AC, McArdell CS, Golet EM, Ibric S, Molnar E, Nipales NS, et al. Occurrence and fate of flouroquinolone, macrolide, and sulfanamide antibiotics during wastewater treatment and in ambient waters in Switzerland. Symposium Series. Washington: American Chemical Society; 2001. 791, p. 56–69.

15. Pedersen J, Yeager M, Suffet I. Xenobiotic organic compounds in runoff from fields irrigated with treated wastewater. J Agric Food Chem. 2003;51:1360–72.

16. Hamscher G, Sczesny S, Ho¨per H, Nau H. Determination of persistent tetracycline residues in soil fertilized with liquid manure by high-performance liquid chromatography with lectrospray ionization tandem mass spectrometry. Anal Chem. 2002;74:1509–18.

17. Samuelsen OB, Torsvik V, Ervik A. Long-range changes in oxytetracycline concentration and bacterial resistance towards oxytetracycline in a fish farm sediment after medication. Sci Total Environ. 1992;114:25–36.

18. Batt AL, Snow DD, Aga DS. Occurrence of sulphonamide antimicrobials in private water wells in Washington County, Idaho, USA. Chemosphere. 2006;64:1963–71.

19. Karthikeyan KG, Meyer T. Occurrence of antibiotics in wastewater treatment facilities in Wisconsin, USA. Sci Total Environ. 2006;361:196–207.

20. Miao XS, Bishay F, Chen M, Metcalfe CD. Occurrence of antimicrobials in the final effluents of wastewater treatment plants in Canada. Environ Sci Technol. 2004;38:3533–41.

21. Rooklidge SJ. Environmental antimicrobial contamination from terraccumulation and diffuse pollution pathways. Sci Total Environ. 2004;325:1–13.

22. Tolls J. Sorption of veterinary pharmaceuticals in soils: a review. Environ Sci Technol. 2001;35:3397–406.

14

Monitoring of pesticides water pollution-
The Egyptian River Nile

Hesham Dahshan[*], Ayman Mohamed Megahed, Amr Mohamed Mohamed Abd-Elall,
Mahdy Abdel-Goad Abd-El-Kader, Ehab Nabawy and Mariam Hassan Elbana

Abstract

Background: Persistent organic pollutants represent about 95 % of the industrial sector effluents in Egypt. Contamination of the River Nile water with various pesticides poses a hazardous risk to both human and environmental compartments. Therefore, a large scale monitoring study was carried on pesticides pollution in three geographical main regions along the River Nil water stream, Egypt.

Methods: Organochlorine and organophosphorus pesticides were extracted by liquid-liquid extraction and analyzed by GC-ECD.

Results: Organochlorine pesticides mean concentrations along the River Nile water samples were 0.403, 1.081, 1.209, 3.22, and 1.192 µg L^{-1} for endrin, dieldrin, p, p′-DDD, p, p′-DDT, and p, p′-DDE, respectively. Dieldrin, p, p′-DDT, and p, p′-DDE were above the standard guidelines of the World Health Organization. Detected organophosphorus pesticides were Triazophos (2.601 µg L^{-1}), Quinalphos (1.91 µg L^{-1}), fenitrothion (1.222 µg L^{-1}), Ethoprophos (1.076 µg L^{-1}), chlorpyrifos (0.578 µg L^{-1}), ethion (0.263 µg L^{-1}), Fenamiphos (0.111 µg L^{-1}), and pirimiphos-methyl (0.04 µg L^{-1}). Toxicity characterization of organophosphorus pesticides according to water quality guidelines indicated the hazardous risk of detected chemicals to the public and to the different environmental compartments. The spatial distribution patterns of detected pesticides reflected the reverse relationship between regional temperature and organochlorine pesticides distribution. However, organophosphorus was distributed according to the local inputs of pollutant compounds.

Conclusions: Toxicological and water quality standards data revealed the hazardous risk of detected pesticides in the Egyptian River Nile water to human and aquatic life. Thus, our monitoring data will provide viewpoints by which stricter legislation and regulatory controls can be admitted to avoid River Nile pesticide water pollution.

Keywords: Monitoring, Organochlorine pesticides, Organophosphorus pesticides, River Nile, Human hazardous risk

Background

Over fifty years pesticides were used in African countries for combating and controlling agricultural pests [1]. Among African countries, Egypt is one of the intensive pesticide use areas. Thus, the main water supply (River Nile) is loaded with various types of persistent organic pollutants (POPs).

Nowadays, POPs represent about 95 % of the major industrial sectors in Egypt as raw and fabricated metals, vehicles, pharmaceuticals, textiles, pesticides, fertilizers, petrochemicals, cement, paper and pulp, and food processing [2].

Organochlorine pesticides represent an important group of POPs, which are believed to be possible carcinogens as well as endocrine disruptors [3]. Due to its hazardous risk, the United Nations Environmental Program (UNEP) has initiated a prospective for reducing these threatening chemicals worldwide as agricultural sectors have forced to be shifted towards organophosphorus pesticides instead of organochlorine. However, these compounds are more toxic to vertebrates than other classes of insecticides [4].

In Egypt, the River Nile ecosystem is of particular interest since river water is the main source of drinking

* Correspondence: Dahshanhesham@gmail.com
Department of Veterinary Public Health, Faculty of Veterinary Medicine, Zagazig University, Zagazig, Sharkia governorate, Egypt

for about 90 million citizens and also the main way for irrigation of agricultural lands. The monitoring and assessment of pesticide water pollution have been well studied in North America, Japan and many Europe countries [5]. However, studies on freshwater aquatic environments in Egypt are scanty especially the large-scale monitoring studies. A former study was carried in 1995 on freshwater aquatic environments along the River Nile revealed that DDT, HCH, and PCBs were detected [6]. Furthermore, inconstancy of organochlorine residues in Nile water during the period from 1982 to 1998 was reported in literatures [7, 8]. For our point of view, no data about the River Nile pesticides water pollution is available since 1998.

As monitoring of pesticide water pollution is an substantial source of information describing the current state of environmental pollution and reflecting the effectiveness of environmental legislation policies, our study was conducted to obtain a large scale monitoring data on spatial distribution of selected organochlorine and organophosphorus pesticides in water samples collected at 20 sampling sites along the River Nile stream and the major delta lakes of Egypt.

Methods

Study area

A large-scale monitoring study was conducted on organochlorine and organophosphorus residue levels in water samples collected at 20 sampling sites along the River Nile, Egypt (Fig. 1). Sampling regions were selected according to the locations of major agricultural and industrial activities. Three geographical regions along the River Nile stream were selected as follows; Greater Cairo, in which about 50 % of all industrial activity is concentrated & Nile Delta, the majority of agricultural lands are located and the remaining industrial activity rests (vehicles, textiles, fertilizers, food, detergents) and & Nile estuaries at Damietta and Rosetta, in which textiles, furniture, pesticides, food factories are the main national income. From Greater Cairo, seven sampling sites were selected. Meanwhile, from Nile Delta, Nile estuaries seven and six sampling sites were selected, respectively.

Sample collection

During the summer of 2013, a total of 60 water sample were collected from the River Nile sampling sites (3 samples each). Water samples were collected using 2.5 L amber glass bottle at 50 cm below water surface. Water samples were filtered through 0.45 μm fiber glass filters to remove sand and debris (WHATMAN) [9, 10].

Reagents and standards

Reagents used are; solvents including, n-hexane, aceto-nitrile, ethanol, and dichloromethane (all solvents were pesticide residue (PR) grade and were purchased from Alliance Bio, USA. Alliance Bio, USA); florisil 60–100 mesh (Sigma, USA); and sodium sulfate anhydrate

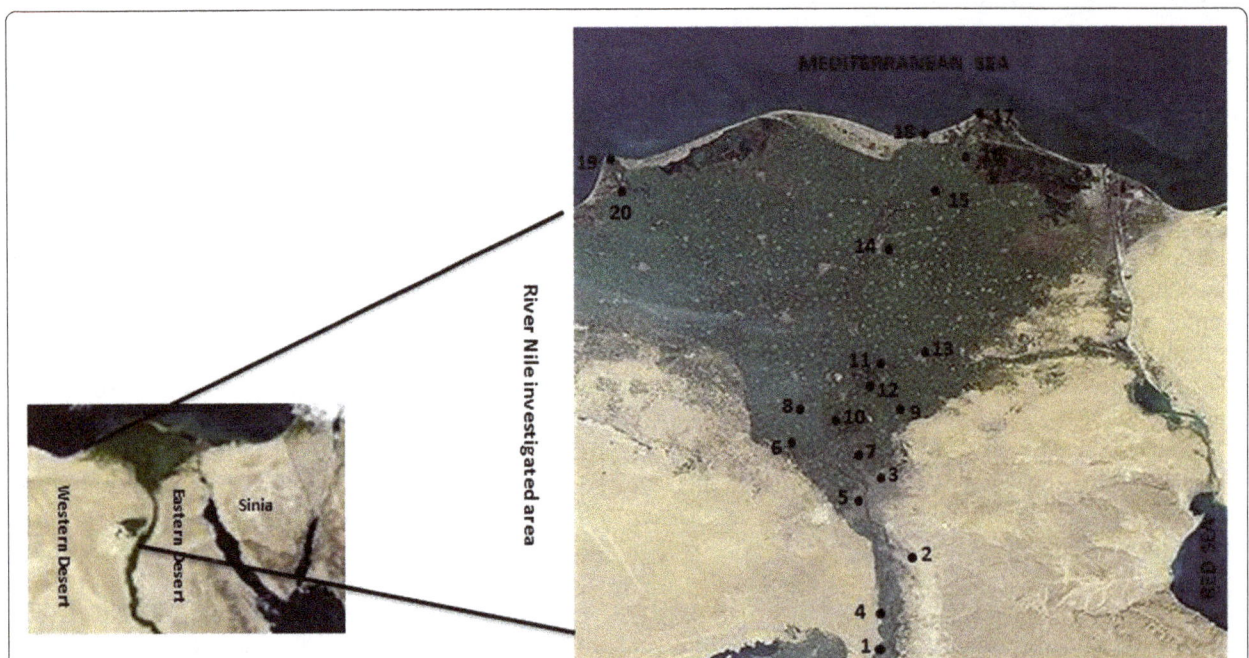

Fig. 1 Map of the River Nile showing sites of water sampling: Greater Cairo; 1. Alwasta-Beni Sweif, 2. Helwan, 3. Cairo, 4. Alaeat-Giza, 5. Giza town, 6. Alknater-Giza, 7. Qalubiya& Nile Delta; 8. Monofea, 9. Belbas, 10. Benha, 11. Alazezea-Menya Elkamh, 12. Menya Elkamh town, 13. Zagazig, 14. Mansoura& Nile estuaries; 15. Fraskour-Damietta, 16. Damietta town, 17. Ras El-bar, 18. Gamasa, 19. Rosetta town, 20. Edfina-Rosetta

(El Nasr Pharmaceutical Chemical Co, Egypt). The individual reference standards used for quantification and identification of organochlorine and organophosphorus residues were obtained from Dr. Ehrenstorfer GmbH (Augsburg, Germany).

Analytical procedures
Extraction
Liquid-liquid extraction was used according to procedures described by APHA [11]. Water sample was extracted twice. In each, A 60 mL volume of 15 % methylene chloride in n-hexane was introduced into a 2 L separating funnel containing 1 L of filtered water and shaken vigorously for 5 min. The combined extracts were dried over anhydrous sodium sulfate and concentrated to about 1 mL in a rotary evaporator.

Clean up
Water extract were cleaned and fractionated on florisil; 20 g of 0.5 % activated florisil was poured into a column topped with 1 g anhydrous sodium sulfate to remove remaining water from the sample. The florisil column was washed with 50 mL n-hexane before the sample loaded. To recover p, p'-DDE, the column was eluted with 60 mL of 30 % methylene chloride in n-hexane (First fraction). The second fraction was achieved by column elution with 35 mL of 30 % dichloromethane in hexane and after that with 45 mL of 50 % dichloromethane in hexane to elute all organochlorine pesticide residues in the samples. Each fraction was evaporated in the rotary vacuum evaporator until the volume reached 2–3 mL [12, 13]. However, to determine the residues of organophosphorus pesticide, the water sample extract was injected in gas chromatography for analysis without clean up.

Quantitative determination
Quantitative analysis of pesticides was carried out at Residue Analysis Department, Central Agri. Pesticides Lab, Dokki, Egypt using an Agilent gas chromatograph 6890 coupled with a HP-5MS (Agilent, Folsom, CA) capillary column of 30 m length × 0.25 mm internal diameter × 0.25 μm film thickness, Agilent). Chemstation software was used for instrument control. A 63Ni-ECD detector was used for analysis. The GC system was operated in a splitless mode. The column oven temperature was programmed as follows; the oven temperature was programmed from an initial temperature 180 °C (2 min hold) to 220 °C (1 min hold) at a rate of 5 °C/min, then finally to 280 °C at a rate of 9 °C/min. the oven was maintained at 280 °C for 30 min. The temperature of the injector operating in splitless mode was held at 260 °C while the detector temperature was 320 °C. The carrier gas was ultra pure nitrogen at flow rate of 4 mL/min.

The target compounds were identified on the basis of the retention times of individual authentic standards.

Quality assurance and quality control
The quality of organochlorine and organophosphorus pesticides was assured through the analysis of solvent blanks, procedure blanks and triplicate samples. LOD and LOQ data in the GC-ECD was presented in Table 1. Sample of each series was analyzed in triplicates.

Results and discussion
Pollution of the River Nile water by pesticides
The quantities of pesticides used in Egypt based on Environmental Affairs agency, Egypt; January 2009, is about 600 ton/annually. Therefore, the spatial distribution of ten organochlorine and twelve organophosphorus pesticide residues in the main water source for Egyptian (River Nile) was investigated.

Occurrence of organochlorine pesticide residues
Water samples taken from three studied regions (Greater Cairo, Nile Delta, Nile estuaries at Damietta and Rosetta) at River Nile were analyzed. The recoveries of organochlorine pesticides ranged between 82 and 98.6 %. The mean concentration values are presented in Table 2. Organochlorine pesticide residues were mainly

Table 1 Analyzed pesticides, limits of detection and limits of quantification data in the GC-ECD ($\mu g/L^{-1}$)

Pesticides	LOD	LOQ
α-HCH	0.005	0.015
γ-HCH	0.004	0.012
Aldrin	0.003	0.009
Heptachlor	0.003	0.009
Endrin	0.003	0.009
Heptachlor epoxide	0.003	0.009
P, P'-DDE	0.003	0.009
Dieldrin	0.002	0.006
P, P'-DDD	0.003	0.009
P, P'-DDT	0.004	0.012
Ethoprophos	0.005	0.015
Phorate Diazinon	0.003	0.009
Dimethoate	0.005	0.015
Pirimiphos-methyl	0.005	0.015
Chlorpyrifos	0.005	0.015
Fenitrothion	0.004	0.012
Quinalphos	0.005	0.015
Prothiofos Ethion	0.005	0.015
Triazophos	0.004	0.012
Fenamiphos	0.005	0.015

Table 2 Mean concentration of organochlorine pesticides ($\mu g/L^{-1}$) detected in water samples from the River Nile, Egypt

Site No.	Site Name	Region	α-HCH	γ-HCH	Aldrin	Heptachlor	Endrin (2 $\mu g/L^{-1}$)[a]	Heptachlor epoxide	P, P'-DDE (2 $\mu g/L^{-1}$)[a]	Dieldrin (0.03 $\mu g/L^{-1}$)[a]	P, P'-DDD (2 $\mu g/L^{-1}$)[a]	P, P'-DDT (2 $\mu g/L^{-1}$)[a]	Total organochlorine pesticides
1	Alwasta-Beni Sweif	Greater Cairo	ND	ND	ND	ND	ND	ND	0.21	ND	ND	ND	0.21
2	Helwan												
3	Cairo												
4	Alaeat-Giza												
5	Giza town												
6	Alknater-Giza												
7	Qalubiya												
8	Monofea	Nile Delta	ND	ND	ND	ND	ND	ND	0.982	ND	ND	0.952	1.934
9	Belbas												
10	Benha												
11	Alazezea-Menya Elkamh												
12	Menya Elkamh town												
13	Zagazig												
14	Mansoura												
15	Fraskour-Damietta	Nile Estuaries	ND	ND	ND	ND	0.403	ND	ND	**1.081**	1.209	**2.268**	4.961
16	Damietta town												
17	Ras El-bar												
18	Gamasa												
19	Rosetta town												
20	Edfina-Rosetta												7.105[a]

Bold numbers: Values above the standard guidelines of World Health Organization
[a]Organochlorine pesticide concentration ($\mu g/L^{-1}$) along the River Nile sampling sites
ND not detectable, Number of samples = 60 (3/each sampling site)

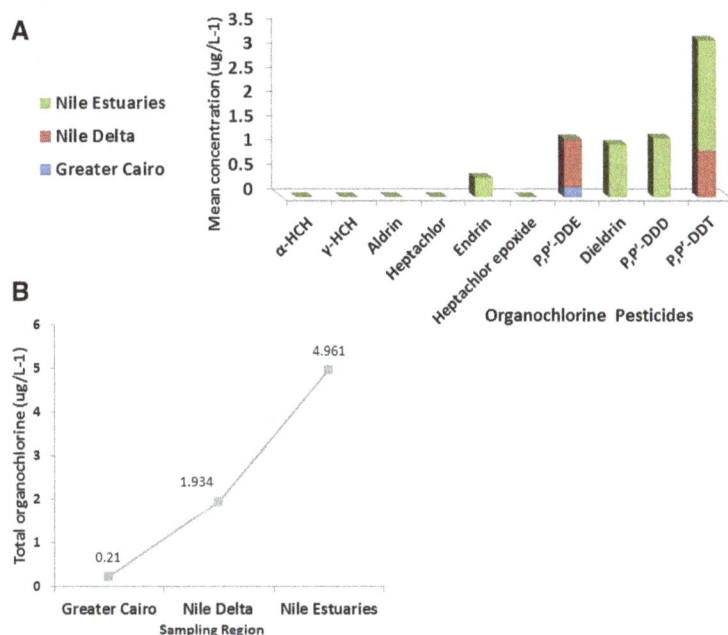

Fig. 2 a Mean organochlorine pesticide concentrations; **b** Spatial distribution of total organochlorine pesticides in water samples collected from three sampling regions along the River Nile, Egypt

detected in the downstream of the river as follows; endrin, dieldrin, p, p'-DDD, and p, p'-DDT at a rate of 0.403, 1.081, 1.209, and 2.268 µg L^{-1}, respectively. The levels of DDTs in this study were higher than those in the Pearl River, the Haihe River, Qiantang River and the Huaihe River [14–17]. However, the concentration is lower than the concentration obtained from water sample collected in Begumganj, Bangladesh [18]. The high concentration level of total organochlorine pesticides at the Nile estuaries (Fig. 2b) could be attributed to the Delta agricultural lands wash off. Further investigations are clearly needed to reveal the sources and patterns of organochlorine pesticides contamination in river water.

It is surprising to note that at Greater Cairo and Nile Delta region, various organochlorine pesticides were not detected although major industrial and agricultural activities are concentrated there. This could be due to the pesticides evaporation in tropical countries (Egypt), pesticide residues dilution or adsorption. p, p'-DDE was detected only at Greater Cairo in a low concentration, (0.21 µg L^{-1}). However at Nile Delta region, p, p'-DDE and p, p'-DDT was estimated in a concentration of 0.982 and 0.952 µg L^{-1}, respectively Table 2.

Along the investigated River Nile region sites, the most frequently detected organochlorine pesticide was endrin. Followed by, dieldrin, p, p'-DDE, p, p'-DDD, and p, p'-DDT. However, α-HCH, γ-HCH, aldrin, heptachlor, and heptachlor epoxide were not detected in the water samples (Fig. 2a). In spite of, p, p'-DDT and its metabolites (p, p'-DDE and p, p'-DDD), endrin and dieldrin

have been officially prohibited since 1980 and in 1996 a Ministerial Decree prohibited the import and use of 80 pesticides including dieldrin, endrin, and DDT [19]. Nonetheless, our study indicates that above mentioned organochlorine pesticides are still sold in Egyptian markets.

Occurrence of organophosphorus pesticide residues
Amongst 12 organophosphorus pesticides analyzed, eight were detected. The recoveries of organophosphorus pesticides were in-between 82.5 and 100 %. The most frequently detected was triazophos, followed by quinalphos, then, fenitrothion, ethoprophos, chlorpyrifos, ethion, fenamiphos, and pirimiphos-methyl. However, prothiofos, dimethoate, diazinon, and phorate were not detected (Fig. 3a). For the Nile estuaries, the highest concentration of organophosphorus pesticide detected in water was 1.488 µg L^{-1} for triazophos. In our monitoring study levels of triazophos are generally higher than those reported in rivers and lakes of Greece [20], River Ravi of Pakistan [21], potable and irrigated water of Brazil [22]. Our results are in concert with a study conducted in Jiulong River in South China [23] as triazophos was the main organophosphorus pesticides detected in the estuary river water. In general, studying of organophosphorus River Nile water pollution is still in its initial stage, and further research is increasingly needed to establish a frame network data about its contamination degree.

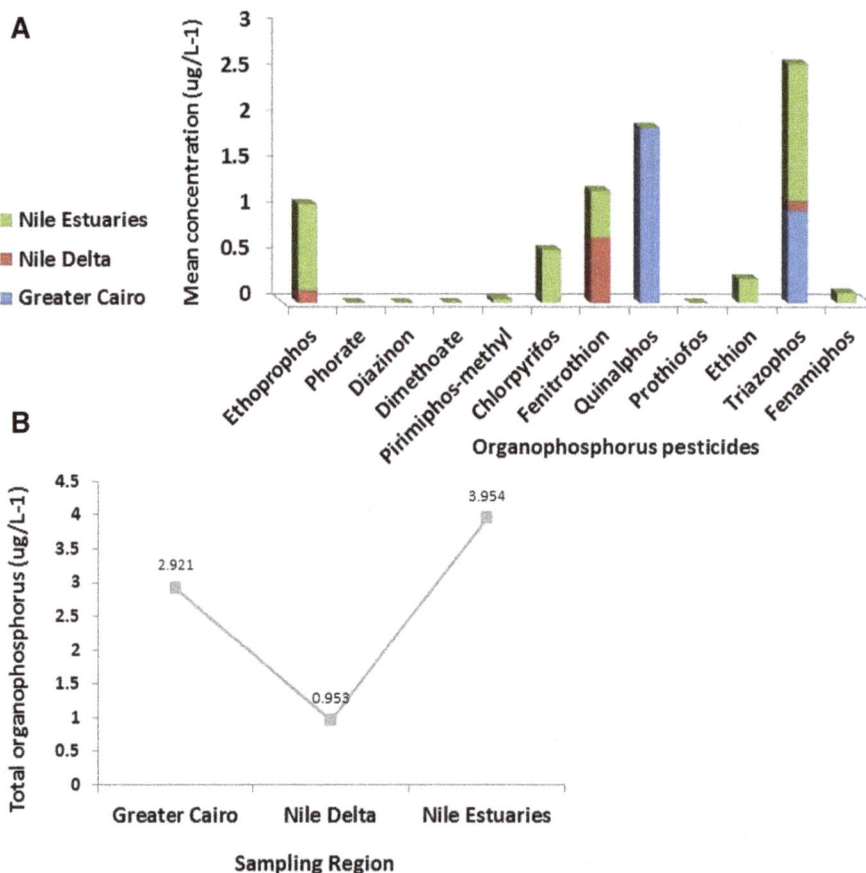

Fig. 3 a Mean organophosphorus pesticide concentrations; **b** Spatial distribution of organophosphorus pesticides in water samples collected from three sampling regions along the River Nile, Egypt

In Greater Cairo and Nile Delta sampling regions the higher concentrations were 1.91 and 0.711 µg L^{-1} for, quinalphos, and fenitrothion, respectively Table 3. Accordingly, our results revealed that organophosphorus pesticide concentrations in the River Nile water, Egypt exceeded the EEC Council Directive 98/83/EC for water quality standard [24]. This could be attributed to the substitution of persistent organochlorine pesticides with organophosphate pesticides in the treatment of scattered cotton fields in Egypt as organophosphates and carbamates are the dominate insecticide used there [25, 26], resulting in serious hazards to the freshwater aquatic environments and adverse harmful effects to wildlife and humans.

Spatial distribution of pesticides in the River Nile water samples
Organochlorine pesticides
Organochlorine pesticides water pollution showed a gradual increase in total organochlorine concentrations from Nile upstream at Greater Cairo in which total organochlorine pesticides were 0.21 µg L^{-1} to the Nile

estuaries in which total organochlorine pesticides were 4.961 µg L^{-1}. (Fig. 2b). In this context, we can expect the reverse relationship between the spatial organochlorine pesticides distribution and sampling regions temperature as organochlorine pesticides volatilize at warm temperatures (Nile upstream) and condense at cooler temperatures, reaching their highest concentrations in the cooler regions (Nile estuaries) [27].

Organophosphorus pesticides
Residues of total organophosphorus pesticides along the River Nile water sampling regions, showing the following spatial distribution pattern: River Nile estuaries > Greater Cairo > Nile Delta (Fig. 3b). Each sampling region was highly contaminated by special organophosphorus compound (Fig. 3a). The Fluxes in organophosphorus levels along the River Nile indicate contaminants local inputs. No cumulative effect toward the river downstream as Greater Cairo water samples were more contaminated by organophosphorus pesticides than Nile Delta samples in spite of its geographical location toward the river upstream Fig. 1.

Table 3 Mean concentration of organophosphorus pesticides ($\mu g/L^{-1}$) detected in water samples from the River Nile, Egypt

Site No.	Site Name	Region	Ethoprophos	Phorate	Diazinon	Dimethoate	Pirimiphos-methyl	Chlorpyrifos	Fenitrothion	Quinalphos	Prothiofos	Ethion	Triazophos	Fenamiphos	Total organophosphorus pesticides
1	Alwasta-Beni Sweif	Greater Cairo	ND	ND	ND	ND	ND	ND	ND	1.91	ND	ND	1.011	ND	2.921
2	Helwan														
3	Cairo														
4	Alaeat-Giza														
5	Giza town														
6	Alknater-Giza														
7	Qalubiya														
8	Monofea	Nile Delta	0.14	ND	ND	ND	ND	ND	0.711	ND	ND	ND	0.102	ND	0.953
9	Belbas														
10	Benha														
11	Alazezea-Menya Elkamh														
12	Menya Elkamh town														
13	Zagazig														
14	Mansoura														
15	Fraskour-Damietta	Nile Estuaries	0.936	ND	ND	ND	0.04	0.578	0.511	ND	ND	0.263	1.488	0.111	3.954
16	Damietta town														
17	Ras El-bar														
18	Gamasa														
19	Rosetta town														
20	Edfina-Rosetta														7.828[a]

[a]Organophosphorus pesticide concentration ($\mu g/L^{-1}$) along the River Nile sampling sites
ND not detectable; Number of samples = 60 (3/each sampling site)

Human hazardous risks

Human exposure to pesticide residues could be through water, food and air. Residue levels vary according to the type of exposure and the individual's daily intake [28]. Therefore, the assessing of human hazardous risks due to the intake of pesticides polluted water is important.

Organochlorine pesticides

The hazardous risk of organochlorine pesticides was evaluated according to water quality guidelines set by the World Health Organization (WHO), which specifies limits for endrin, p, p'-DDE, dieldrin, p, p'-DDD, and p, p'-DDT as 2, 2, 0.03, 2, and 2 µg L^{-1}, respectively [29]. Our results showed that dieldrin and p, p'-DDT residues in some sampling sites were above the standard guidelines of WHO Table 2. Thus, water from the River Nile generally possessed an environmental and human health hazard as dieldrin is highly toxic to the central nervous system [30] and eating DDT contaminated fish over a short time would most likely affect the nervous system [31].

Organophosphorus pesticides

Of the organophosphorus pesticides detected in water, Ethoprophos, Triazophos, and Fenamiphos are considered highly hazardous to fish and other aquatic organisms, while others are considered moderately to slightly toxic. Water quality standards and toxicological data for human and aquatic organisms in relation to the detected organophosphorus pesticides are listed in Table 4. Toxicity characterization based on the Pesticide Action Network databases, WHO, Canadian Water Quality Guidelines, and U.S. National Drinking Water Standards and Health Criteria, revealed that all the detected organophosphorus pesticides are related to at least one health effect [32]. Thus, new tools and policies with greater reliability than those already existing by the Egyptian Environmental Affairs Agency (EEAA) of the Ministry of State for Environmental Affairs are needed to prevent or reduce the use of these harmful chemicals in industrial and agricultural sectors.

Conclusions

Organochlorine and organophosphorus pesticides River Nile water pollution was investigated. Organochlorine pesticides detected were dieldrin; endrin; p, p'-DDE; p, p'-DDD; and p, p'-DDT. While, organophosphorus pesticides detected were triazophos, ethoprophos, quinalphos, chlorpyrifos, fenitrothion, ethion, fenamiphos, and pirimiphos-methyl. Spatial distribution of detected pesticides showed the reverse relationship between sampling regions temperature and organochlorine pesticides distribution. Meanwhile, organophosphorus pesticides were distributed according to the local inputs of pollutant compounds. Toxicological and water quality standards data revealed the hazardous risk of detected chemicals to human and aquatic life. We expect our results will provide viewpoints by which stricter legislation and regulatory controls can be admitted to avoid River Nile pesticide water pollution.

Table 4 Hazardous risks of detected organophosphorus pesticides in the River Nile, Egypt

Detected Organophosphorus compound	Total Concentration along the River Nile µg/L^{-1}	PAN Bad Actors[b]	WHO Acute Hazard	Carcinogen	WHO Water Quality Criteria µg/L^{-1}	Canadian Water Quality Guidelines for the Protection of Aquatic Life µg/L^{-1}
Ethoprophos	1.076	Yes	Ia, Extremely hazardous	Yes	No water quality standard.	No water quality guidelines but induce mortality.
Pirimiphos-methyl	0.04	Yes	III, Slightly hazardous	Unclassifiable	Not recommended for direct application to drinking water.	No water quality guidelines but (Moderate to high toxicity)
Chlorpyrifos	0.578	Yes	II, Moderately hazardous	Not likely	30.0	0.0035
Fenitrothion	1.222	Yes	II, Moderately hazardous	Not likely	Occurs at concentrations below toxic effects.	Moderately toxic
Quinalphos	1.91	Yes	II, Moderately hazardous	Not likely	No water quality standard.	No water quality guidelines but induce mortality.
Ethion	0.263	Yes	II, Moderately hazardous	Not likely	No water quality standard.	No water quality guidelines (Moderate to high toxicity)
Triazophos	2.601	Yes	Ib, Highly hazardous	Not likely	Unlikely to occur.	Unlikely to occur, but induce mortality.
Fenamiphos	0.111	Yes	Ib, Highly hazardous	Not likely	3.50[a]	Unlikely to occur, but induce mortality.

Data of Hazardous risk were presented according to (Kegley et al., 2014) [32]
[a]U.S. Drinking Water Equivalent Level cited by U.S. National Drinking Water Standards and Health Criteria
[b]Pan Bad Actors: are chemicals that are highly acutely toxic, cholinesterase inhibitor

Acknowledgments
This study was supported by the Zagazig University Research Projects, Egypt. The authors would like to thank Dr. Hend Mahmoud "Residue Analysis Department, Central Agriculture Pesticides Lab, Dokki, Egypt" for her technical assistance.

Funding
This study was funded under The Project of Environmental Monitoring Programmes, Zagazig University Research Projects, Egypt.

Authors' contributions
HD, AMM, AMMA have participated in the study conception and design, acquisition of data, and analysis& interpretation of data. MAA and EN participated in the intellectual helping in different stages of the study. HD and MHE participated in drafting of manuscript and preparation of critical version. All Authors have read the manuscript and have agreed to submit it in its current form for consideration for publication. All read and approved the final manuscript.

Competing interests
The authors declare that they have no competing interests.

References
1. Williamson S, Ball A, Pretty J. Trends in pesticide use and drivers for safer pest management in four African countries. Crop Prot. 2008;27:1327–34.
2. Mansour SA. Persistent organic pollutants (POPs) in Africa: Egyptian scenario. Hum Exp Toxicol. 2009;28:531–66.
3. Peter O, Lin KC, Karen P, Joe A. Persistent Organic Pollutants (POPs) and Human Health. Washington: World Federation of Public Health Association Publications; 2002. p. 1–35.
4. Chambers HW, Boone JS, Carr RL, Chanbers JE. Chemistry of organophosphorus insecticides. In: Robert IK, editor. Handbook of Pesticide Toxicology. secondth ed. California: Academic; 2001. p. 913–7.
5. Yamaguchi N, Gazzard D, Scholey G, Macdonald DW. Concentration and hazard assessment of PCBs, organochlorine pesticides and mercury in fish species from the upper Thames River pollution and its potential effects on top predators. Chemosphere. 2003;50:265–73.
6. Wahaab RA, Badawy MI. Water quality assessment of the River Nile system: an overview. Biomed Environ Sci. 2004;17:87–100.
7. El-Dib MA, Badawy MI. Organo-chlorine insecticides and PCBs in River Nile water, Egypt. Bull Environ Contam Toxicol. 1985;34:126–33.
8. Abou-Arab AAK, Gomaa MNE, Badawy A, Naguib K. Distribution of organochlorine pesticides in the Egyptian aquatic ecosystem. Food Chem. 1995;54:141–6.
9. APHA. Standard methods for the examination of water and waste water) (10th ed., pp. 391–448. Washington: American Public Health Association; 1985.
10. Wilde FD. National Field Manual for the Collection of Water-Quality Data, Chapter A1. Preparations for Water Sampling. In: Handbooks for Water-Resources Investigations. USA: U.S. Geological Survey TWRI Book 9; 2005.
11. APHA. Standard method for examination of water and wastewater. 14th ed. Washington: AWWA/WPCE; 1975.
12. UNEP. Determination of DDTs and PCBs by capillary gas chromatography and electron capture detectors. Reference method for marine pollution studies no.4. 1988.
13. Khaled A, El Nemr A, Said TO, El-Sikaily A, Abd- Allah AMA. Polychlorinated biphenyls and chlorinated pesticides in mussels from the Egyptian red sea coast. Chemosphere. 2004;54:1407–12.
14. Guan YF, Wang JZ, Ni HG, Zeng EY. Organochlorine pesticides and polychlorinated biphenyls in riverine runoff of the Pearl River Delta, China: assessment of mass loading, input source and environmental fate. Environ Pollut. 2009;157:618–24.
15. Wang T, Zhang ZL, Huang J, Hu HY, Yu G, Li FS. Occurrence of dissolved polychlorinated biphenyls and organic chlorinated pesticides in the surface water of Haihe River and Bohai Bay, China. Environ Sci. 2007;28:730–5.
16. Zhou R, Zhu L, Kong Q. Levels and distribution of organochlorine pesticides in shellfish from Qiantang River, China. J Hazard Mater. 2008;152:1192–200.
17. Feng J, Zhai M, Liu Q, Sun J, Guo J. Residues of organochlorine pesticides (OCPs) in upper reach of the Huaihe River, East China. Ecotoxicol Environ Saf. 2011;74:2252–9.
18. Matin M, Malek M, Amin M, Rahman S, Khatoon J, Rahman M, Aminuddin M, Mian A. Organochlorine insecticide residues in surface and underground water from different regions of Bangladesh. Agric Ecosyst Environ. 1998;69:11–5.
19. Sallam KI, Morshedy AMA. Organochlorine pesticide residues in camel, cattle and sheep carcasses slaughtered in Sharkia Province, Egypt. Food Chem. 2008;108:154–64.
20. Konstantinou IK, Hela DG, Albanis TA. The status of pesticide pollution in surface waters (rivers and lakes) of Greece. Part I. Review on occurrence and levels. Environ Pollut. 2006;141:555–70.
21. Mahboob S, Niazi F, Sultana S, Ahmad Z. Assessment of pesticide residues in water, sediments and muscles of Cyprinus Carpio from Head Balloki in the River Ravi. Life Sci J. 2013;10:11s.
22. Milhome MAL, Sousa PLR, Lima FAF, Nascimento RF. Influence the USE of pesticides in the Quality of surface and groundwater located in irrigated areas of Jaguaribe, Ceara, Brazil. Int J Environ Res. 2015;9:255–62.
23. Zheng S, Chen B, Qiu X, Chen M, Ma Z, Yu X. Distribution and risk assessment of 82 pesticides in Jiulong River and estuary in South China. Chemosphere. 2016;144:1177–92.
24. EECD. European Economic Community Directive, EEC Council Directive 98/83/EC. Off J Eur Communities. 1998;L330:42.
25. WRI. World Resources Institute in Collaboration with the UN Environmental Programme, World resources 1994–1995. Washington: World Resources Institute; 1996.
26. El-Sebae AH, Abou Zeid M, Saleh MA. Status and environmental impact of toxaphene in the third world-a case study of African agriculture. Chemosphere. 1993;27:2063–72.
27. Anonymous. Persistent organic pollutants and the Stockholm Convention: a resource guide, A Report Prepared by Resource Futures International for the World Bank and CIDA. 2001. p. 22.
28. Jonsson V, Liu GJK, Armbruster J, Kettelhut LL, Drucker B. Chlorohydrocarbon pesticide residues in human milk in Greater St. Louis, Missouri. Am J Clin Nutr. 1997;30:1106–9.
29. Hamilton DJ, Ambrus A', Dieterle RM, Felsot AS, Harris CA, Holland PT, Katayama A, Kurihara N, Linders J, Unsworth J, Wong SS. Regulatory limits for pesticide residues in water (IUPAC technical report). Pure Appl Chem. 2003;75:1123–55.
30. WHO. Aldrin and dieldrin, Geneva, World Health Organization, International Programme on Chemical Safety (Environmental Health Criteria 91). 1989.
31. ATSDR, Agency for Toxic Substances and Disease Registry. Toxicological Profile for DDT, DDE, DDD. Atlanta: U.S. Department of Health and Human Services, Public Health Service; 2002.
32. Kegley SE, Hill BR, Orme S, Choi AH. Pan Pesticide Database, Pesticide Action Network, North America (Oakland, CA). 2014. http://www.pesticideinfo.org/.

15

A new bioindicator, shell of *Trachycardium lacunosum*, and sediment samples to monitors metals (Al, Zn, Fe, Mn, Ni, V, Co, Cr and Cu) in marine environment: The Persian Gulf as a case

Vahid Noroozi Karbasdehi[1], Sina Dobaradaran[1,2,3*], Iraj Nabipour[4], Afshin Ostovar[4], Amir Vazirizadeh[5], Masoumeh Ravanipour[1], Shahrokh Nazmara[6], Mozgan Keshtkar[1], Roghayeh Mirahmadi[1] and Mohsen Noorinezhad[7]

Abstract

Background: The present work was designed to detect heavy metal contents of Al, Zn, Fe, Mn, Ni, V, Co, Cr and Cu in sediments and shells of the *Trachycardium lacunosum* collected in polluted and unpolluted areas along the Persian Gulf.

Methods: The samples were taken from surface sediments (0-10 cm) and shells of *Trachycardium lacunosum* in two separated areas (polluted and unpolluted) in northern part of the Persian Gulf, Asaluyeh Bay, during summer 2013. The prepared samples were analyzed by inductively coupled plasma-optical emission spectrometry (ICP-OES).

Results: Based on the results, all measured metals including Al, Zn, Fe, Mn, Ni, V, Co, Cr and Cu were meaningfully higher in the sediment samples of polluted area compared to unpolluted area and the order of metal concentrations in the sediment samples were Cr > Co > V > Ni > Zn > Cu > Fe > Al > Mn in polluted area. In the case of shell samples of *Trachycardium lacunosum*, polluted area contained significantly higher contents of Al, Zn, Fe, Mn, Ni, Co, Cr and Cu compared to unpolluted area and the order of metal concentrations in the shell samples were Fe > Zn > Al > Mn > Cu > Cr > Ni > Co in the polluted area.

Conclusion: It was concluded that shells of the *Trachycardium lacunosum* can be used as a suitable bioindicator for heavy metals in the aquatic environment. Results confirmed that due to the possible contaminations by oil and gas activities near the polluted area perennial monitoring and mitigation measures is extremely necessary.

Keywords: Aquatic Organisms, Environmental Monitoring, Geologic Sediments, Metals, Persian Gulf, *Trachycardium lacunosum*, Toxicology

* Correspondence: sina_dobaradaran@yahoo.com;
s.dobaradaran@bpums.ac.ir
[1]Department of Environmental Health Engineering, Faculty of Health, Bushehr University of Medical Sciences, Bushehr, Iran
[2]The Persian Gulf Marine Biotechnology Research Center, The Persian Gulf Biomedical Sciences Research Institute, Bushehr University of Medical Sciences, Boostan 19 Alley, Imam Khomeini Street, Bushehr, Iran
Full list of author information is available at the end of the article

Background

Environment protection needs awareness of the circumstance of the environments and the way in which they change. Hence, deterioration due to human and industrial activities and change in environments are the principal topics of monitoring studies [1]. The data attained in monitoring studies may use as a basic for managers and policy makers for evaluation and enhancement of environment condition by imposing proper actions to protect the environment. Coastline areas are subject to suffer from different negative environmental impacts due to industrial and human activities. Chemical pollution associated with industrial production is the main concern in the marine environment [2]. Heavy metals are considered as one of the most critical contaminants in the marine environment due to their bioaccumulation and biomagnification throughout the trophic chain [3, 4]. Heavy metals toxicity in marine organisms, long residence time within trophic chains, as well as the probable risk of human exposure to heavy metals, makes it essential to evaluate the concentrations of them in the aquatic environment and organisms [5]. Heavy metals may also induce sublethal effect in marine organisms, such as disruption of homeostasis, and impairment at cellular and molecular levels [6]. Additionally, these impacts may seriously decrease the persistence capacity of the organism by enhancing susceptibility to diseases and impairment [7]. Sediments act as a reservoir for various pollutants such as heavy metals and while many bivalves existing inside sediment accumulate elevated concentration levels of metals with regard to their bioaccessibility [8]. The ecological significance of bivalves, their simplicity of applying, their vast distribution and numerous abundance, and their relative to polluted sediments make them suitable species for toxicity testing of sediment [9]. Metals accumulate differentially in the shells and soft tissues of bivalves [10] however there is no particular position on if the use of shells or soft tissues alone is preferred in evaluating of metal [11]. But soft tissues have received further consideration amongst researchists for metals monitoring mostly because of agreement with the US coastal mussel watch monitoring scheme [12]. However, shell can provide a more precise symptom of pollution and environmental change [13]; they give minor variation than the living organism's tissue also present a historic record of metal level all over the organism's life cycle. This record still preserved after organism death [14]. High levels of different metals in sediments and organisms of marine environment are a well-documented environment concern [15]. But there are a few comprehensive studies in the Persian Gulf region especially on evaluation of metal contents in the bivalve shells of *Trachycardium lacunosum* with its connection to metal contents in the sediments. *Trachycardium lacunosum* is a marine and infaunal bivalve as well as a filter feeder pelecypod that belongs to the Cardiidae family. This bivalve has a white-rimmed shell, with the characteristic pink, brown, and purple spots overt. The average *Trachycardium lacunosum* length is about 25–35 mm. *Trachycardium lacunosum* is native to intertidal zone and sandy substrates of the Persian Gulf [16]. Due to the high dispersion of this bivalve in Nayband Bay and Lavar-e-Saheli, in this study we used *Trachycardium lacunosum* to evaluate its efficiency as a suitable bioindicator for metals.

The Persian Gulf is one of the oldest sea passageways in the world, and nearly 45 % of natural gas and 57–66 % of known oil reserves of the world lie in the region of the Persian Gulf. The presence of large amounts of natural gas and oil has made the Persian Gulf as one of the most strategic waterway in the world. The Persian Gulf has been the main waterway for oil transport in the last decades and during our time has also suffered from repeated oil spills to its marine environment.

To the best of our knowledge there is no report on the concentrations of heavy metals in the shells of *Trachycardium lacunosum* also there is no detailed study on heavy metal contents in the northern part of the Persian Gulf. So in this study for the first time in the offshore South Pars, the northern part of the Persian Gulf, we aimed to (1) measure the contents of Al, Zn, Fe, Mn, Ni, V, Co, Cr, as well as Cu in the shells of *Trachycardium lacunosum* and sediments simultaneously in two separated areas (polluted and unpolluted) (2) comparison between the metal contents of sediments in the polluted and unpolluted areas as well as shells (3) determine the interrelationships between metal contents in the shells of Tracycardium lacunosum as well as the sediments in both polluted and unpolluted areas.

Methods

Study area description

The South Pars/North Dome is the world's biggest gas field, shared between Iran and Qatar, and situated in the Persian Gulf. This natural gas field covers a space of 9700 km² and the name of this field in Iranian territorial is South Pars. Closest land point to this gas field in the northern part of the Persian Gulf is Asaluyeh. It was chosen as the site for all facilities related to this gas field in Iranian territorial. Asaluyeh is situated on the shore of the Persian Gulf in southeast of Bushehr province. Two different areas were selected in the Asaluyeh as sampling points including polluted area (Nayband Bay) and unpolluted area (Lavar-e-Saheli) (Fig. 1 and Table 1). The surface sediment textures of both polluted and unpolluted areas are silt-clay.

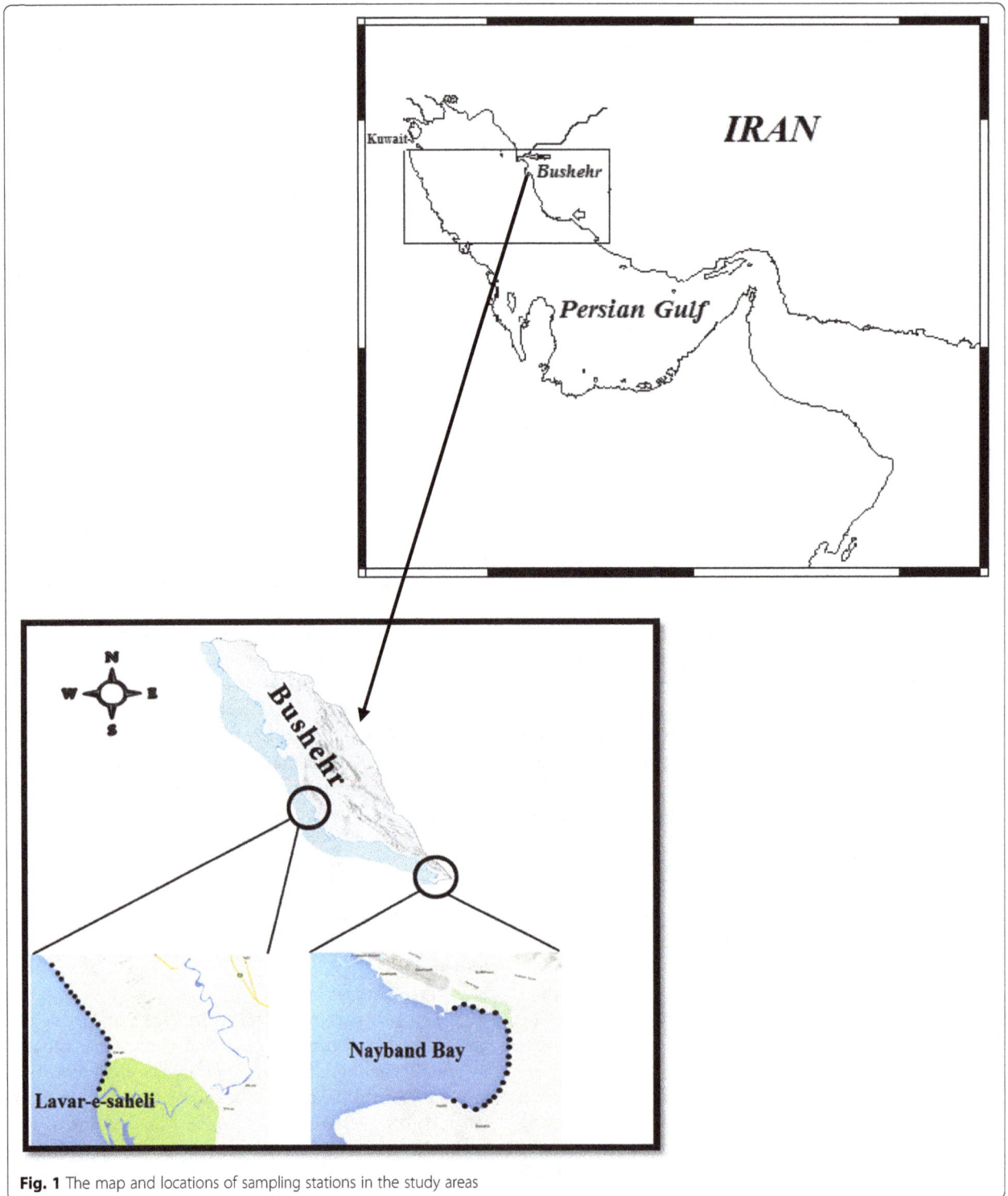

Fig. 1 The map and locations of sampling stations in the study areas

Sample collection

Composite samplings based on area (3 different locations for each sample) were performed at low tide times from the tidal area along the Persian Gulf coastal. Samples were collected from surface sediments (0-10 cm) and shells of *Trachycardium lacunosum* in both polluted and unpolluted areas during summer 2013 as fallow:

a) In polluted area: 20 sediment samples and 18 shell samples

Table 1 Geographical coordinates of the stations studied

Stations of unpolluted area	Number	E
1	28°13'45.59"N	51°17'12.51"E
2	28°13'42.77"N	51°17'13.12"E
3	28°13'40.38"N	51°17'13.37"E
4	28°13'38.14"N	51°17'13.72"E
5	28°13'36.13"N	51°17'13.94"E
6	28°13'34.10"N	51°17'14.12"E
7	28°13'33.77"N	51°17'14.45"E
8	28°13'32.27"N	51°17'14.67"E
9	28°13'30.01"N	51°17'15.02"E
10	28°13'27.49"N	51°17'15.31"E
11	28°13'24.97"N	51°17'15.76"E
12	28°13'22.36"N	51°17'16.18"E
13	28°13'19.04"N	51°17'16.46"E
14	28°13'16.31"N	51°17'17.74"E
15	28°13'14.09"N	51°17'17.81"E
16	28°13'12.08"N	51°17'17.68"E
17	28°13'5.35" N	51°17'17.71"E
18	28°13'2.74" N	51°17'17.86"E
19	28°12'59.71"N	51°17'17.64"E
Stations of polluted area	Number	E
20	27°26'39.57"N	52°40'32.36"E
21	27°26'21.06"N	52°40'34.43"E
22	27°26'2.91" N	52°40'36.37"E
23	27°25'48.69"N	52°40'35.29"E
24	27°25'33.86"N	52°40'35.21"E
25	27°25'21.54"N	52°40'33.93"E
26	27°25'11.54"N	52°40'32.18"E
27	27°25'1.77" N	52°40'29.43"E
28	27°24'52.85"N	52°40'25.78"E
29	27°24'45.36"N	52°40'22.32"E
30	27°24'36.78"N	52°40'18.48"E
31	27°24'27.10"N	52°40'14.73"E
32	27°24'19.29"N	52°40'9.73" E
33	27°24'11.30"N	52°40'5.49" E
34	27°24'4.09" N	52°40'0.34" E
35	27°23'57.01"N	52°39'52.23"E
36	27°23'50.64"N	52°39'41.26"E
37	27°23'49.45"N	52°39'4.93" E
38	27°23'46.16"N	52°39'15.78"E
39	27°23'43.78"N	52°39'27.46"E

b) In unpolluted area: 19 sediment samples and 13 shell samples

After transporting the collected surface sediments to the laboratory, the samples were dried at 105 °C for 24 h, homogenized, and packed in polyethylene bags and kept at -20 °C before analysis. The shell samples washed under a jet of water to liminate algae, sand, clay as well as other impurities, and then dried at 105 °C for 24 h and kept at -20 °C before analysis.

Reagents
All the employed oxidants and mineral acids (HNO_3, H_2O_2, HF, and HCl) were of suprapure quality (Merck, Germany). All plastic and glassware were cleaned by drenching overnight in a 10 % (w/v) HNO_3 solution and afterward washed with deionized water before use. All solutions were prepared by ultrapure water (18.2 MΩ cm).

Digestion and analytical procedures
The sediment samples (0.5 g) were digested with 6 ml hydrochloric acid (37 %), 2 ml nitric acid (65 %) in a microwave digestion system for 30 min and then diluted to 25 ml with ultrapure water and stored in polyethylene bottle until analysis. 0.5 g of powdered shell was fully digested in a Teflon cup using a mixture of conc. HNO_3, $HClO_4$ and HF with the ratio 3:2:1 respectively. Acids were added to dried sample and left overnight prior to further process. After that the samples were heated at 200°C then left to cool and filtered. The filtered solution was justified to a volume of 25 ml. It should be noted here that shell samples with similar shell length were selected for analysis in each sample point to minimize effects of body weight [17]. The bivalve length was measured by using a caliper with an accuracy of 0.02 mm. Blank digest was similarly performed. Metals analysis of Al, Zn, Fe, Mn, Ni, V, Co, Cr as well as Cu was performed by inductively coupled plasma optical spectrometry (ICP-OES). In Table 2, specifications of the instrumental operating circumstances are shown. All metal levels were represented as $\mu g\ g^{-1}$ dry wet (dw). Statistical analysis of data was performed with the SPSS, Version 21 and Mann-Whitney U test as well as the Spearman's rho correlation coefficient were used for statistical significant differences. Differences in mean values were accepted as being significant if $P < 0.05$.

Result and discussion
Content of metals in sediments and shells
The concentration levels of examined metals (Al, Zn, Fe, Mn, Ni, V, Co, Cr and Cu) in sediment samples of polluted (Nayband Bay) and unpolluted (Lavar-e-Saheli) areas are shown in Table 3.

Table 2 ICP-OES instrumental operating details

Parameters	
Company, model	SPECTRO (Germany), Spectro arcos
RF generator power (W)	1400
Frequency of RF generator (MHz)	27.12 MHz
Type of detector	Charge coupled devices (CCD)
Torch type	Flared-end EOP torch 2.5 mm
Plasma, auxiliary, and nebulizer gas	High purity (99.99 %) argon
Plasma gas flow rate (l/min)	14.5
Auxiliary gas flow rate (l/min)	0.9
Nebulizer gas flow rate (l/min)	0.85
Sample uptake time (s)	240 total
Delay time of (s)	-
Rinse time of (s)	45
Initial stabilization time (s)	Preflush: 45
Time between replicate analysis (s)	-
Measurement replicate	3
Pump rate	30 RPM
Element (λ/nm)	Al 396.152; Cu 324.754; Fe 259.941 Mn 257.611; Ni 231.604; Zn 268.416 Cr 205.618; Co 228.616; V 292.402

The orders of metal concentration levels in the sediment samples were Cr > Co > V > Ni > Zn > Cu > Fe > Al > Mn in the polluted area (Nayband Bay) and Co > Cr > V > Zn > Ni > Cu > Al > Fe > Mn in the unpolluted area (Lavar-e-Saheli).

In the unpolluted area the contents of Al, Zn, Fe, Mn, Ni, V, Co, Cr and Cu ranged from 0.074–0.811 (Mean: 0.3005), 3.2–9.2 (Mean: 5.737), 0.065–0.482 (Mean: 0.246), 0.004–0.024 (Mean: 0.01), 2–9.4 (Mean: 3.79), 2.8–13.4 (Mean: 6.57), 16-25 (Mean: 22.85), 1–74.7 (Mean: 16.57), and 0.1–4.5 (Mean: 2.47) $\mu g\ g^{-1}$ respectively. In the polluted area the contents of Al, Zn, Fe, Mn, Ni, V, Co, Cr as well as Cu in the sediment samples ranged from 0.161–1.543 (Mean: 0.960), 5.4–15.6 (Mean: 10.37), 0.171–1.532 (Mean: 1.108), 0.007–0.054 (Mean: 0.048), 5.9-20.7 (Mean: 15.490), 5.6–27.2 (Mean: 19.38), 21–160 (Mean: 34.3), 41.3–438.3 (Mean:104.16), and 1.6–8.3 (Mean: 3.38) $\mu g\ g^{-1}$ respectively. Ismail and Safahieh measured the content levels of Cu and Zn in the sediment samples collected from intertidal areas in the Lukut River. They have reported that Cu and Zn in the surface sediments were within the range of 37 to 100 $\mu g\ g^{-1}$ and 100 to 210 $\mu g\ g^{-1}$ respectively [18]. According to Usero et al. report, the concentrations of Cr, Cu, Pb, Zn, As and Hg in the sediments of Atlantic coast in southern Spain ranged from 10–33, 3–13, 0.26–0.72, 2–46, 18–460, 3.5–102 and 0.11-0.41 mg kg^{-1} dry mass respectively [19]. In another study, Palpandi and

Kesavan measured concentration levels of heavy metals including Zn, Mn, Cu, Al, Cr and Ni in the sediment samples of Velar estuary, Southeast coast of India. They reported that the mean concentration levels of Cu, Fe and Zn ranged from 39.28 ± 0.6, 178.28 ± 1.12, 16.28 ± 1.24, 542.00 ± 487.58, 9.44 ± 3.11 and 1.64 ± 1.20 $\mu g\ g^{-1}$ respectively [20].

Statistical analysis of Mann-Whitney U test showed that sediment samples in the polluted area contained significantly higher concentrations ($P < 0.05$) of all measured metals (Al, Zn, Fe, Mn, Ni, V, Co, Cr and Cu) compared to unpolluted area (Table 4). The comparison between metal concentrations in polluted and unpolluted areas are shown in Fig. 2.

Sediments act as both sinks and carriers for pollutants in the marine environments. Heavy metals are among the most usual marine contaminants and their occurrence in the marine environment indicates the presence of natural or anthropogenic source. Many studies have illustrated that heavy metal concentration in sediments can be sensitive indicators of pollutants in the marine environment [21, 22]. High concentration levels of trace metals in marine environments due to human activities have been recorded since old times. But elevated releases of toxic metals in to the municipal areas and the related health consequences just become clear in the 1960s [23]. Our study showed higher contents of Al, Zn, Fe, Mn, Ni, V, Co, Cr, as well as Cu in the Nayband Bay (polluted area) compare to unpolluted area mainly due to the activities of all related industries to gas and oil field in the region, boat repairing platform, shipping activities and discharge of effluents from the domestic sources nearby. The activities of industries after a while can release a diversity of poisonous sand possibly poisonous contaminants into the environment [24]. In a recent study in Jade Bay in NW Germany, the trace metal pollution in surface sediment and suspended particulate substance was described. Various metals including As, Cd, Cu, Ni, Pb, Sn and Zn were increased in the surface sediments. The potential metal sources in the region were the harbor area, floodgates and dumped harbor sludge in different parts of the region [25]. In a study in the Montenegrin coastal area, the overall trend for the concentration levels of measured metals in sediment samples was Fe > Mn > Cr > Ni > Zn > Cu > Co. The result of this study showed the anthropogenic impacts on the metal concentration levels in the Montenegrin beach zone [26]. In another study at Vellar estuary, Southeast coast of India the order of metal accumulation was Fe > Al > Mg > Mn > Cd > Cu > Cr > Zn > Ni > Pb. It was reported that higher level of metals could be due to effluents from municipal, domestic and agricultural wastes [20]. The contents of Al, Zn, Fe, Mn, Ni, V, Co, Cr as well as Cu in the shell samples of *Trachycardium lacunosum* in polluted

Table 3 Concentration of heavy metals (µg g^{-1} dw) in sediment samples at polluted & unpolluted areas

Area	Station	Al	Zn	Fe	Mn	Ni	V	Co	Cr	Cu
	1	0.136	5.4	0.162	0.006	2.1	6.1	16	1.5	2.2
	2	0.160	5.7	0.176	0.007	2	4.4	25	7.3	4.3
	3	0.201	6	0.273	0.010	3.1	8.8	22	64.9	1.2
	4	0.208	5.2	0.224	0.009	4.2	7.1	23	1.3	2.5
	5	0.251	6.2	0.278	0.010	3.7	7	22	3.5	2
	6	0.336	6	0.302	0.012	2.8	9.8	22	12.6	0.9
	7	0.526	6.1	0.411	0.018	6.4	4.4	22	70.7	0.1
	8	0.704	7.9	0.482	0.024	7.8	6.4	25	1.5	2.6
	9	0.811	9.2	0.461	0.023	9.4	6.1	22	58.6	2.3
Unpolluted area	10	0.799	8.7	0.445	0.023	8.1	13.4	23	74.7	3.8
	11	0.466	6.4	0.340	0.016	2.1	10.1	24	2.4	3.7
	12	0.109	3.2	0.095	0.005	2.2	7	25	1.5	1.9
	13	0.141	4.4	0.150	0.006	2.6	5.4	23	1	4.5
	14	0.074	4.7	0.065	0.004	2.2	4.1	25	1.5	3.3
	15	0.171	7.3	0.129	0.005	4.7	6.8	21	1.3	1.3
	16	0.145	4.6	0.169	0.007	2.1	4.1	24	6.3	2.1
	17	0.199	3.9	0.201	0.001	2.1	8	23	1.5	3.9
	18	0.144	3.8	0.170	0.007	2.2	2.8	23	1.3	2.7
	19	0.129	4.3	0.139	0.006	2.2	3.1	24	1.4	1.6
Mean ± SD		0.3005 ± 0.24	5.737 ± 1.65	0.246 ± 0.13	0.01 ± .006	3.79 ± 2.38	6.574 ± 2.66	22.85 ± 2.06	16.57 ± 27.1	2.47 ± 1.2
	20	0.161	5.4	0.171	0.007	5.9	5.6	29	170	1.9
	21	1.543	13.5	1.256	0.051	19.3	23.6	22	148.3	4.1
	22	1.488	15.6	1.532	0.053	19.4	27.2	26	438.3	1.6
	23	1.233	10.6	1.183	0.051	16	23.1	27	103.9	3.3
	24	1.108	10.3	1.150	0.049	17.4	21.7	21	93.8	8.3
	25	0.903	8.6	1.066	0.049	14.8	17	34	102.3	2.6
	26	0.928	10.1	1.067	0.048	16.9	21.6	35	92.3	1.8
	27	0.648	9.2	0.939	0.049	15.3	17.4	45	49.4	2.2
	28	0.831	9.6	0.997	0.048	15.3	13.3	24	48.9	4.2
Polluted area	29	0.988	11.1	1.188	0.050	16.4	19.1	34	71.5	3.8
	30	0.986	9.6	1.205	0.051	17	21.7	160	68.7	4.5
	31	1.167	11.5	1.243	0.049	18.4	26.4	21	96.1	2.5
	32	1.374	12	1.333	0.052	20.7	27	25	147.1	3.5
	33	0.843	9.9	1.144	0.048	16.4	13.4	25	53.6	2.9
	34	0.778	9.1	1.083	0.047	16.2	17.9	26	41.3	5.1
	35	0.712	11.2	1.016	0.049	13.6	19.4	35	44.9	2.7
	36	1.046	10.7	1.230	0.051	17.8	15.1	27	97	4
	37	0.955	11.2	1.211	0.054	14.8	23.5	22	76.6	3.1
	38	0.784	9.9	1.097	0.050	15.3	19.9	23	74.2	2.8
	39	0.739	8.2	1.051	0.048	16.2	13.7	25	65	2.7
Mean ± SD		0.960 ± 0.315	10.37 ± 2.05	1.108 ± 0.26	0.048 ± 0.009	15.490 ± 2.3	19.38 ± 5.4	34.3 ± 30.2	104.16 ± 86.4	3.38 ± 1.5

Table 4 The differences between the metal concentrations of samples in polluted and unpolluted areas

Heavy metals	P-value sediments	P-value shells
Al	0.000	0.006
Co	0.009	0.000
Cr	0.000	0.000
Cu	0.021	0.001
Fe	0.000	0.000
Mn	0.000	0.000
Ni	0.000	0.009
V	0.000	-
Zn	0.000	0.000

(Nayband Bay) and unpolluted (Lavar-e-Saheli) areas are given in Table 5.

The orders of metal concentration levels in shell samples were Fe > Zn > Al > Mn > Cu > Cr > Ni > Co in the polluted area (Nayband Bay) and Fe > Al > Zn > Mn > Cu > Co > Ni > Cr in the unpolluted area (Lavar-e-Saheli). In the polluted area the contents of Al, Zn, Fe, Mn, Ni, Co, Cr, and Cu in the shell samples ranged from 0.139–5.36 (Mean: 0.995), 0.335–6.915 (Mean: 1.385), 0.645–6.85 (Mean: 3.170), 0.234–1.269 (Mean: 0.565), 0.003–0.234 (Mean: 0.063), 0.012–0.022 (Mean: 0.016), 0.001–0.242 (Mean: 0.075), and 0.003-1.677 (Mean: 0.285) $\mu g\ g^{-1}$ respectively. In the unpolluted area the concentration levels of Al, Zn, Fe, Mn, Ni, Co, Cr, and Cu ranged from 0–0.758 (Mean: 0.408), 0.003–0.756 (Mean: 0.3805), 0.526–1.564 (Mean: 1.029), 0.07–0. 242 (Mean: 0.176), 0.002–0.02 (Mean: 0.006), 0–0.15 (Mean: 0.009), 0.001–0.001 (Mean: 0.001), 0.003–0.061 (Mean: 0.016) $\mu g\ g^{-1}$ respectively. In a study in Pantai Lido, west coast of Peninsular Malaysia, mean concentrations of

Cu, Cd, Fe, Ni, Pb and Zn in the shell samples of *Perna viridis* were 8.41, 6.67, 48.3, 40.4, 59.4, and 5.96 $\mu g\ g^{-1}$ respectively [27]. Ravera et al also determined the heavy metal levels in the shell samples of *Uniopictorium mancus* from shallow Bay located in Ranco, Italy. They reported that the mean values Al, Cu, Zn, Fe and Mn were found to be (80.86 ± 100.48), (3.53 ± 3.29), (24.00 ± 14.63), (211.20 ± 273.71) and (461.52 ± 252.67) $\mu g\ g^{-1}$ respectively [28]. In a study in Tersakan River, south-west Turkey, mean concentration of Cd, Co, Cr, Cu, Fe, Mn, Ni, Pb and Zn in the shell samples of *Unio sp.* ranged from 0.382 ± 0.06, 1.155 ± 0.08, 7.403 ± 0.54, 15.902 ± 1.24, 671.182 ± 55.05, 268.291 ± 18.24, 20.821 ± 1.77, 4.157 ± 0.21 and 8.475 ± 2.48 $\mu g\ g^{-1}$ respectively [29]. Statistical analysis of Mann-Whitney U test showed that Shell samples of *Trachycardium lacunosum* in polluted area contained significantly higher concentrations ($P < 0.05$) of all measured metals (Al, Zn, Fe, Mn, Ni, Co, Cr and Cu) compared with unpolluted area (Table 4). The comparison between metal concentrations in the polluted and unpolluted areas are shown in Fig. 3.

Beside sediment that may be good indicators of long and medium term of metal loads, bivalve shell is also an indicator of metal contamination since it is sessile and sedentary and reflects the metal level of the special region [30]. In the marine environments, metals discharged from sewage or industrial effluents may be quickly transported from water column to the sediment [31]. The accessibility of various metals in sediments provides a chance for marine organisms to biomagnify these metal and later remobilized them via the food chain. The metal concentrations in the shell samples of *Trachycardium lacunosum* in polluted area were higher than those of the samples taken from the unpolluted area. This indicated that the polluted area had higher pollution and bioaccessibilities of heavy

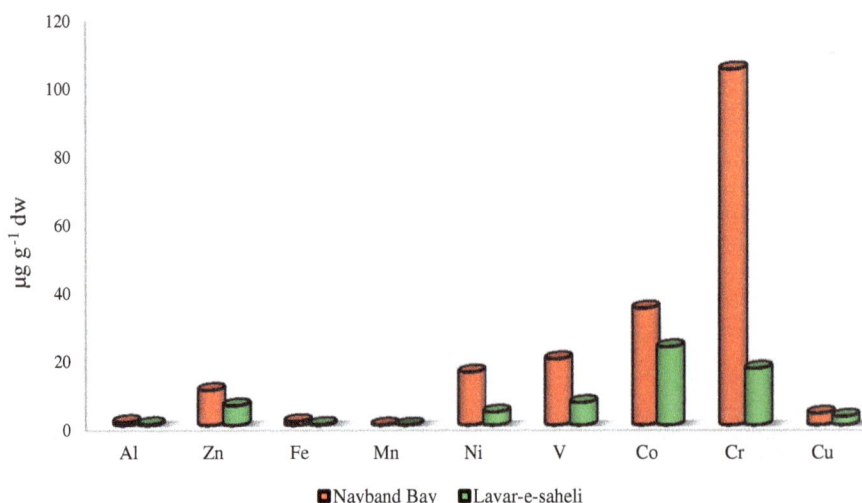

Fig. 2 Comparison of heavy metal concentration levels in the sediment samples at polluted and unpolluted areas

Table 5 Concentration of heavy metals ($\mu g\ g^{-1}$ dw) in the shell samples at polluted and unpolluted areas

Area	Station	Al	Zn	Fe	Mn	Ni	Co	Cr	Cu
	1	0.360	0.465	0.935	0.106	0.003	0.012	0.001	0.061
	2	0.409	0.528	0.809	0.155	0.005	0.012	0.001	0.043
	3	0.081	0.003	1.564	0.220	0.007	0.010	0.001	0.004
	4	0.586	0.756	0.526	0.070	0.020	0.015	0.001	0.043
	5	0.699	0.404	1.265	0.176	0.006	0.003	0.001	0.003
	12	0.682	0.274	1.307	0.242	0.005	0.000	0.001	0.008
Unpolluted area	13	0.516	0.666	1.102	0.140	0.002	0.015	0.001	0.003
	14	0.000	0.179	1.295	0.219	0.003	0.000	0.001	0.010
	15	0.257	0.332	0.942	0.176	0.003	0.004	0.001	0.009
	16	0.463	0.598	1.294	0.226	0.007	0.013	0.001	0.014
	17	0.180	0.232	0.694	0.179	0.003	0.015	0.001	0.003
	18	0.758	0.104	0.802	0.186	0.005	0.006	0.001	0.003
	19	0.314	0.405	0.838	0.199	0.003	0.013	0.001	0.003
Mean ± SD		0.408 ± 0.24	0.3805 ± 0.22	1.029 ± 0.3	0.176 ± 0.05	0.006 ± 0.004	0.009 ± 0.006	0.001 ± 0	0.016 ± 0.02
	20	0.260	0.335	0.645	0.234	0.003	0.013	0.001	0.003
	22	0.727	0.938	4.514	0.980	0.008	0.020	0.057	0.028
	23	0.624	0.805	2.415	0.743	0.005	0.015	0.019	0.049
	24	0.641	0.827	3.024	1.022	0.007	0.016	0.055	0.116
	25	0.213	0.880	1.623	0.299	0.021	0.013	0.118	0.303
	26	0.139	0.902	1.868	0.388	0.003	0.017	0.083	0.930
	27	0.678	0.977	3.208	0.285	0.227	0.019	0.071	0.484
	28	0.640	0.826	2.340	0.645	0.121	0.016	0.001	0.006
Polluted area	29	0.557	0.719	3.265	0.673	0.055	0.012	0.001	0.014
	31	0.608	0.784	1.998	0.550	0.020	0.015	0.146	0.046
	32	1.114	1.437	6.850	1.269	0.055	0.022	0.068	0.141
	33	1.249	1.611	1.911	0.264	0.076	0.014	0.104	0.600
	34	1.722	2.221	6.126	0.630	0.065	0.015	0.001	0.003
	35	5.360	6.915	1.828	0.416	0.003	0.013	0.013	0.003
	36	0.956	1.233	3.747	0.800	0.003	0.017	0.048	0.036
	37	1.335	1.722	2.333	0.375	0.225	0.017	0.157	0.339
	38	0.699	0.875	6.543	0.246	0.003	0.019	0.162	0.349
	39	0.383	0.901	2.790	0.350	0.234	0.015	0.242	1.677
Mean ± SD		0.995 ± 1.16	1.385 ± 1.45	3.170 ± 1.76	0.565 ± 0.3	0.063 ± 0.083	0.016 ± 0.002	0.075 ± 0.07	0.285 ± 0.43

metals. These results are in accordance with the fact that there are different anthropogenic activities, such as petrochemical plants and harbor activities in the Nayband Bay. Use of bivalve shells for metal contamination monitoring in the aquatic environments has various advantages over that of soft tissues. The shells are simple to keep and handle and become clear to be sensitive to environmental metals over the long period. As shells growth occurs incrementally they can provide an indication over a distinct time period, unlike the soft tissues which are good accumulator of various metals and integrate the chemical pollution indication over the living of the marine organisms [32].

The findings of this study showed that *Trachycardium lacunosum* is a good biological indicator for all examined metals except V in the Persian Gulf coastal areas due to its capability in bioaccumulating of metals from the sediment. In a study, Palpandi and Kesavan measured the levels of metals in sediment, shell and soft tissues of mangrove gastropod Nerita Crepidularia. They have reported that the order of metal accumulation in shell and soft tissues of Nerita Crepidularia was Fe > Al > Mg > Mn > Cd > Cu > Cr > Zn > Ni > Pb. They concluded that the higher levels of metals could be due to the heavy inflow of freshwater, which brought lot of effluents

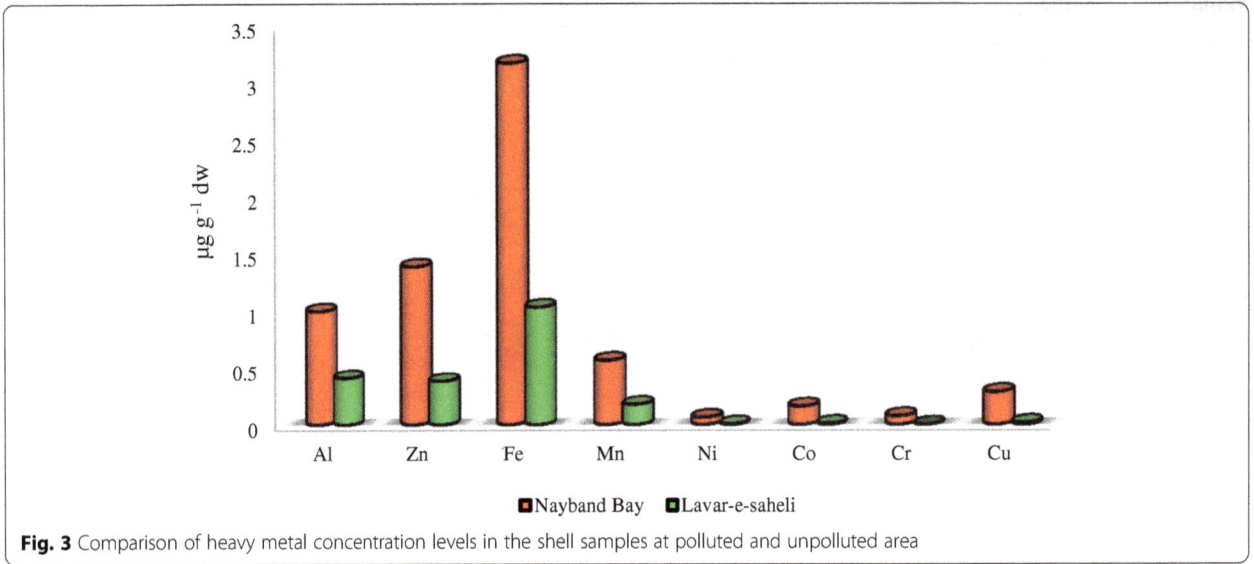

Fig. 3 Comparison of heavy metal concentration levels in the shell samples at polluted and unpolluted area

from municipal drainage and irrigation channels [20]. In another study it has been reported that between measured metals, Zn had the highest concentration level in the shell samples of *Perna viridis* and *Modiolus metcalfei* in Vellar Estuary, South East shoreline of India [33]. In another study in the Egyptian Red Sea shoreline, significant spatial differences in the metal concentration levels in *Tridacna maxima* were observed. The concentrations of most investigated metals in the *Tridacna*

maxima shells and sediments were higher in the anthropogenic areas compare with unpolluted areas [34].

Identification of metal interrelationships

The Spearman's rho correlation coefficients were calculated to assess the association of metals in the sediment (Table 6) and shell samples (Table 7) in polluted and unpolluted areas.

Table 6 The Spearman's rho correlations between metal concentrations in the sediments in polluted and unpolluted areas

		Al	Co	Cr	Cu	Fe	Mn	Ni	V	Zn
Unpolluted area	Al	1.000	−0.306	0.505[b]	−0.069	0.947[a]	0.928[a]	0.632[a]	0.558[b]	0.813[a]
	Co		1.000	0.015	0.358	−0.196	−0.163	−0.230	−0.269	−0.338
	Cr			1.000	−0.293	0.549[b]	0.505[b]	0.309	0.255	0.486[b]
	Cu				1.000	0.009	0.038	−0.323	−0.011	−0.124
	Fe					1.000	0.989[a]	0.513[b]	0.474[b]	0.710[a]
	Mn						1.000	0.486[b]	0.444[b]	0.654[a]
	Ni							1.000	0.221	0.616[a]
	V								1.000	0.432
	Zn									1.000
polluted area	Al	1.000	−0.328	0.867[a]	0.173	0.886[a]	0.689[a]	0.757[a]	0.722[a]	0.728[a]
	Co		1.000	−0.246	−0.232	−0.343	0.018	−0.238	−0.256	−0.304
	Cr			1.000	−0.154	0.735[a]	0.632[a]	0.585[a]	0.656[a]	0.594[a]
	Cu				1.000	0.141	0.203	0.143	−0.105	−0.081
	Fe					1.000	0.767[a]	0.759[a]	0.740[a]	0.765[a]
	Mn						1.000	0.411	0.477[a]	0.606[a]
	Ni							1.000	0.510[b]	0.517[b]
	V								1.000	0.751[a]
	Zn									1.000

[a]Correlation is significant at the 0.01 level
[b]Correlation is significant at the 0.05 level

Table 7 The Spearman's rho correlations between metal concentrations in the shells in polluted and unpolluted areas

		Al	Co	Cr	Cu	Fe	Mn	Ni	Zn
Unpolluted area	Al	1.000	−0.042	.	−0.092	−0.130	−0.187	0.515	0.371
	Co		1.000	.	−0.142	−0.604[b]	−0.376	−0.034	0.578[b]
	Cr			.	.	.	.	.	.
	Cu				1.000	0.098	−0.313	0.312	0.352
	Fe					1.000	0.420	0.232	−0.301
	Mn						1.000	−0.092	−0.534[b]
	Ni							1.000	0.268
	Zn								1.000
polluted area	Al	1.000	0.158	−0.224	−0.352	0.282	0.083	0.016	0.779[a]
	Co		1.000	0.304	0.276	0.551[b]	0.191	−0.062	0.129
	Cr			1.000	0.809[a]	−0.184	−0.602[b]	0.161	0.012
	Cu				1.000	−0.170	−0.567[b]	0.230	0.006
	Fe					1.000	0.400	0.021	0.017
	Mn						1.000	−0.248	−0.150
	Ni							1.000	0.101
	Zn								1.000

[a]Correlation is significant at the 0.01 level
[b]Correlation is significant at the 0.05 level

As shown in Table 4, most metals in the sediment samples in the polluted area are well correlated. Fe had remarkable positive correlations ($P < 0.01$) with Mn ($r = 0.767$), Ni ($r = 0.759$), V ($r = 0.740$), and Zn ($r = 0.765$). Cr had also noticeable correlations ($P < 0.01$) with Fe ($r = 0.735$), Mn ($r = 0.632$), Ni ($r = 0.585$), V ($r = 0.656$), and Zn ($r = 0.594$). In the case of Mn remarkable positive correlations ($P < 0.01$) were observed vs V and Zn. The significant correlation between Al and other metals (except Cu, Co) in both polluted and unpolluted areas confirms that these metals are associated with alumina silicate minerals. Similar significant positive correlations between metals in the sediment samples have been reported in different areas [25, 35]. As seen in Table 4, in the shell samples of *Tracycardium lacunosum* in the polluted area there are correlations for Al vs Zn ($r = 0.779$), Cr vs Cu ($r = 0.809$) and in the cases of Co vs Fe ($r = 0.557$), Cr vs Mn ($r = -0.602$), and Fe vs Mn ($r = -0.567$). The correlations were significant at the level of 0.05 in the polluted area. The significant correlations found between heavy metals could be due to several reasons such as differences in the biological half-life and biochemical behaviors of metals found in the sediments and shells [36–38].

Conclusion

In this work, the levels of metals including Al, Zn, Fe, Mn, Ni, V, Co, Cr and Cu were determined in the sediment and shell samples of the bivalve *Tracycardium lacunosum* from two areas (polluted and unpolluted) of Asaluyeh Bay, northern part of the Persian Gulf. This study was the first effort to consider shell of *Tracycardium lacunosum* as a bioandicator for monitoring of heavy metals. Results of this study indicated that all measured metals including Al, Zn, Fe, Mn, Ni, V, Co, Cr and Cu were significantly higher in the sediment samples of polluted area compared with unpolluted area. In the case of shell samples of *Trachycardium lacunosum*, polluted area contained significantly higher concentrations of Al, Zn, Fe, Mn, Ni, Co, Cr and Cu compared to unpolluted area. It was concluded that shells of the *Trachycardium lacunosum* can be applied as a suitable bioandicator for heavy metals in the marine environment. Results confirmed that due to the possible pollution by oil and gas activities near the polluted area continuing and permanent evaluating as well as mitigation measures in this area is highly necessary.

Abbreviation
ICP-OES: Inductively Coupled Plasma-Optical Emission Spectrometry

Acknowledgements
The authors are grateful to the Bushehr University of Medical Sciences for their financial support and the laboratory staff of the Environmental Health Engineering Department for their cooperation. This project was partly supported by Iran National Science Foundation (Research Chair Award No. 95/INSF/44913).

Funding
This study was performed as a master thesis in Environmental Health Engineering founded by Bushehr University of Medical Science

Authors' contributions

VNK was the main investigator, collected the data, and wrote the first draft of the manuscript. SD was the supervisor of study in all steps, edited and polished the final version of manuscript. IN has guided and collected the samples. AO performed the statistical analysis. AV, MN, MR and SN were advisors of the study. MK and RM conducted the experiments. All authors read and approved the final version of manuscript.

Competing interests

The authors declare that they have no competing interests.

Author details

[1]Department of Environmental Health Engineering, Faculty of Health, Bushehr University of Medical Sciences, Bushehr, Iran. [2]The Persian Gulf Marine Biotechnology Research Center, The Persian Gulf Biomedical Sciences Research Institute, Bushehr University of Medical Sciences, Boostan 19 Alley, Imam Khomeini Street, Bushehr, Iran. [3]Systems Environmental Health, Oil, Gas and Energy Research Center, The Persian Gulf Biomedical Sciences Research Institute, Bushehr University of Medical Sciences, Bushehr, Iran. [4]The Persian Gulf Tropical Medicine Research Center, The Persian Gulf Biomedical Sciences Research Institute, Bushehr University of Medical Sciences, Bushehr, Iran. [5]The Persian Gulf Studies and Researches Center Marine Biotechnology Department, Persian Gulf University, Bushehr, Iran. [6]School of Public Health, Tehran University of Medical Sciences, Tehran, Iran. [7]Ecology Department, Iranian Shrimp Research Institute, Bushehr, Iran.

References

1. Dobaradaran S, Mahvi AH, Nabizadeh R, Mesdaghinia A, Naddafi K, Yunesian M, Rastkari N, Nazmara S. Hazardous organic compounds in groundwater near Tehran automobile industry. Bull Environ Contam Toxicol. 2010;85(5):530–3.

2. Arfaeinia H, Nabipour I, Ostovar A, Asadgol Z, Abuee E, Keshtkar M, Dobaradaran S. Assessment of sediment quality based on acid-volatile sulfide and simultaneously extracted metals in heavily industrialized area of Asaluyeh, Persian Gulf: concentrations, spatial distributions, and sediment bioavailability/toxicity. Environ Sci Pollut Res Int. 2016;9:1–20.

3. Dobaradaran S, Naddafi K, Nazmara S, Ghaedi H. Heavy metals (Cd, Cu, Ni and Pb) content in two fish species of Persian Gulf in Bushehr Port, Iran. Afr J Biotechnol. 2013;9(37):6191–3.

4. Abadi DR, Dobaradaran S, Nabipour I, Lamani X, Ravanipour M, Tahmasebi R, Nazmara S. Comparative investigation of heavy metal, trace, and macro element contents in commercially valuable fish species harvested off from the Persian Gulf. Environ Sci Pollut Res Int. 2015;22(9):6670–8.

5. Giarratano E, Amin OA. Heavy metals monitoring in the southernmost mussel farm of the world (Beagle Channel, Argentina). Ecotoxicol Environ Saf. 2010;73(6):1378–84.

6. Tsangaris C, Kormas K, Strogyloudi E, Hatzianestis I, Neofitou C, Andral B, Galgani F. Multiple biomarkers of pollution effects in caged mussels on the Greek coastline. Comp Biochem Physiol C Toxicol Pharmacol. 2010;151(3):369–78.

7. Chandurvelan R, Marsden ID, Glover CN, Gaw S. Assessment of a mussel as a metal bioindicator of coastal contamination: relationships between metal bioaccumulation and multiple biomarker responses. Sci Total Environ. 2015;511:663–75.

8. Marsden ID, Smith BD, Rainbow PS. Effects of environmental and physiological variables on the accumulated concentrations of trace metals in the New Zealand cockle Austrovenus stutchburyi. Sci Total Environ. 2014;470:324–39.

9. Riba I, Casado-Martínez C, Forja JM, Valls ÁD. Sediment quality in the Atlantic coast of Spain. Environ Toxicol Chem. 2004;23(2):271–82.

10. Szefer P, Szefer K. Metals in molluscs and associated bottom sediments of the southern Baltic. Helgol Mar Res. 1990;44(3-4):411–24.

11. Zuykov M, Pelletier E, Harper DA. Bivalve mollusks in metal pollution studies: from bioaccumulation to biomonitoring. Chemosphere. 2013;93(2):201–8.

12. Goldberg ED, Bowen VT, Farrington JW, Harvey G, Martin JH, Parker PL, Risebrough RW, Robertson W, Schneider E, Gamble E. The mussel watch. Environ Conserv. 1978;02:101–25.

13. Yap CK, Ismail A, Tan SG, Rahim IA. Can the shell of the green-lipped mussel Perna viridis from the west coast of Peninsular Malaysia be a potential biomonitoring material for Cd, Pb and Zn? Estuar Coast Shelf Sci. 2003;57(4):623–30.

14. Huanxin W, Lejun Z, Presley BJ. Bioaccumulation of heavy metals in oyster (Crassostrea virginica) tissue and shell. Environ Geol. 2000;39(11):1216–26.

15. Falcó G, Llobet JM, Bocio A, Domingo JL. Daily intake of arsenic, cadmium, mercury, and lead by consumption of edible marine species. J Agric Food Chem. 2006;54(16):6106–12.

16. Abbott RT, Morris PA. A field guide to shells: Atlantic and Gulf Coasts and the West Indies. California: Houghton Mifflin Harcourt; 2001.

17. Marina M, Enzo O. Variability of zinc and manganese concentrations in relation to sex and season in the bivalve Donax trunculus. Mar Pollut Bull. 1983;14(9):342–6.

18. Ismail A, Safahieh A. Copper and zinc in intertidal surface sediment and Telescopium telescopium from Lukut River, Malaysia. Coast Mar Sci. 2005;29(2):111–5.

19. Usero J, Morillo J, Gracia I. Heavy metal concentrations in molluscs from the Atlantic coast of southern Spain. Chemosphere. 2005;59(8):1175–81.

20. Palpandi C, Kesavan K. Heavy metal monitoring using Nerita crepidularia-mangrove mollusc from the Vellar estuary, Southeast coast of India. Asian Pac J Trop Biomed. 2012;2(1):S358–67.

21. Jain CK. Metal fractionation study on bed sediments of River Yamuna, India. Water Res. 2004;38(3):569–78.

22. Idris AM, Eltayeb MA, Potgieter-Vermaak SS, Van Grieken R, Potgieter JH. Assessment of heavy metals pollution in Sudanese harbours along the Red Sea Coast. Microchem J. 2000;87(2):104–12.

23. Kazemi A, Riyahi Bakhtiari A, Mohammad Karami A, Haidari B, Kheirabadi N. Bioavailability and variability of Cd, Pb, Zn, and Cu pollution in soft tissues and shell of Saccostrea cucullata collected from the coast of Qeshm Island, Persian Gulf, Iran. Iranian J Toxicol. 2013;7(21):836–41.

24. Daskalakis KD. Variability of metal concentrations in oyster tissue and implications to biomonitoring. Mar Pollut Bull. 1996;32(11):794–801.

25. Beck M, Böning P, Schückel U, Stiehl T, Schnetger B, Rullkötter J, Brumsack HJ. Consistent assessment of trace metal contamination in surface sediments and suspended particulate matter: A case study from the Jade Bay in NW Germany. Mar Pollut Bull. 2013;70(1):100–11.

26. Joksimovic D, Tomic I, Stankovic AR, Jovic M, Stankovic S. Trace metal concentrations in Mediterranean blue mussel and surface sediments and evaluation of the mussels quality and possible risks of high human consumption. Food Chem. 2011;127(2):632–7.

27. Edward FB, Yap CK, Ismail A, Tan SG. Interspecific variation of heavy metal concentrations in the different parts of tropical intertidal bivalves. Water Air Soil Pollut. 2009;196(1-4):297–309.

28. Ravera O, Cenci R, Beone GM, Dantas M, Lodigiani P. Trace element concentrations in freshwater mussels and macrophytes as related to those in their environment. J Limnol. 2003;62(1):61–70.

29. Genç TO, Yilmaz F, İnanan BE, Yorulmaz B, Ütük G. Application of multi-metal bioaccumulation index and bioavailability of heavy metals in Unio sp. (Unionidae) collected from Tersakan River, Muğla, South-West Turkey. Fresen Environ Bull. 2015;24(1a):208–15.

30. Brügmann L. Heavy metals in the Baltic Sea. Mar Pollut Bull. 1981;12(6):214–8.

31. Lo CK, Fung YS. Heavy metal pollution profiles of dated sediment cores from Hebe Haven, Hong Kong. Water Res. 1992;26(12):1605–19.

32. Phillips DJ. Quantitative aquatic biological indicators; their use to monitor trace metal and organochlorine pollution. 1980.

33. Ponnusamy K, Sivaperumal P, Suresh M, Arularasan S, Munilkumar S, Pal AK. Heavy metal concentration from biologically important edible species of bivalves (Perna viridis and Modiolus metcalfei) from vellar estuary, south east coast of India. J Aquac Res Development. 2014;5(5):1–5.

34. Madkour HA. Distribution and relationships of heavy metals in the giant clam (Tridacna maxima) and associated sediments from different sites in the Egyptian Red Sea coast. 2005.

35. Schöne BR, Zhang Z, Radermacher P, Thébault J, Jacob DE, Nunn EV, Maurer AF. Sr/Ca and Mg/Ca ratios of ontogenetically old, long-lived bivalve shells (Arctica islandica) and their function as paleotemperature proxies. Palaeogeogr Palaeoclimatol Palaeoecol. 2011;302(1):52–64.

16

Optimization of interpolation method for nitrate pollution in groundwater and assessing vulnerability with IPNOA and IPNOC method in Qazvin plain

Elham Kazemi[1,2], Hamid Karyab[1,2*] and Mohammad-Mehdi Emamjome[1,2]

Abstract

Background: The presence of nitrate is one of the factors limiting the quality of groundwater resources, particularly in arid and semi-arid climates. Therefore, the knowledge about the distribution of nitrate in groundwater and its source has an effective role in protecting health. The study aimed to optimize an interpolation method to predict the nitrate concentration and assessment of aquifer vulnerability in Qazvin plain.

Methods: One hundred sixty-two deep wells in Qazvin plain aquifer were randomly selected and nitrate concentration was analyzed in four different lands including agricultural, residential, steppe and mixed-use areas. Interpolation was done by IDW, Spline, Kriging and National neighbor methods using ArcGIS software. To select the best interpolation method, errors of predicted values were determined by Mean Relative Error (RME) and Root Mean Square Error (RMSE). For analysis of potential vulnerability of aquifer to nitrate pollution due to agricultural activity and sewage leaks, hazard factors and control factors were used for identification of hazard indexes (HI) using IPNOA and IPNOC model.

Results: The results showed that in 8.82% and 18.52% of samples in agricultural and residential areas, the detected nitrate was above the acceptable level at 50 mg/L. National neighbor method with the lowest RME and Spline method with the lowest RMSE were provided the most accurate estimates of nitrates in the aquifer. The highest hazard was obtained in agricultural areas (HI = 6.11). Also, the most influential parameters on aquifer vulnerability were mineral fertilizer ($HF_f = 3$), organic fertilizers ($HF_m = 3$), irrigation systems ($CF_i = 1.04$) and tillage patterns ($CF_{ap} = 1.04$).

Conclusions: According to the results, National neighbor with the lowest RME was preferable than the other spatial interpolation methods for prediction of nitrate concentration in the aquifer. This method provided similar spatial distribution maps of nitrate in groundwater and that was an efficient method for assessing water quality. Hazard index as a result of agricultural activities (IPNOA) was ranged from "very low" to "low" which was in accordance with detected and predicted nitrate concentration in the aquifer. In addition he hazard of nitrate contamination from household (IPNOC) was in very low (class 2).

Keywords: Groundwater, Nitrate, Spatial interpolation, Vulnerability, IPNOA, IPNOC, Qazvin plain

Background

Due to its wide distribution and ease access, groundwater is considered as one of the most important sources of drinking water [1]. Thanks to high quality, these resources often do not require an advanced treatment [2]. But in recent years, various reasons including population growth, urbanization, heavy use of nitrogenous fertilizers in cropping agriculture systems [3, 4], not having a proper treatment for municipal and industrial wastewater have increased nitrate, nitrite and other chemicals concentration in groundwater [5–7]. Nitrogen is one of the most contaminations considered as a common problem in many parts of the world which can exist in two forms: organic and inorganic (including ammonium, ammonium compounds, nitrite and nitrate). Nitrates is

* Correspondence: hkaryab@gmail.com
[1]Department of Environmental Health Engineering, School of Health, Qazvin University of Medical Sciences, Qazvin, Iran
[2]Bahonar Blvd, College of Medical Sciences, Qazvin, Iran

considered as an important indicator of chemical in water and health, so that the World Health Organization has established a maximum contaminant level of 50 mg/L for drinking water [8]. Nitrate in drinking water can cause some serious diseases such as blue baby in infants and saliva, stomach, colon and bladder cancers in chronic exposures in adults [8–10].

Nowadays, due to increased human activities and negligence in chemical usage, an increase in nitrate concentration in residential areas groundwater seems natural. It can be attributed to the use of fertilizers, septic systems and leaking sewage systems [11, 12]. Many studies have shown high concentration of nitrate in areas with septic tank. It was illustrated that the groundwater resources were under strong anthropogenic pressure posed by the city [13]. Ouedraogo and Vanclooster (2016) showed nitrate problems in many metropolis in Africa [14]. In some cases, the increase in population has also considered as an influential factor in increasing nitrate concentration. Nas and Berktay showed that the average nitrate concentration increased from 2.2 to 16.1 mg/L during 1998 to 2001 in Konya, Turkey [15].

Today, efforts have been made to identify the predicting systems for water quality assessing as the best way to prevent pollution and investigate the quality of groundwater [16]. GIS (Geographic Information System) is considered as one of the most powerful technologies in this field to identify, analyze, interpret and make inferences about data [17]. Its capabilities for spatial interpolation have improved through integrating advanced methods as well as linking GIS to a system designed for modeling, analyzing, and visualizing a continuous field [18]. This system gathers data from a determined geographic location in order to store, collect and analyze data which is a great step to make a huge source of spatial and descriptive data accessible in a short time [9]. Numerous studies have shown that GIS used as explored spatial analyzes, interpolation and mapping all over the world. Also in science and health services, it can provide users and authorities with useful information [10, 19].

It is demonstrated that Kriging, IDW and Spline are efficient for spatial interpolation of nitrate concentrations in flat areas water resources [9, 18, 20]. Uyan and cay (2010) showed that Universal kriging, a type of geostatistical technique, can be applied to distribute the groundwater nitrate concentration data [8]. In statistics, mainly in geo statistics, it is a powerful method of interpolation. The basic idea of Kriging is to predict the value of a function at a given point by computing a weighted average of the known values neighboring the function. Mathematically, the method is closely related to regression analysis. Semi-variogram plays a central role in the analysis of the geo statistical data using kriging method [21, 22]. It is showed that kriging method

was the most suitable technique for mapping the bathymetry of the Yucatan submerged platform [23]. Spline method estimates values using a mathematical function that minimizes overall surface curvature. This results in a smooth surface passing exactly through the input point [24]. Inverse Distance Weight (IDW) is based on the extent of similarity of cells used in order to determine the depth and spatial variability of groundwater quality in areas which are not flat [17, 25, 26]. Azpurua and Dos Ramos showed that it is most likely to produce the best estimation in interpolation [27]. Natural Neighbor is based on a discrete set of spatial points. The value of an interpolation point is estimated using weighted values of the closest surrounding points in a triangulation [28]. One of the important keys in interpolation method is errors determination. There are a lot ways to determine the interpolation errors such as the Mean Bias Error and Root Mean Square Error [8, 17, 25]. The IPNOA is a parametric model which assesses the potential hazard of nitrate contamination originating from agriculture in aquifers. The method integrates the hazard factors (HF) and the control factors (CF). The HFs represents all farming activities that cause an impact on soil quality including application of fertilizers, livestock, poultry manure, industry wastewater and urban sludge. In addition the CFs assesses the characteristics of geographical location, climatic conditions and agronomic actions. IPNOA method has been used in several studies to assess the vulnerability of aquifers to nitrate [29–31].

In this article, after monitoring the nitrate concentration in selected groundwater resources, optimized spatial interpolation method and IPNOA index were implemented to present the groundwater vulnerability to nitrate in saline aquifer in Qazvin plain.

Methods

The study area and sampling stations

The research was done in Qazvin plain including two semi-arid and arid cold climates with an area of about 74,737 km^2 in Qazvin province. This area was located in Saline Aquifer with an area of 9502 km^2 in the northern basin. It was geographically situated between $49^\circ - 50^\circ$ and $17' - 32'$ at east longitude and $35^\circ - 36^\circ$ and $39' - 21'$ at north latitude including five cities, 14 sections, 30 districts, 18 towns, and 289 villages. The most important water resources in the study area were in Kharrud, a seasonal river, 3 permanent rivers, 189 springs and 23 aqueducts. In addition, more than 1200 semi-deep and deep wells were drilled to supply water for agricultural, industrial and residential destinations. The rain fall was 62 and 345 mm/year in the arid and semi-arid climates, respectively, with the average of 141 mm/year in the plain. Furthermore, the soil types in surface layers of arid climate and semi-arid climate were saline, sodium (alkaline) and gypsum.

Considering approximately 1000 underground wells, setting the confidence level at 95% and the average standard deviation at 1 mg/L, 162 wells were selected as the sample size in the arid and semi-arid climates. Different climates were specified by Dommartin method based on temperature and annual rainfall in the study area. In other studies, this categorization has been confirmed [32]. Since a vast area of the study was located in semi-arid region, the greater proportion of the samples was chosen from this climate. Figure 1 displayed the distribution of sampling points in different climates of the study area. As shown in Fig. 2, sampling stations were chosen in flat areas. In this figure, DEM (Digital Elevation Model) raster reveals that the study area was located at an altitude of 234 to 418 m above sea the level. Sampling stations were located in agricultural (41%), steppe (17%), residential (18%) and mixed- use (24%) areas. There was no station in industrial areas. Mixed-use was a type of area that blends residential, commercial, industrial and agricultural uses [33].

Calculated fixed radial (CFR) method based on saturated thickness (m), aquifer porosity (%), pumping rate (m^3/year) and pumping time (year) was used to estimate groundwater quality protection zone [34]. Accordingly, it was approximately conducted 500 m around the water wells. Then, in this radius, factors affecting water pollution to nitrate including septic sewage discharges,

fertilization, industrial activities, agronomic practices, irrigation systems and tillage were investigated by visiting around randomly selected water wells.

Nitrate monitoring, quality control and data analysis

A cross-sectional study was done to monitor the nitrate concentration in the randomly selected groundwater resources in the study area during two seasons, spring and summer. In each stage, one liter water sample was taken from selected wells in a glass bottle. The grab samples were transported to a laboratory under controlled temperature conditions. Nitrate analysis was performed as soon as possible in accordance with the standard methods of water and wastewater examination [35] with a spectrophotometer DR6000 (HACH) at a wavelength of 220 nm.

In this study, precision, accuracy, representativeness and sensitivity were used as the quality control indicators. *Precision* was the degree of similarity among measurements taken from three repeated samples. It was calculated as the relative standard deviation (RSD). In general, data should be viewed with caution, when the RSD for triplicates is at 18%. It was calculated with Eq. 1, where s was standard deviation of the nitrate concentration [36].

$$RSD\ (\%) = \frac{sd}{mean} \times 100 \tag{1}$$

Accuracy is an agreement measure for a variable value in a sample with a known or "true" value. In this study,

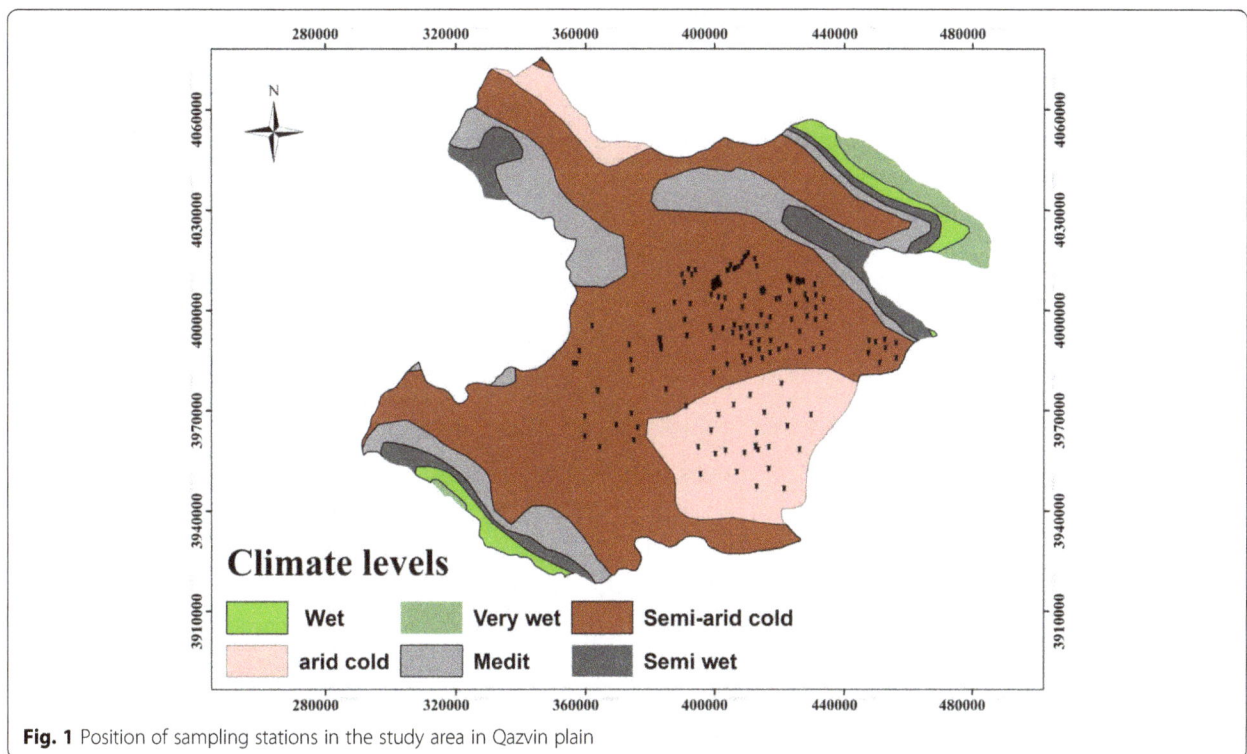

Fig. 1 Position of sampling stations in the study area in Qazvin plain

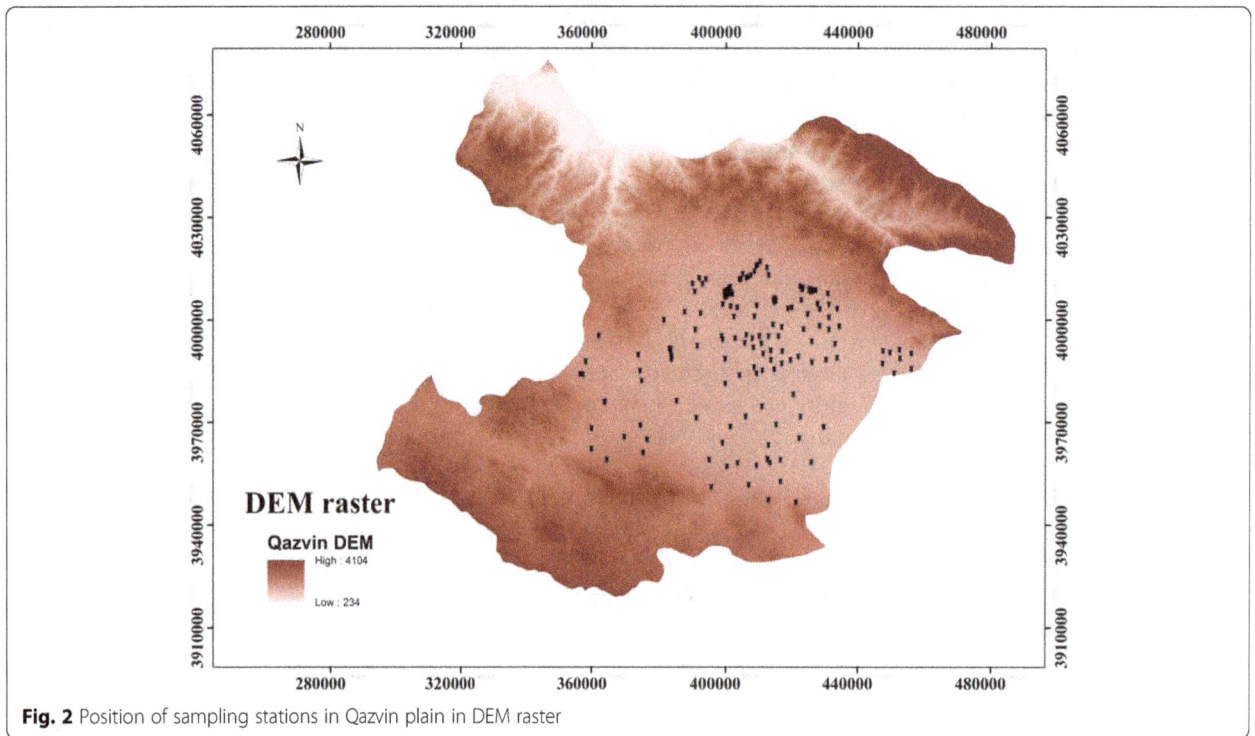

Fig. 2 Position of sampling stations in Qazvin plain in DEM raster

spiked samples were used to provide an estimate of accuracy. Levels in the reference samples were selected within the range of values examined in the water body being assessed. For this purpose, certified reference samples of 5, 25 and 50 mg/l on nitrate with three repetitions were chosen to yield the accuracy. Finally, the results were presented as % of recovery using Eq. 2.

$$\text{Re covery } (\%) = \frac{A\text{-}B}{A} \times 100 \qquad (2)$$

Where A was true value and B was measured nitrate concentration in the laboratory. The lab quality control samples must meet 85% - 115% recovery level. If the internal standard has a recovery of <70%, the results will be flagged and may be higher than what has been

Table 1 The hazard and control factors scores in assessment of vulnerability by IPNIA model

Relative hazard classes of different fertilization types					
Inorganic (kg/ha)	HF_f	Organic (kg/ha)	HF_m	Sludge (kg/ha)	HF_s
0	1	0	1	0	1
1–25	2	1–150	2	1–150	2
26–100	3	151–300	3	151–500	3
100–180	4	301–500	4	501–1500	4
>180	5	>500	5	>1500	5

Control factors									
Soil nitrogen content (%)	CF_n	Irrigation system	CF_i	Rainfall (mm/year)	temperature (°C)	CF_c	Tillage	Type of fertilization	CF_{ap}
>0.5	1.04	Basin	1.06	>1200	6–15	1.01	Traditional	Fertirrigation	1.04
0.22–0.5	1.02	Border	1.04	1050–1150	13	1.08		Total surface	1.00
0.15–0.22	1.00	Sprinkler	1.02	950–1100	14–16	1.06		Through leaves	0.98
0.1–0.15	0.98	No irrigation	1.00	800–1000	12	1.04	Minimum	Localized	0.96
<0.1	0.96			600–1000	15–16	1.02	No tillage		0.94
				600–800	12–13	1.00			
				< 600	15–30	0.98			

Table 2 Hazard index and relative classification [29]

Hazard index	Hazard level	Classification
2.54–3.18	1	Unlikely
3.19–5.88	2	Very low
5.89–7.42	3	low
7.43–9.31	4	Moderate
9.32–11.10	5	High
11.11–17.66	6	Very high

reported [37]. *Representativeness* is a qualitative term expressing how well the data which reflect the true environmental condition are sampled. It was a method to minimize the variability to ensure discriminate uniformity. It was visible in sample transportation and analyzing and using certified glass in lab [38]. *Sensitivity* is the capability of a method or instrument to discriminate between measurement responses for different levels of the variable of interest. It was investigated through detecting the minimum amount of nitrate concentration measured with spectrophotometer and considered 10 times lower than the water quality objectives for nitrate concentration in drinking water. Hallock and Ehinger (2003) express 20%, 7% and 0.01 mg/l, respectively, for accuracy, precision and detection limit in nitrate monitoring in water bodies [39–41].

Two sample t-tests using STATS.12 software was used for analysis of obtained data in arid and semi arid climates. In addition one-way analysis of variance was used for analysis of obtained data in four different land usages including agriculture, residential, steppe and mixed-use areas.

Spatial interpolation and errors

GIS was used for interpolating the nitrate concentration using specific explanatory variables. Interpolation methods such as kiriging, spline, IDW and natural neighbor are powerful tools for data estimation based on the structure of a building [42]. Kriging assumed that the distance between sample points reflects a spatial correlation which can be used to explain variations. It is presented in Eq. 3. In this equation, F(x, y) were the estimated values of the index at the point with coordinates x and y. W_i was weight factor of points (i), and f_i was the index values in

measuring points. In this equation, the amount of weight was provided through Variogram model [43].

$$F(x, y) = \sum_{i=1}^{n} W_i \times f_i \qquad (3)$$

Inverse Distance Weighting (IDW) is a method to determine multivariate interpolation with a known scattered set of points. The assigned values to unknown points are calculated with a weighted average of the values available at the known points. In this method, the weight factor is determined based on points far from each other. Nearby points of samples were assigned with more weights. The IDW is presented in Eq. 4, where D_i was the proportion of the observed distance points to estimated point, π was potential distance weighting and n was the number of adjacent points. With the increasing power of the IDW, the amount of RMSE increases, as well [8].

$$y_i = \frac{D^{-\pi}}{\sum_{i=1}^{n} D_i^{-\pi}} \qquad (4)$$

The Spline method uses polynomials and is based on sample data fit a polynomial function from which values of unknown points are estimated. The key feature of spline is that there is no sudden change in level. According to the impact of degree to data interpolation, whether the higher degree is selected, the result will be smoother, but significantly the accuracy of the model is reduced. It is calculated by Eq. 5, where N was the number of sampling point, λ_j was factor solution of linear equations and r_j was the distance from "j" sample points. Depending on the type of option $T_{(x, y)}$, $R(r_j)$ was set by the user [44].

$$Z_{(x,y)} = T_{(x,y)} + \sum_{j=1}^{N} \lambda_j R(r_j) \qquad (5)$$

Natural neighbor is another weighted-average method that has many positive features. It can be used for both interpolation and extrapolation. The basic equation used in natural neighbor interpolation is identical to the one used in IDW and can efficiently handle large input point databases [17]. In this study, to determine the interpolation errors, the values of RMSE, %RMSE and MRE were assessed by eqs 6, 7 and 8, respectively. The values of RMSE and MRE, the more approximating zero,

Table 3 The precision and accuracy of nitrate analysis in water samples

Concentration (mg/L)	n	Mean (mg/L)	Std. Dev.	RSD	Recovery	Std. Err. of mean	95% Conf. Interval of mean
5	3	5.64	0.02	0.01	113%	0.01	5.59–5.69
25	3	26.64	0.14	0.01	106.56%	0.08	26.29–26.99
50	3	51.65	2.03	0.04	103.3%	1.17	46.52–56.60

Table 4 Nitrate concentration in water resources in different climates in Qazvin aquifer

Climates	n	Nitrate concentration (mg/L)				
		Range	mean	Std. Dev.	Std. Err. of mean	95% Conf. Interval of mean
Semi- arid	121	5.45–76.55	26.25	14.04	1.28	23.73–28.79
Arid	41	10.03–43.98	19.97	7.56	1.18	17.59–22.36
Total	162	5.45–76.55	24.66	12.98	1.02	22.64–26.68

indicate a better estimate of model used to assess the unknown parameters [2, 5].

$$RMSE = \sqrt{\frac{\sum_{j=1}^{n}\left(x(P)_j - x(m)_j\right)^2}{n}} \quad (6)$$

$$\%RMSE = \frac{RMSE}{\overline{X}} \times 100 \quad (7)$$

$$MRE = \frac{1}{n}\sum_{i=1}^{n}\left|\frac{z^*(x_i) - z(x_i)}{z(x_i)}\right| \quad (8)$$

Where x(p) was the estimated value of each component, x(m) was the measurement of water quality component, n was sample number, $\overline{X}$ was the average of a measured component, $z(x_i)$ was the observed value at location i, $z^*(x_i)$ was the interpolated value at location i, and n was the sample size. Since RMSE is sensitive to outliers, %RMSE can be used to calculate the percent [45].

Nitrate hazard index

Since 40.7% and 18.5% groundwater in the study were located in agriculture and residential areas, IPNOA and IPNOC were used to evaluate the hazard level of nitrate due to fertilization and sewage leakage. The IPNOC method assesses the Hazard Index from sewage leaks and sinkholes, whiles IPNOA gives a hazard potential from farming. Natural nitrogen content in soil, the climate, agronomic practices and the irrigation systems were assessed to obtain nitrate hazard index due to agricultural activities [28–30]. The relative hazard classes of different fertilization types and control factors are shown in Table 1. The hazard index (HI) for nitrate contamination was obtained by multiplying the different hazard factors (HF) by the control classes (CF) as shown in the Eq. 9. Where f was fertilizers, m was manure, s was sludge, n was soil nitrogen content, c was climate, ap was agronomic practices and i was irrigation system.

$$HI = \left(HF_f + HF_m + HF_s\right) \times CF_n \times CF_c \\ \times CF_{ap} \times CF_i \quad (9)$$

Nitrogen leaking from the sewer pipes was assessed to obtain IPNOC index. Due to lack of information, nitrogen leakage from houses which not connected to sewerage systems was not considered. The annual value of nitrogen

leaking from the sewer pipes (N_y) was calculated with Eq. 10 which L_i was length of pipe (m), L_c was adjusted wastewater leakage (%) and N_i was average annual nitrogen load in transit (kg/year). The adjusted leakage percentage in each pipe (L_c) was calculated based on its material and age using Eq. 11 which L_e was leakage related to the average life time of pipe (%).

$$Ny = \sum L_c \times N_i \times L_i \quad (10)$$

$$L_c = L_e(age/average\ life) \quad (11)$$

Information required in this study was obtained from two sources. The first source was obtained from site visit and implementation of prepared checklist by researchers in the field including land-use, fertilizer types, fertilizer consumption, irrigation systems and tillage patterns in the study area. The second sources of information were obtained from government offices such as agriculture and water authority and meteorological organization. This information was included rainfall pattern, temperature, fertilization, soil nitrogen contents and general information of aquifer in the Qazvin plain. Finally hazard levels and vulnerability classification was obtained from Table 2. The prepared information from different sources were pipe material: concrete; average life time: 15 years; age of sewerage: 5 year; leakage: 7%; nitrogen production: 4.38 kg/capita.year; infiltration of nitrogen: 85%; number of inhabitants in each zone: 9812 person; area in each zone: 0.785 km^2; mean of sewerage length: 21.23 m/ha;

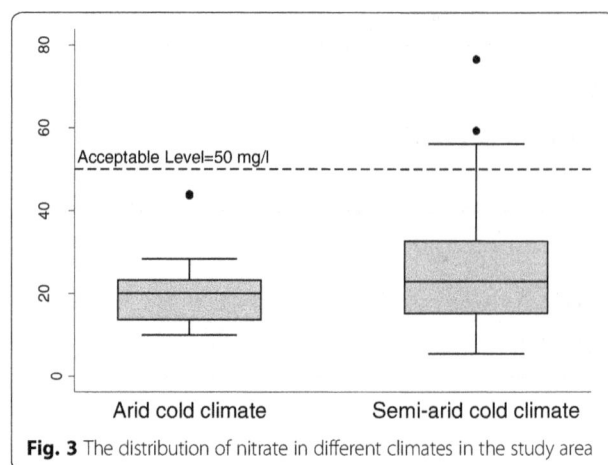

Fig. 3 The distribution of nitrate in different climates in the study area

Table 5 Nitrate concentration in water resources in different land usages in Qazvin aquifer

Land usages	n	Range	mean	Std. Dev.		
Agricultural	66	5.45–76.55	22.45	9.51		
Steppe	27	9.79–32.56	21.05	11.64		
Residential	30	6.67–56.17	26.58	15.44		
Mixed-use	39	8.30–76.55	29.43	15.54		
Source	SS	Df	MS	F	Prob > F	Prob > chi2
Between groups	1670.22	3	556.74	3.45	0.018	0.002

Rainfall: 62 mm/year in the arid and 345 mm/year in semi-arid climates; temperature: 13 °C; Soil nitrogen content: less than 0.1%; inorganic fertilization: 70–120 kg/ha in agricultural lands; Irrigation system: Sprinkler and border; Tillage: Traditional (no irrigation and no tillage was observed in steppe and residential area).

Result and discussion
Nitrate in different climates
As shown in Table 3, the recovery of triple samples was ranged in accepted levels of 85–115%. In addition,

calculated RSD values were ranged between 0.01 to 0.04% that were lower than the acceptable value of 18%.

The means of nitrate concentration in semi-arid and arid cold climates were 25.66 and 19.97 mg/l (Table 4). The mean concentration of nitrate in semi-arid climate was 19.93% more than arid climate. In addition the results showed that nitrate concentration was significantly different in two climates ($p < 0.05$). The results indicated that nitrate concentration in both climates was significantly lower than acceptable level of 50 mg/l ($p < 0.05$). As presented in Fig. 3, the results showed that among 10

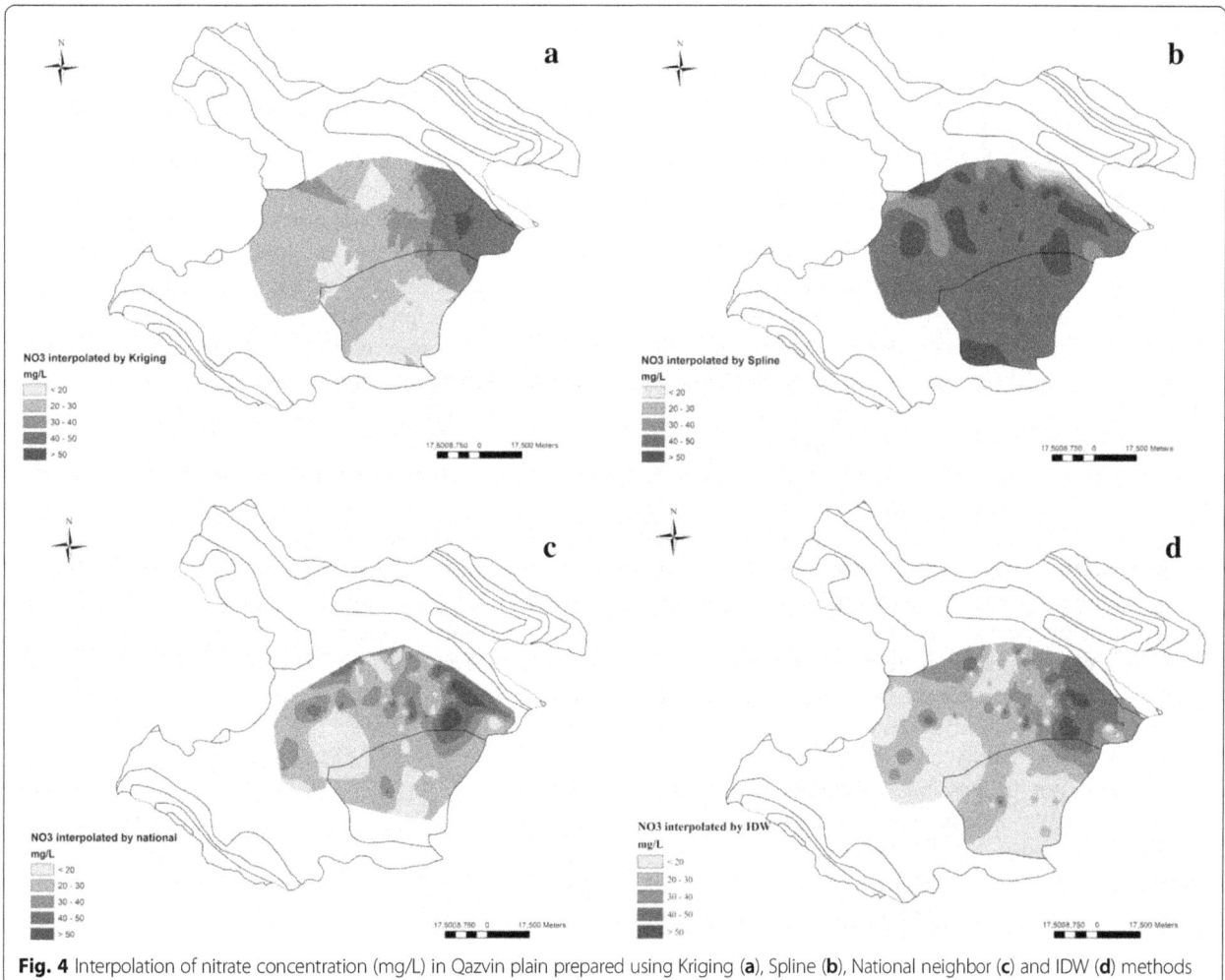

Fig. 4 Interpolation of nitrate concentration (mg/L) in Qazvin plain prepared using Kriging (**a**), Spline (**b**), National neighbor (**c**) and IDW (**d**) methods

samples (4.6%) in semi-arid climate, the nitrate concentration was above the acceptable level of 50 mg/L in drinking water [46]. In addition, no unacceptable level was identified in water samples in arid climate. Williams and et al. (2016) in the assessment of California watershed concluded that nitrate in 42% of samples were highly more than the American standard. They concluded that nitrate was attributed to human activities [13].

Nitrate in different land usages

The results of nitrate concentration in different land-use activities in the study area are shown in Table 5. The results showed that the mean of nitrate concentration has a significant difference in different land usages (F = 3.45, p < 0.05). The detected nitrate concentrations in different areas were ranked as mixed-use > residential > agricultural > steppe. In addition, in 8.82%, 18.52% and 20.08% of samples in agricultural, residential and mixed-use areas, the detected nitrate was above the acceptable level at 50 mg/L. The main reason for the nitrate detected in the residential areas was explained for the neighboring residential wells with cesspools. Also, the consumption of fertilizers in agricultural land was shown in studies as the main reason for the contamination of nitrate. The excessive use of nitrogen fertilizers not only increases groundwater pollution but also causes spatial spread of pollution. It is demonstrated that point and non-point sources, such as leaching chemical fertilizer in agriculture usage and leaching from septic and sewage discharges in residential usage can affect the nitrate concentration in groundwater [47]. Same results were reported by Schaider et al. (2016) which emphasized septic systems were as major sources for the contamination of aquifer with various pollutants such as nitrate [48]. Inverse relation between fertilizers application rate and groundwater nitrate concentration was reported by Mahvi et al. (2005). They explained that the reason was soil characteristics in different regions [49].

The results showed that 27 sample stations were located in residential areas at a distance of 500 m between water wells and residential septic based on CFR method. According to the obtain results, it can be concluded that sample stations located in residential area were more contaminated than agricultural areas. Because the different density of settlements around water wells and lack of access to precision distance between water wells and septic systems in residential land-use, relationship between population density and nitrate concentration in aquifer could not be possible. Schneider et al. (2016) showed that septic systems were a major source of nitrate in groundwater; however, they could not show a significant relationship between density and nitrate concentration in aquifer [48]. The obtained results did not correspond with the study carried out by Kristin et al. (2005) that introduced agricultural land-uses as the main

Table 6 The prediction accuracy of interpolation methods

Methods	Mean relative error	Root mean square error	% RMSE
Kriging	0.21	11.93	0.483
Natural neighbor	0.01	12.47	0.505
IDW	0.10	28.49	1.155
Spline	0.21	5.25	0.212

reason for the groundwater contamination in Nantucket Island, in Massachusetts, America [15]. These results emphasize that the contamination of water resources was caused by man-made factors more than natural factors in Qazvin plain. In site visits, it was revealed that cesspools were the major man-made factor that can affect water quality in residential lands in Qazvin plain.

Agricultural activities were an important factor affecting the quality of water resource, especially in arid and semi-arid areas in the study area, in the same line, more than 40% of sample stations which were located in the agricultural lands had the mean nitrate concentration of 25.27 mg/L. Too much agricultural activities, using fertilizers especially nitrogen ones and high pumping of groundwater were the most effective factors observed in the site visit in the study area. In other studies, the role of agricultural activities in the contamination of groundwater with nitrates was proven (Jafari Malak Abadi. 2002). The highest nitrate concentrations were detected in mixed-use land, which blends residential, commercial, industrial and agricultural uses. Increasing nitrate concentration in this area can be attributed to the synergistic agents. Also, some studies have showed that industrial activities may highly increase nitrate concentration in groundwater resources [50]. The lowest concentration of nitrates was identified in steppe land. This can be due lack of human activities, the soil type and properties such as texture, structure, rainfall, irrigation, evaporation and transpiration [51].

Spatial interpolation of nitrate

Figure 4 shows the spatial distribution of nitrate concentration in the study area. Patterns a, b, c and d were obtained by interpolating the satisfaction levels using Kriging, Spline, Natural Neighbor and IDW methods,

Table 7 Average of hazard indexes using IPNOA method in the study area

Land usage	HI$_i$	Hazard level	Hazard classification
Agricultural	6.11	3	Low
Steppe	3.05	2	Very low
Residential	3.05	2	Very low
Mixed	5.09	2	Very low
Average	4.08	2	Very low

respectively. Their outputs showed that in most of the study areas nitrate concentration was variable between 20 to 30 mg/L. One of the problems in spatial interpolation is errors that can derivate from estimation of unknown values. As presented in Table 6, cross validation in results showed that the interpolation by National Neighbor had the lowest amount of MRE, while the Spline method presented the lowest RMSE. The obtained results were not consistent with other studies that emphasized the kriging method was the most suitable technique for mapping [23]. Acceptable amount for errors is not provided because it depends on the quality of the map being geo-referenced, the quality of the target (base) map and the purpose of the geo referencing.

Potential risk of nitrate

The obtained data from inorganic and organic fertilizers (kg/ha), sludge usage (kg/ha), Soil nitrogen content (%), irrigation systems, rainfall (mm/year), temperature (°C) and tillage patterns were converted into scores including HF_f, HF_m, HF_s, CF_n, CF_i, CF_c, CF_{ap} and HI_i. As presented in Table 7, the highest nitrate hazard due to farming and fertilization was obtained in agricultural areas with HI of 6.11 whereas HI of nitrogen leaking from sewerage in residential areas was 7.12. Obtained results showed that hazard index, as a result of agricultural activities, was ranged from "very low" to "low" which was in accordance with detected and interpolated nitrate concentration in the aquifer. In addition he hazard of nitrate contamination from household (IPNOC) was in very low (class 2). In assessment of IPNOA index, the maximum aquifer vulnerability was appertained to mineral fertilizer ($HF_f = 3$), whereas sewage sludge have no major effect on hazard levels ($HF_m = 1$). Despite of limitations in considering soil characteristics and hydrological structure of subsoil, IPNOA and IPNOC methods can give an evaluation of potential risk of groundwater contamination.

Conclusion

The research was developed to monitoring and optimization an interpolation method to predict the nitrate concentration and assessment of aquifer vulnerability. The mean concentration of nitrate in semi-arid climate was 19.93% more than arid climate and in 4.6% samples the detected nitrate was above the acceptable level of 50 mg/L. In addition the results showed that nitrate concentration was significantly different in different climates and land usages ($p < 0.05$). In addition in 8.82%, 18.52% and 20.08% of samples in agricultural, residential and mixed-use areas, the detected nitrate was above the acceptable level at 50 mg/L. Application of IDW, kriging, National Neighbor and Spline methods represents that National neighbor with the lowest RME and Spline with

the lowest RMSE provide the most accurate estimates of nitrates in the aquifer. The results of parametric models of IPNOA and IPNOC showed that the quality of Qazvin aquifer is mainly influenced by the fertilization in agriculture land and wastewater leakage in residential areas. Obtained results emphasize that conserving practices are highly important since the groundwater resources are limited in arid and semi-arid climates in aquifer of Qazvin plain.

Abbreviations
CF: control factor; GIS: Geographic Information System; HF: hazard factor; HI: hazard index; IDW: Inverse Distance Weighting; RME: Mean Relative Error; RMSE: Root Mean Square Error; RSD: relative standard deviation

Acknowledgements
The authors thank the sponsor of the project.

Funding
This paper was part of a Ms. degree thesis of the first author and has been funded by the Vice Chancellor for Research of Qazvin University of Medical Science under Project No. 93–12-20 32,578.

Authors' contributions
EK was the main investigator, analyzed nitrate in lab and drafted the manuscript. HK supervised and analyzed the data. MME was advisors of the study. All of authors contributed in all steps of study and read and approved the final manuscript.

Competing interests
The authors declare that they have no competing interests.

References
1. Nwobodo TN, Anikwe MAN, Chukwu KE. Assessment of Spatio-temporal variation of groundwater quality in Udi-Ezeagu watershed. Enugu Area Southeastern Nigeria. 2015;3(4):210–7.
2. Luczaj J, Masarik K. Groundwater quantity and quality issues in a water-rich region: examples from Wisconsin, USA. Resources. 2015;5(1):10–3390. https://doi.org/10.3390/resources4020323.
3. Buvaneshwari S, Riotte J, Ruiz L, Sekhar M, Mohan Kumar MS, Sharma AK, et al. High spatial variability of nitrate in the hard rock aquifer of an irrigated catchment: Implications for water resource assessment and vulnerability. Geophysical Research Abstracts, EGU General Assembly, Conference Abstracts. 2016. http://meetingorganizer.copernicus.org/EGU2016/EGU2016-5430-4.pdf.
4. Suthar S, Bishnoi P, Singh S, Mutiyar PK, Nema AK, Patil NS. Nitrate contamination in groundwater of some rural areas of Rajasthan, India. J hazar. Mater. 2009;171(1):189–99.
5. Tizro AT, Voudouris K, Vahedi S. Spatial variation of groundwater quality parameters: a case study from a semiarid region of Iran. Int bull water Resour & Dev. 2014;1:3.
6. Azadi NA, Fallahzadeh RA, Sadeghi S. Dairy wastewater treatment plant in removal of organic pollution: a case study in Sanandaj, Iran. Environ Health Engin and Manag J. 2015;2(2):73–7.
7. Atafar Z, Mesdaghinia A, Nouri J, Homaee M, Yunesian M, Ahmadimoghaddam M, Mahvi AH. Effect of fertilizer application on soil heavy metal concentration. Environ Monit Assess. 2010;160(1):83–9.
8. Uyan M, Cay T, editors. Geostatistical methods for mapping groundwater nitrate concentrations. 3rd International conference on cartography and GIS. Nessebar, Bulgaria. 2010. https://www.cartography-gis.com/pdf/20_Mevlut_Uyan_Turkey_paper.pdf.
9. Anselin L, Getis A. Spatial Statistical Analysis and Geographic Information Systems. In: Anselin L, Rey S. Perspectives on Spatial Data Analysis. Advances in Spatial Science. Berlin: Springer; 2010:35–47.

10. Sahoo S, Jha M. Analysis of spatial variation of groundwater depths using geostatistical modeling. Inter J Applied Engin Res. 2014;9(3):317–22.

11. Shaffer MJ, Delgado JA. Essentials of a national nitrate leaching index assessment tool. J Soil Water Conserv. 2002;57(6):327–35.

12. Pontius RG, Cornell JD, Hall CA. Modeling the spatial pattern of land-use change with GEOMOD2: application and validation for Costa Rica. Agric Ecosyst Environ. 2001;85(1):191–203.

13. Ouedraogo I, Vanclooster MA. Meta-analysis of groundwater contamination by nitrates at the African scale. Hydrol earth. Syst. 2016; doi:10.5194/hess-20-2353-2016.

14. Ouedraogo I, Vanclooster M. A meta-analysis and statistical modelling of nitrates in groundwater at the African scale. Hydrol Earth Syst Sc. 2016;20(6):2353–81.

15. Nas B, Berktay A. Groundwater contamination by nitrates in the city of Konya,(Turkey): a GIS perspective. J Environ Manag. 2006;79(1):30–7.

16. Fallahzadeh RA, Almodaresi SA, Dashti MM, Fattahi A, Sadeghnia M, Eslami H, et al. Zoning of nitrite and nitrate concentration in groundwater using Geografic information system (GIS), case study: drinking water wells in Yazd City. J Geosci Environ Protect. 2016;4(03):91.

17. Childs C. Interpolating surfaces in ArcGIS spatial analyst. ESRI Education Services. http://webapps.fundp.ac.be/geotp/SIG/interpolating.pdf. ArcUser, July-September 2004.

18. Mitas L, Mitasova H, Spatial interpolation. Geographical information systems: principles, techniques, management and applications, vol. 1; 1999. p. 481–92.

19. Merwade V. Effect of spatial trends on interpolation of river bathymetry. J Hydrol. 2009;371(1):169–81.

20. Curtarelli M, Leão J, Ogashawara I, Lorenzzetti J, Stech J. Assessment of spatial interpolation methods to map the bathymetry of an Amazonian hydroelectric reservoir to aid in decision making for water management. Int J Geo-Inf. 2015;4(1):220–35.

21. Okobiah O, Mohanty SP, Kougianos E. Geostatistical-inspired fast layout optimisation of a nano-CMOS thermal sensor. IET Circuits, Devices & Systems. 2013;7(5):253–62. doi:10.1049/iet-cds.2012.0358.

22. Koziel S, Bandler JW. Accurate modeling of microwave devices using kriging-corrected space mapping surrogates. Int J Nume Model: Electron Networks, Devices and Fields. 2012;25(1):1–14.

23. Merwade VM, Maidment DR, Goff JA. Anisotropic considerations while interpolating river channel bathymetry. J Hydrol. 2006;331(3):731–41.

24. Parker SJ, Butler AP, Jackson CR. Seasonal and interannual behaviour of groundwater catchment boundaries in a chalk aquifer. Hydrol Process. 2016;30(1):3–11.

25. Naoum S, Tsanis I. Ranking spatial interpolation techniques using a GIS-based DSS. Global Nest Journal. 2004;6(1):1–20.

26. Kay PJ. Applying geoscience to Australia's most important challenges. Portland GSA Annual Meeting, paper No. 2009:161–11.

27. Azpurua MA, Ramos KD. A comparison of spatial interpolation methods for estimation of average electromagnetic field magnitude. Progress In Electromagnetics Research M. 2010;14:135–45.

28. Bannister R, Kennelly P. Incorporating stream features into groundwater contouring tools within GIS. Groundwater. 2016;54(2):286–90. doi:10.1111/gwat.12332.

29. Capri E, Civita M, Corniello A, Cusimano G, De Maio M, Ducci D, et al. Assessment of nitrate contamination risk: the Italian experience. J Geochem Explor. 2009;102(2):71–86.

30. Sacco D, Offi M, DeMario M, Grignani C. Groundwater nitrate contamination risk assessment: a comparison of parametric systems and simulation modelling. Am J Environ Sci. 2007;3:117–25.

31. Murshed AY, Asmaat WR, Nasher G. Determining nitrates hazard in agricultural area using IPNOA index, in the lower part of Wadi Siham, al-Hodeidah, Yemen. Int J Adv Sci Tech Res. 2016;3(6):267–79.

32. Raziei T, Pereira LS. Estimation of ET_o with Hargreaves–Samani and FAO-PM temperature methods for a wide range of climates in Iran. Agric Water Manag. 2013;121:1–18.

33. Luck M, Wu JA. Gradient analysis of urban landscape pattern: a case study from the phoenix metropolitan region, Arizona, USA. Landsc Ecol. 2002;17(4):327–39.

34. Kresic N. Hydrogeology and groundwater modeling, second edition, CRC press: Taylor and francis group; 2006:333–34.

35. APHA, Standard methods for the examination of water and wastewater: American Public Health Association. https://www.standardmethods.org, 2012.

36. Klesta EJ, Bartz JK, Sparks D, Page A, Helmke P, Loeppert R, et al. Quality assurance and quality control. Methods of soil analysis Part 3-chemical methods. 1996:19–48.

37. Evans JR, Lindsay WM. The management and control of quality. Cincinnati: South-Western; 2002.

38. Liu L, Özsu T. Encyclopedia of Database Systems. Springer US; 2009. p. 3247–51.

39. Hallock D, Ehinger W. Quality assurance monitoring plan. In: Stream ambient water quality monitoring, Environmental Assessment Program, vol. 28. Olympia, Washington: Washington State Departament of Ecology; 2003.

40. Hubbard KG, You J. Sensitivity analysis of quality assurance using the spatial regression approach:a case study of the maximum/minimum air temperature. J Atmos Ocean Tech. 2005;22(10):1520–30.

41. Mitchell P. Guidelines for quality assurance and quality control in surface water quality programs in Alberta. Alberta, Environment; 2006.

42. Li J, Heap AD. A Review of spatial interpolation methods for environmental scientists. Ecological Informatics. 2011;6(3):228–41.

43. Barca E, Passarella G. Spatial evaluation of the risk of groundwater quality degradation. A comparison between disjunctive kriging and geostatistical simulation. Environ Monit Assess. 2008;137(1–3):261–73.

44. Chen X, Ralescu DA. B-spline method of uncertain statistics with applications to estimate travel distance. Journal of Uncertain Systems. 2012;6(4):256–62.

45. Liu X, Wang H, Guo J, Wei J, Ren Z, Zhang J, et al. Spatially-explicit modelling of grassland classes–an improved method of integrating a climate-based classification model with interpolated climate surfaces. Rangel J. 2014;36(2):175–83.

46. WHO, Guidelines for Drinking-water Quality, fourth edition, World Health Organization, 2011. http://www.who.int/water_sanitation_health/publications/2011/dwq_guidelines/en/.

47. Goss M, Barry D, Rudolph D. Contamination in Ontario farmstead domestic wells and its association with agriculture results from drinking water wells. J Contam Hydrol. 1998;32(3):267–93.

48. Schaider LA, Ackerman JM, Rudel RA. Septic systems as sources of organic wastewater compounds in domestic drinking water wells in a shallow sand and gravel aquifer. Sci Total Environ. 2016;547:470–81.

49. Mahvi AH, Nouri J, Babaei A, Nabizadeh R. Agricultural activities impact on groundwater nitrate pollution. Int J Environ Sci Technol. 2005;2(1):41–7.

50. Power J, Schepers J. Nitrate contamination of groundwater in North America. Agric Ecosyst Environ. 1989;26(3–4):165–87.

51. Scanlon BR, Reedy RC, Stonestrom DA, Prudic DE, Dennehy KF. Impact of land use and land cover change on groundwater recharge and quality in the southwestern US. Glob Change Biol. 2005;11(10):1577–93.

17

Indicator bacteria community in seawater and coastal sediment: the Persian Gulf as a case

Vahid Noroozi Karbasdehi[1], Sina Dobaradaran[1,2,3*], Iraj Nabipour[4], Afshin Ostovar[4], Hossein Arfaeinia[5], Amir Vazirizadeh[6], Roghayeh Mirahmadi[1], Mozhgan Keshtkar[1], Fatemeh Faraji Ghasemi[1] and Farzaneh Khalifei[1]

Abstract

Background: The aim of present work was to assess the concentration levels as well as vertical distribution of indicator bacteria including total coliform, fecal coliform, Pseudomonas aeruginosa, and Heterotrophic Plate Count (HPC) in the marine environment (seawater and coastal sediments) and evaluate the correlation between indicator bacteria and some physicochemical parameters of surface sediments as well as seawaters.

Methods: A total number of 48 seawater and sediment samples were taken from 8 stations (each site 6 times with an interval time of 2 weeks) between June and September 2014. Seawater and sediment samples were collected from 30 cm under the surface samples and different sediment depths (0, 4, 7, 10, 15, and 20 cm) respectively, along the Persian Gulf in Bushehr coastal areas.

Results: Based on the results, the average numbers of bacterial indicators including total coliform, fecal coliform, and *Pseudomonas aeruginosa* as well as HPC in seawater samples were 1238.13, 150.87, 8.22 MPN/100 ml and 1742.91 CFU/ml, respectively, and in sediment samples at different depths (from 0-20 cm) varied between 25×10^3 to 51.67×10^3, 5. 63×10^3 to 12.46×10^3, 17.33 to 65 MPN/100 ml, 36×10^3 to 147.5×10^3 CFU/ml, respectively. There were no statistically significant relationships between the indicator organism concentration levels with temperature as well as pH value of seawater. A reverse correlation was found between the level of indicator bacteria and salinity of seawater samples. Also results revealed that the sediment texture influenced abundance of indicators bacteria in sediments. As the concentration levels of indicators bacteria were higher in muddy sediments compare with sandy ones.

Conclusion: Result conducted Bushehr coastal sediments constitute a reservoir of indicator bacteria, therefore, whole of the indicators determined were distinguished to be present in higher levels in sediments than in the overlying seawater. It was concluded that the concentration levels of microbial indicators decreased with depth in sediments. Except total coliform, the numbers of other bacteria including fecal coliform, *Pseudomonas aeruginosa* and HPC bacteria significantly declined in the depth between 10 and 15 cm.

Keywords: Bushehr coastal, Indicator bacteria, Persian Gulf, Sediment texture

* Correspondence: s.dobaradaran@bpums.ac.ir;
sina_dobaradaran@yahoo.com
[1]Department of Environmental Health Engineering, Faculty of Health, Bushehr University of Medical Sciences, Bushehr, Iran
[2]The Persian Gulf Marine Biotechnology Research Center, The Persian Gulf Biomedical Sciences Research Institute, Bushehr University of Medical Sciences, Boostan 19 Alley, Imam Khomeini Street, Bushehr, Iran
Full list of author information is available at the end of the article

Background

Recreational water and beaches are often considered as a place where sensitive individuals may contact with microbial contaminations [1]. These areas are susceptible to fecal contamination from wastewater, septic leachate, farming drainage, livestock and domestic animals, or nonpoint sources of human and animal waste [2]. Fecal contamination in maritime areas can be dangerous to recreational users because feces may contain pathogenic microorganisms that can be ingested and bring intestinal problem [3]. Epidemiological surveys have revealed the positive relationship between fecal contamination at marine beaches and swimming-related diseases [4]. Microbial indicators have been utilized worldwide to show if a water body is contaminated by fecal contamination. Some of these indicators, i.e. fecal coliforms, *E. coli* and *Enterococcus spp.*, are used to monitor the fecal contamination of seawater bodies worldwide [5]. Microbial impairment of drinking, irrigation, or recreational seawaters is generally monitored using concentration levels of fecal indicator bacteria [6]. But other bacteria including *Pseudomonas aeruginosa*, a gram-negative opportunistic human pathogen, and HPC bacteria may also be useful in defining seawater body quality [7]. Exceeding contents of indicator bacteria in seawater and sediments have been related to increased risk of pathogenic microorganism-induced sickness to humans [8]. Various researches have documented an elevated risk of contracting gastrointestinal diseases, skin infections as well as acute respiratory infections after exposure with recreational waters and seawater body with increased concentrations of indicator bacteria [9–12].

Within aquatic systems, it is highlighted that the indicator microorganisms can be highly related to the sediment fraction [13, 14]. This relationship is due to four ecological performances of sediments; 1) provision a place for microbial attachment [15] serves as a favorable organic substance and nutrients for microbes [16], 3) protection from environmental stresses such as sunlight UV [17], protozoan grazing [18], etc., and 4) extracellular polymeric substances (EPS) of bacteria, which enhance sediment flocculation by coagulating and attaching particles together to create a floc matrix and in turn results in an increased downward flux of sediment [19] and accordingly connected bacteria (with potential pathogens) to the sediment [20]. In general, indicator bacteria can stay alive much longer in sediment than in the water column in both freshwater and maritime environments and many studies have confirmed this [21–24]. Pachepsky and Shelton [25] and Brinkmeyer et al. [26] observed significant correlations between fecal indicator bacteria in the seawater column and underlying sediments. They found that the levels of

indicator bacteria in sediments considerably higher than seawaters. Koirala et al. also found that numbers of indicator organisms in sediments are greater than in water samples due to protection behavior of sediments [17]. There are some activities related to sediment resuspension in coastal areas such as commercial or recreational boating and storms that can lead to considerable effects on microbial loads of water [27]. In addition, recreational activity and wave action in the swash zone of the coast can also contribute to resuspension the bacteria from the sediment and consequently may predispose human to health risk [1]. Bushehr province with a long coastline (more than 707 kilometers) along the Persian Gulf and its strategic and geopolitical position, as one of the most important port, is located in southwestern Iran and northern part of the Persian Gulf (Fig. 1). Bushehr as energy capital in Iran is facing with industrial pollution in its marine environment [28] and its region is of special interest for environmental studies [29–33]. The climate is warm and wet in summer and mild in winter. Swimming in the Persian Gulf and playing in the coastline areas are the most important entertainments of people in the Bushehr port. Also, there are sporadic studies on organism's concentrations in the coastal areas of the Persian Gulf but to our best knowledge there is no report yet on comprehensive and baseline data on indicator organisms profile in seawaters and sediments along the Persian Gulf. In this work for the first time in the region of the Persian Gulf, we aimed to (1) assess the concentration levels of indicator bacteria in different depths of surface sediments and seawaters as baseline information in the region (2) mapping and kriging interpolations of the microbial contamination in the surface sediments and seawaters (3) ascertain the correlation between indicator bacteria and some physicochemical parameters of surface sediments as well as seawaters.

Methods
Chemicals and reagents

The employed media including Lactose broth, Brilliant green, EC broth and R$_2$A agar media were prepared from Merck, Germany. The Asparagine broth and Acetamide broth media were prepared from Sigma-Aldrich, USA.

Study area and sediment sample collection

Seawater and sediment sampling were done between June and September 2014. A total number of 48 samples were taken from 8 sampling sites, including TV Park (S1), Skele-Jofreh (S2), Daneshjo Park (S3), Gomrok (S4), Skele-Solh Abad (S5), Skele-Jabri (S6), Bandargah (S7), and Shoghab (S8) (each site 6 times with an interval time of 2 weeks), which were located in the intertidal zone along the Persian Gulf in Bushehr beach zone (Fig. 1). First, seawater samples were collected from 30 cm under the surface (to avoid direct effect of sun ultraviolet radiation on the water surface layer) using a sterile 500 mL glass vessel (for bacterial analyses) and an open-mouthed bottle (for physico-chemical parameters analysis). Next, surface sediments were collected from different depth (including 0, 4, 7, 10, 15 and 20 cm) using an Ekman steel grab sampler $(25 \times 25 \times 25 \text{ cm}^3)$. For each sampling sites, 2–4 sampling points were selected based on the place or shape, and samplings was carried out. The pH of each samples were measured in the place directly after sampling using a U-50 multi-parameter water quality checker (HORIBA, Germany). Samples were placed in a cold box (temperature roughly 4 C and darkness) [34] and directly transported to the lab in coolers on ice within 2 hours.

Media and procedures for bacterial analysis

All microbial indicator analyses including total coliform, fecal coliform, *Pseudomonas aeruginosa* as well as HPC bacteria were done according to standard methods [34]. Lactose broth, EC broth and asparagine broth were employed to determine the most probable number (MPN) per 100 ml of total coliforms, fecal coliforms, as well as *Pseudomonas aeruginosa* respectively, using a five-tube multiple-dilution technique. R_2A agar was used to ascertain the colony forming unit (CFU) per ml of HPC bacteria, using the spread plate technique. In the case of sediment samples, sediments were mixed thoroughly and diluted 1:10 with sterile distilled water (1 g of sediment added into 9 ml of sterile distilled water). This mixture was centrifuged with a speed of 8000 rpm for 1-2 min and then was left to stand for 5-10 min to allow big particles to settle. Sediment suspensions were subsequently processed by the similar procedures as for water samples.

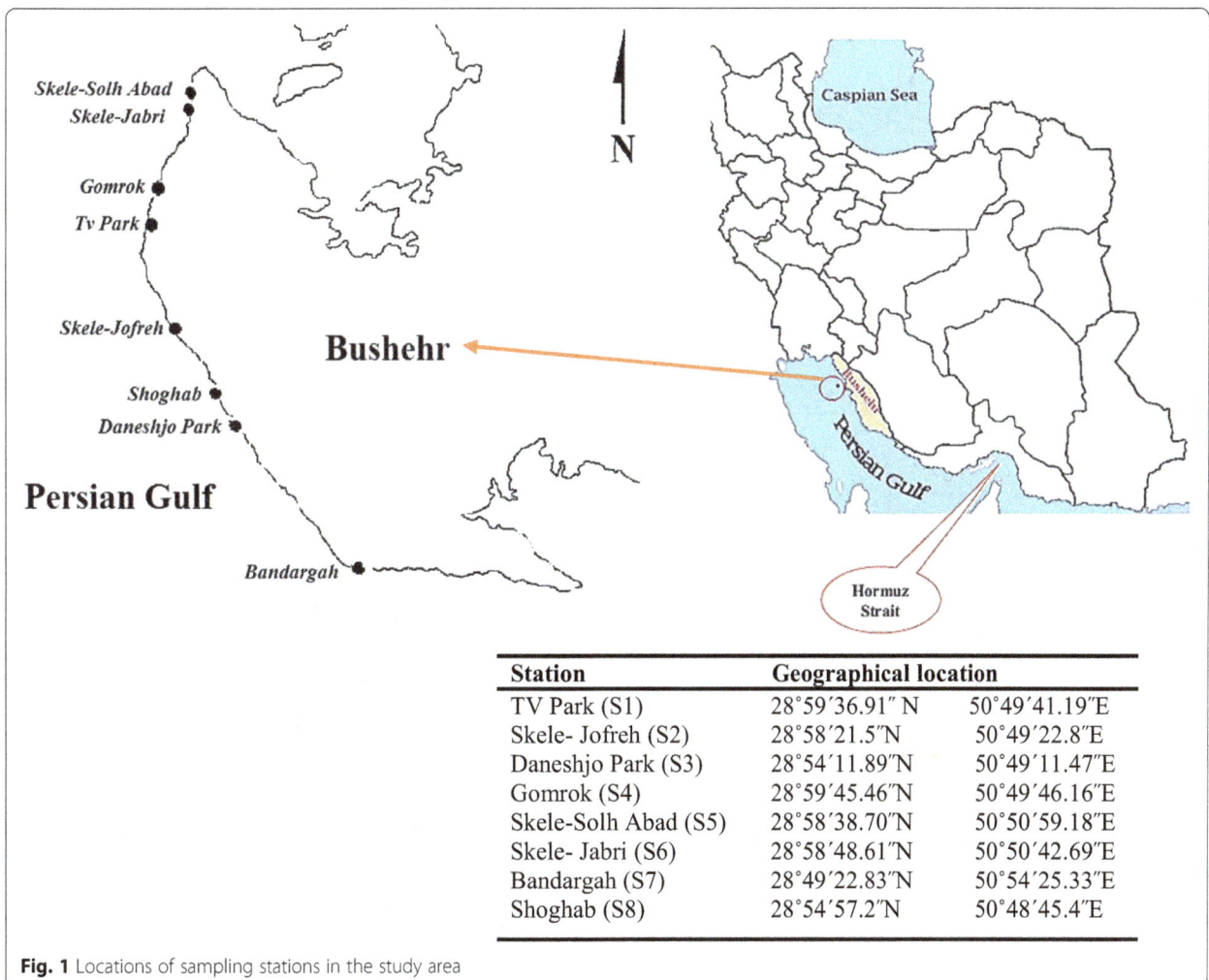

Station	Geographical location	
TV Park (S1)	28°59′36.91″ N	50°49′41.19″E
Skele- Jofreh (S2)	28°58′21.5″N	50°49′22.8″E
Daneshjo Park (S3)	28°54′11.89″N	50°49′11.47″E
Gomrok (S4)	28°59′45.46″N	50°49′46.16″E
Skele-Solh Abad (S5)	28°58′38.70″N	50°50′59.18″E
Skele- Jabri (S6)	28°58′48.61″N	50°50′42.69″E
Bandargah (S7)	28°49′22.83″N	50°54′25.33″E
Shoghab (S8)	28°54′57.2″N	50°48′45.4″E

Fig. 1 Locations of sampling stations in the study area

Grain size analysis of sediment samples

Sediment samples were collected by a grab sampler and coning and quartering technique was used to prepare sediments for grain size analysis [35]. Coning and quartering method involves five steps including: (1) pour the samples onto a flat surface to form a cone (2) flatting the cone (3) divide cone in half (4) divides halves into quarters and discard alternate quarters (5) two quarters are retain and mix together, reform cone and repeats steps until remaining sample be in a correct amount for analysis). After 5 cited steps, sediment sample was kept in a polythene bag labeled with number and location

and transferred to the laboratory by cold box and stored in the freezer at -20 °C until grain size examination according to Buchanan's method [36]. For analysis, sediment dried for 24 hours at 70 °C in Heraeus oven (UT 6420 model). 25 grams of dried sediment of each sample were put in a flask containing 250 ml of distilled water. Then 10 ml of 2.6 grams per liter of sodium hexametaphosphate [Na $(PO_3)_6$] solution was added to the flask contents. After stirring the solution three times, each time for nearly 15 minutes, it was kept in the laboratory for 24 hours. In order to dry, the solution was placed in chines plates and then moved to the oven at 70 °C for

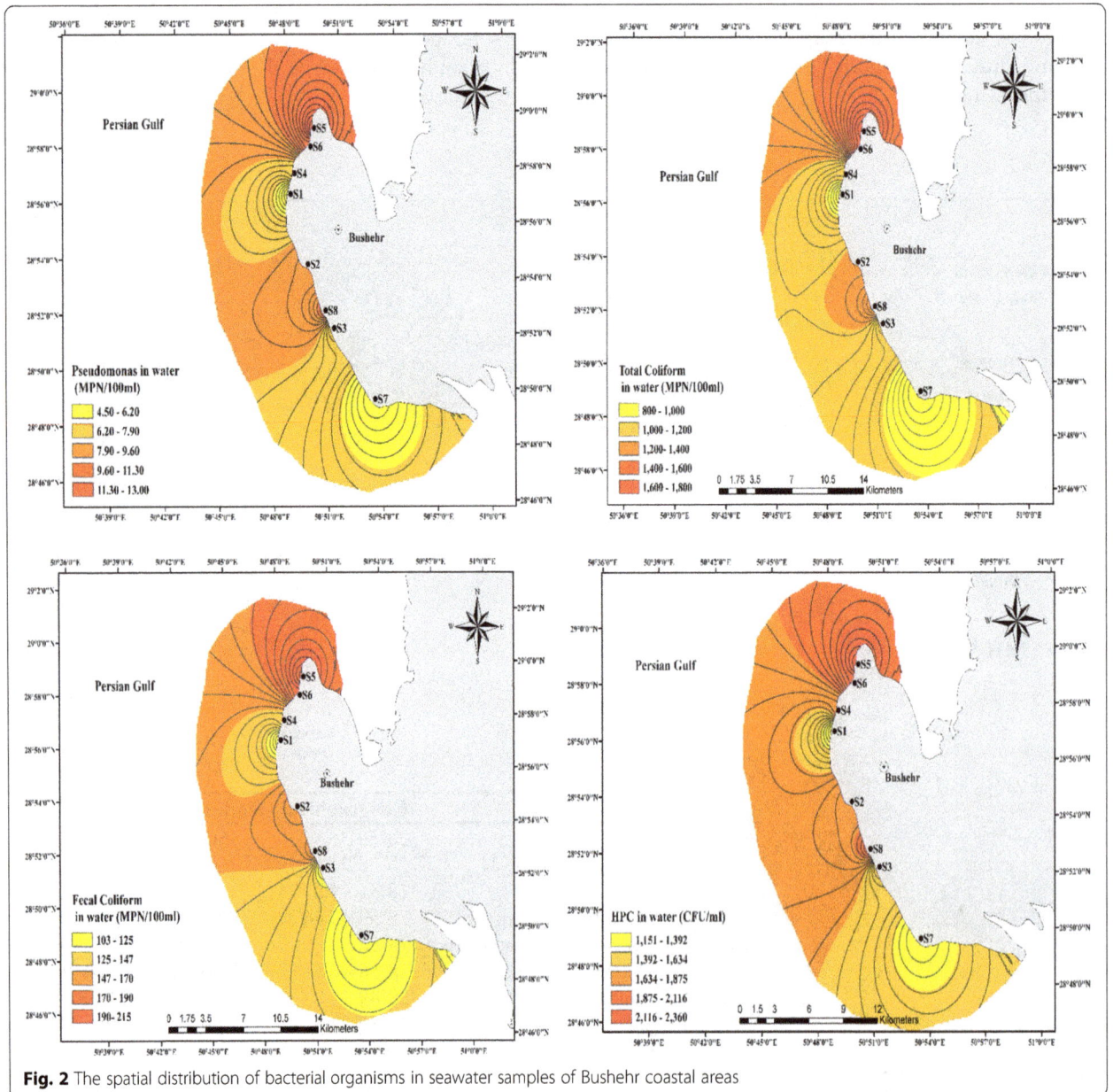

Fig. 2 The spatial distribution of bacterial organisms in seawater samples of Bushehr coastal areas

24 hours. After drying , samples were sieved by shaker Heraeus device (Analysette 3PRO model), and a series of sieves including 4, 2, 1, 0.5, 0.25, 0.125 and 0.0625 mm which climbed on each other, respectively and a

container were placed under them (for weight the particles smaller than 0.0625 mm). Each sample was kept on device for 15 minutes. After that the sediment remaining on each sieve, and sediments of the lower container, weighed carefully with an accuracy of 0.1 mg. By multiplying the weight of each sieve in 4, the percent of its grain size was obtained. Finally as a percentage of dry matter in the sediment, have been reported in 4 different ranges (coarse sand (>500 μm), medium sand (500-250 μm), fine sand (250-125 μm), mud (<125 μm)). So the dominant group determined the types of grain size.

Data analysis
Statistical processing of data was done by using the SPSS version 20 (IBM Corp., USA). The normality of

Table 1 The mean, SD, minimum and maximum values of bacterial indicator levels in seawater samples at different stations (maximum values are expressed as bold italics; minimum values as bold underlined)

Indicator bacteria	Station	N	Mean	Std. Deviation	Minimum	Maximum
Total coliform (MPN/100 ml)	1	6	**800**	112.07	600	920
	2	6	1220	184.93	920	1400
	3	6	1000	161.12	680	1100
	4	6	1050	168.99	750	1200
	5	6	*1800*	328.00	1200	2200
	6	6	1750	339.76	1100	2000
	7	6	885	111.23	660	950
	8	6	1400	247.54	920	1600
Average		48	1238.12	206.7	853.75	1421.25
Fecal coliform (MPN/100 ml)	1	6	**103.67**	15.57	79	120
	2	6	165	8.07	160	180
	3	6	111.17	17.11	94	140
	4	6	143.33	20.15	110	170
	5	6	*213.33*	72.67	110	330
	6	6	193.33	43.29	140	260
	7	6	107.17	16.29	79	130
	8	6	170	38.14	110	230
Average		48	150.875	28.91125	110.25	195
Pseudomonas (MPN/100 ml)	1	6	**4.50**	1.22	3	6
	2	6	8.50	2.07	5	11
	3	6	6.33	2.16	3	9
	4	6	7.33	2.58	4	11
	5	6	*13.00*	2.28	11	17
	6	6	11.00	1.414	9	13
	7	6	4.83	1.169	3	6
	8	6	10.33	1.966	7	12
Average		48	8.227	1.857	5.625	10.625
HPC (CFU/ml)	1	6	**1150.00**	89.443	1000	1250
	2	6	1841.67	724.166	1100	2800
	3	6	1350.00	420.095	940	1850
	4	6	1743.33	406.776	1350	2500
	5	6	*2358.33*	941.497	1500	3500
	6	6	2173.33	344.770	1890	2700
	7	6	1248.33	217.754	970	1610
	8	6	2078.33	302.815	1640	2400
Average		48	1742.91	430.914	1298.75	2326.25

Table 2 The mean, SD, minimum and maximum values of temperature, pH and salinity in seawater samples at various stations (maximum values are expressed as bold italics; minimum values as bold underlined)

	Station	N	Mean	Std. Deviation	Minimum	Maximum
Temperature (°C)	1	6	31.83	0.637	31.04	32.62
	2	6	32.00	0.515	31.34	32.66
	3	6	32.00	0.860	31.02	33.01
	4	6	32.48	0.805	31.62	33.38
	5	6	32.17	0.696	31.38	32.96
	6	6	31.83	0.422	31.04	32.62
	7	6	*32.50*	0.698	31.46	33.16
	8	6	**31.50**	0.724	30.93	32.07
Average		48	32.04	0.67	31.23	32.81
pH	1	6	8.38	0.423	7.79	8.88
	2	6	8.35	0.187	8.10	8.60
	3	6	**8.16**	0.037	8.10	8.20
	4	6	8.26	0.071	8.20	8.40
	5	6	8.41	0.330	7.74	8.60
	6	6	8.21	0.117	7.99	8.30
	7	6	*8.48*	0.179	8.23	8.71
	8	6	8.23	0.267	7.89	8.71
Average		48	8.227	0.201	8.01	8.55
Salinity (ppt)	1	6	*37*	0.648	36.2	38
	2	6	30.1	1.859	28.7	33.6
	3	6	35.1	1.334	33.2	36.5
	4	6	32.4	1.545	31.1	35.4
	5	6	**27.7**	2.126	24.2	29.9
	6	6	28.9	1.898	25.3	30.4
	7	6	35.6	2.459	32.1	37.6
	8	6	29.6	1.283	28.2	31.5
Average		48	32.05	1.644	29.875	34.112

data was checked by the Shapiro-Wilk test before analyzing. Descriptive statistics were applied for presentation of total coliform, fecal coliform, *Pseudomonas aeruginosa* as well as HPC concentration levels in seawater and sediment samples. Parametric Pearson test was applied to establish correlations between association of sediment and seawater characteristics and concentration levels of indicator bacteria. ArcMap 10.2 Geographical Information System (GIS) (ESRI, Redlands, CA) was also used as an appropriate tool for mapping and kriging interpolations of the bacterial concentration levels.

Result and discussion
Seawater
Number of bacterial indicators
The concentration levels of bacterial indicators in seawater samples of Bushehr coastal areas are presented in Fig. 2 and related data is summarized in Table 1. The average numbers of bacterial indicators including total coliform, fecal coliform, and *Pseudomonas aeruginosa* as well as HPC in seawater samples were

1238.13, 150.87, 8.22 MPN/100 ml and 1742.91 CFU/ml, respectively.

The highest average numbers of bacterial indicators in seawater samples between all stations were found in S5 and S6, which were the samples collected from the Solh-Abad and Jabari fishing ports with more frequent anthropogenic activities. Hamilton et al. have previously reported that pollutants from anthropogenic -influenced sources may carry diverse bacteria into the beaches and seawaters [37]. These cited fishing ports were also close to the pisciculture and aquaculture zones for fish and shellfish. Fish and shellfish are effective filter feeders and concentrate high levels of aquatic microorganisms such as fecal coliform and other bacteria in their bodies and may be built up to contents serious to human health. Most incidents of fecal contamination of aquaculture areas are ascribed to anthropogenic origins of such as illegal discharges from boats, inadequately maintained septic systems and run-off from farms [36]. Due to such risks to human health, it is proposed that all fishing ports with high concentration levels of

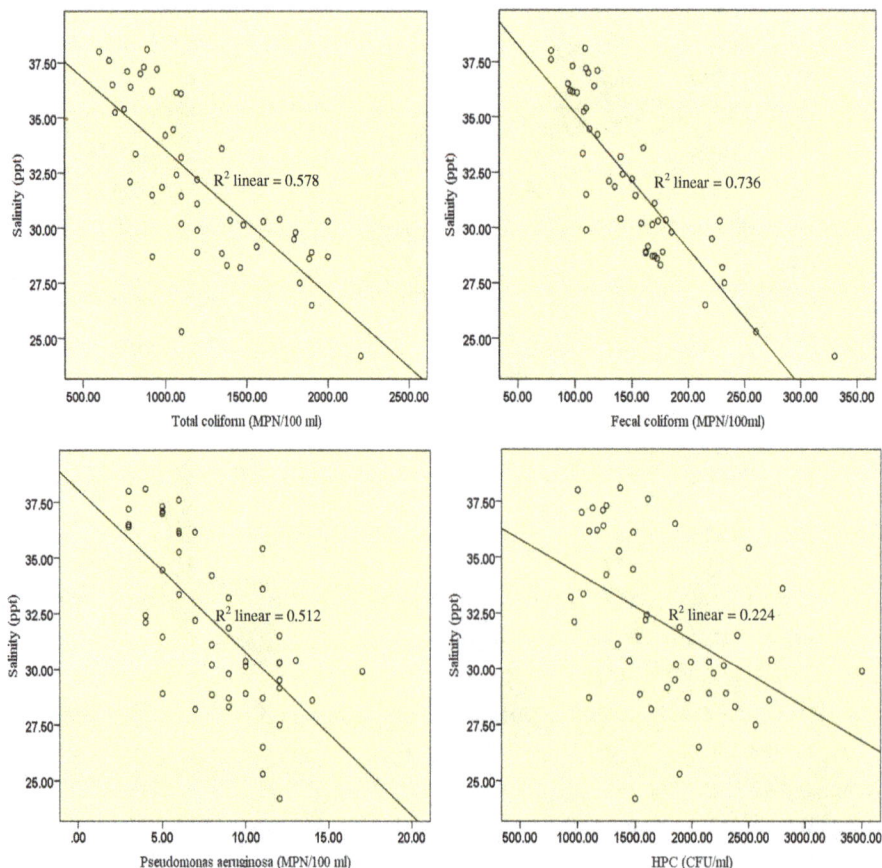

Fig. 3 Pearson correlation between the levels of all examined indicator organisms and salinity of seawater samples

indicators bacteria (like these two stations in Bushehr coastal area) rigidly monitored and categorized by individual state health authorities. Generally, ports 'approved' for fishing must not exceed particular amounts of contamination, especially threshold values of fecal coliforms.

In addition to above mentioned two stations, one of the highest average numbers of bacterial indicators was found in S8 with an average levels of 1400, 170, 10.33 MPN/100 ml and 2078.33 CFU/ml for total coliform, fecal coliform, *Pseudomonas aeruginosa*, as well as HPC respectively. This station is located in coastal area of the Shoghab Park, which is facing wide variety anthropogenic sources such as swimmers feces, throwing up garbage via tourisms, domestic wastewater treatment and disposal practices that may lead to the entrance of high levels of coliform bacteria and enteric human pathogens into the seawater. This area is also frequently utilized for recreational applications such a swimming, recreational fishing and recreational boating. Fecal pollution at swimming

marine area can be dangerous to human health because feces may comprise bacteria, viruses and protozoa and there is possibility of feeding by swimmers which leads to various diseases [1].

Physical and chemical factors

The physical and chemical characteristics of seawater samples from the eight stations were determined and the maximum, minimum, mean and standard deviation values are presented in Table 2. The results of physical parameter measurements in seawater samples showed that these parameters are not tangible changed at different stations. The mean values of temperature and pH in seawater samples were in the range of 31.5-32.5°C and 8.16-8.41 respectively. The salinity in seawater samples was in the range between 27.7 and 37 ppt. Pearson analysis showed that there was a significant inverse correlation between the concentrations levels of total coliforms, fecal coliforms, and *Pseudomonas aeruginosa* and salinity of seawater samples, and a weak inverse correlation between

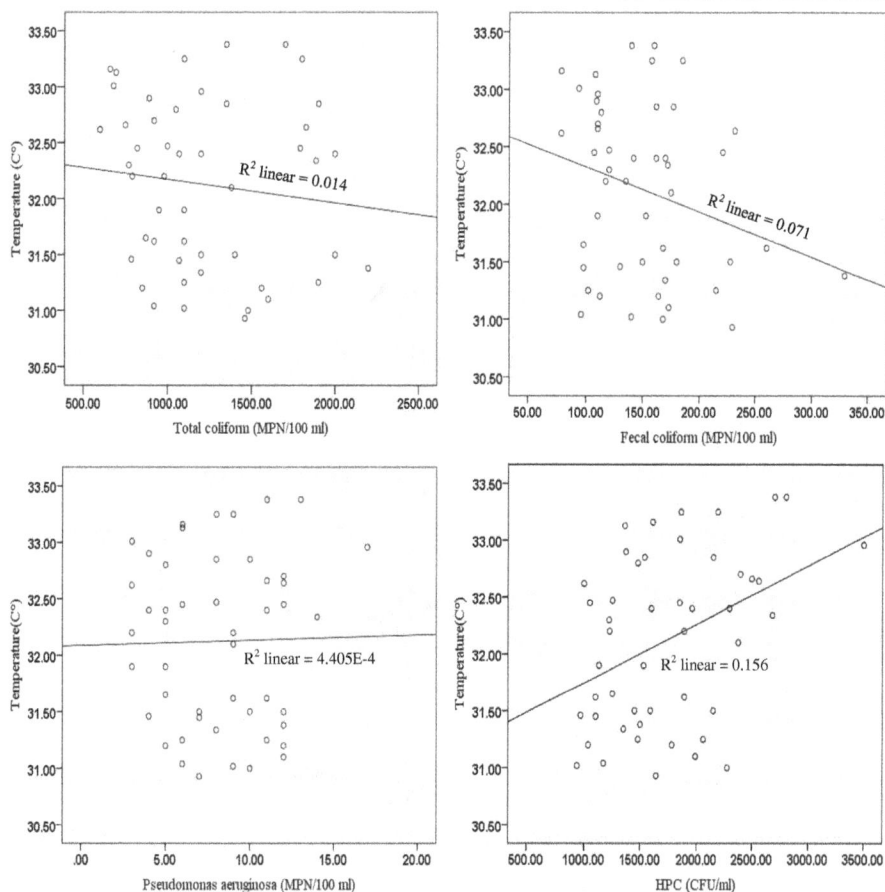

Fig. 4 Pearson correlation between the concentrations of all examined indicator bacteria and temperature of seawater samples

salinity and HPC bacteria (Fig. 3). Therefore, the data revealed that with decreasing salinity, there was a corresponding increase in the values of indicator bacteria in seawater samples. A negative effect of salinity is attributed to specific characteristics of sea water, such as osmotic pressure and the toxicity of inorganic salts [38]. Most studies showed a negative correlation between values of indicator bacteria and salinity. For example, Rozen and Belkini found a reverse correlation between the level of *E. coli* and water salinity in seawater [39]. An inverse correlation between survival of *E. coli* and salinity of water has also been demonstrated by Anderson et al. [40]. In contrast to our studies, Jozić et al showed that there was no statistically significant effect of salinity on the *E. coli* bacteria [41]. Also Pearson analysis showed there were no statistically significant correlations between temperature and pH parameters and indicator organisms (Fig. 4 and Fig. 5). Similarly, Guyal et al. found that the indicator organisms (total coliforms, fecal coliforms, and salmonellae) were no statistically significant relationships with temperature, pH, turbidity, and suspended solids contents of seawater [42].

In another study, Shibata et al. reported that except total coliform, there were no statistically significant relationships between the concentrations levels of enterococci, *Escherichia coli*, fecal coliform, and *C. perfringens* and physical–chemical parameters (rainfall, temperature, pH, and salinity) [43]. But Blaustein et al. [44] and Sampson et al. [45] found that temperature was a major factor in the survival of *E. coli* in surface waters. In another study Placha et al. reported that the survival of *Salmonella typhimurium* and indicator bacteria (coliform and faecal coliform bacteria and faecal streptococci) was considerably affected by temperature [46].

Sediment

Number of bacterial indicators

The concentration levels of bacterial indicators in sediment samples of Bushehr coastal are given in Fig. 6 and associated data is summarized in Table 3. The mean log values of bacterial indicators between eight stations at various depths were compared and are shown in Fig. 7.

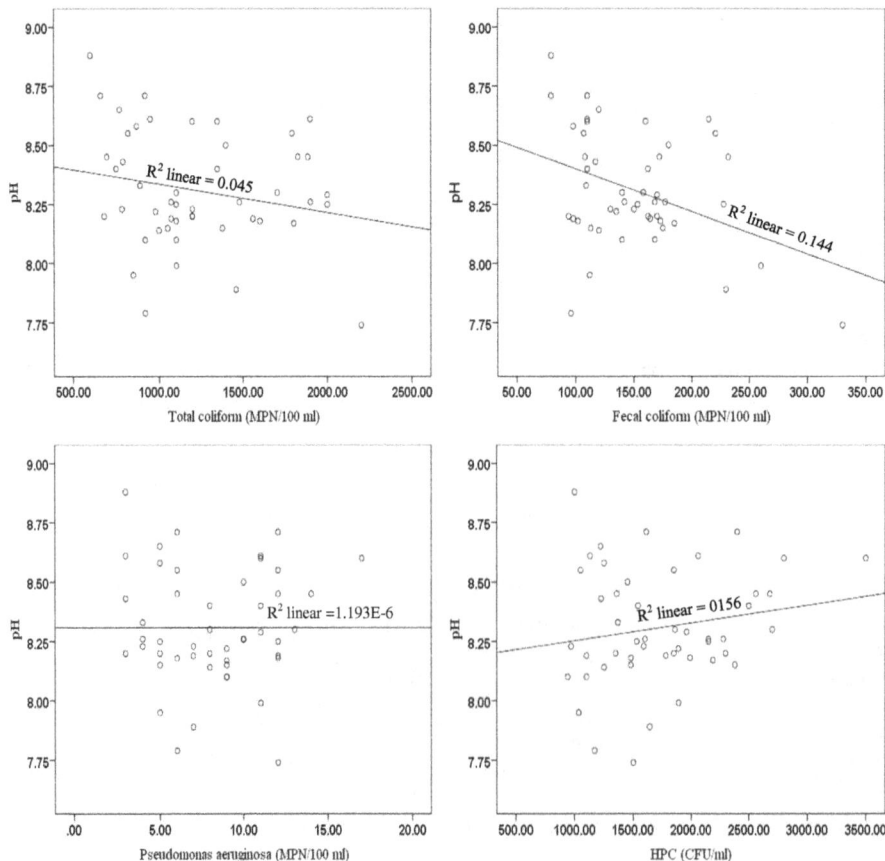

Fig. 5 Pearson correlation between the concentrations of all examined indicator bacteria and pH of seawater samples

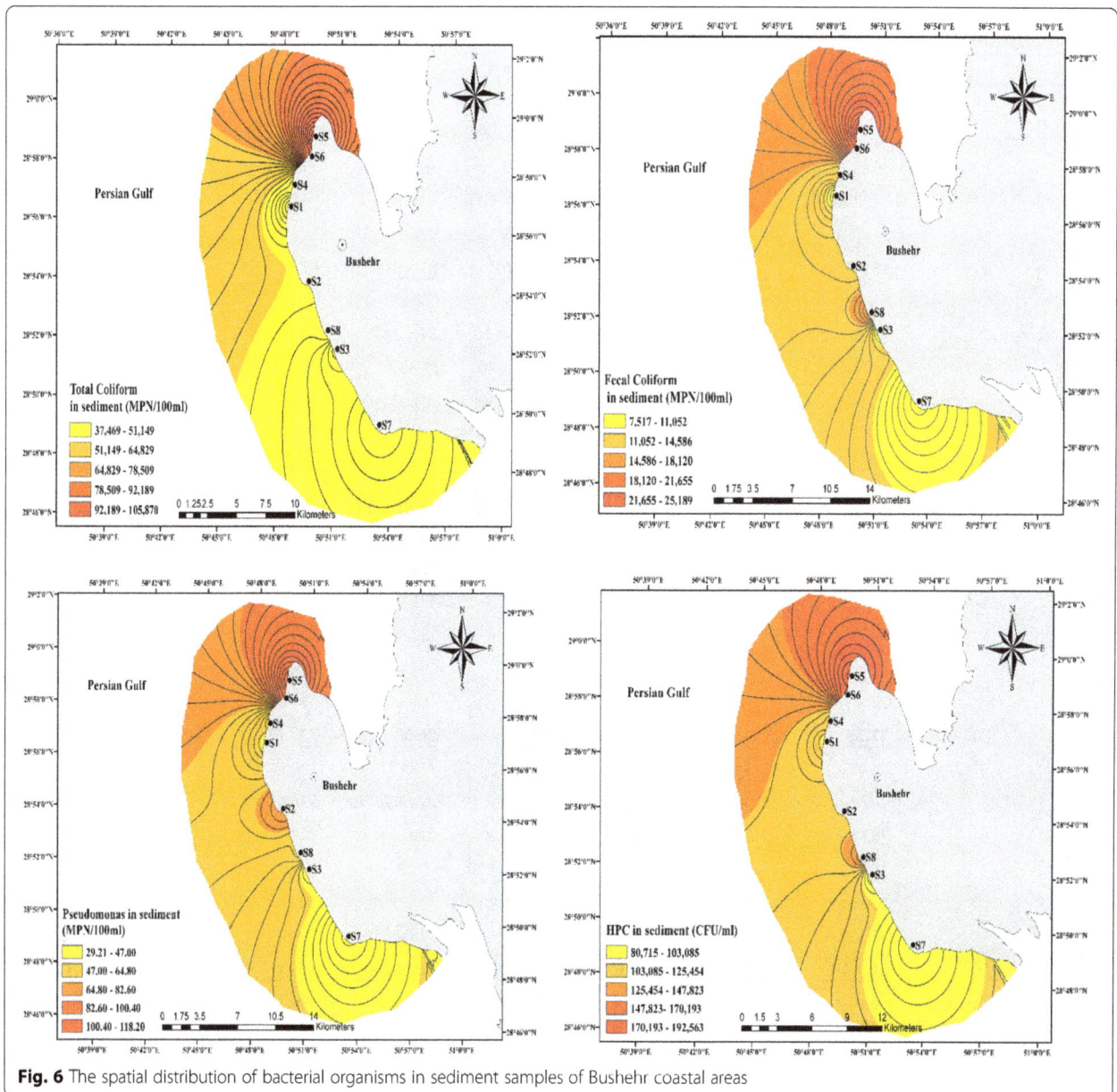

Fig. 6 The spatial distribution of bacterial organisms in sediment samples of Bushehr coastal areas

As seen in Table 3, the levels of total coliforms, fecal coliform, *pseudomonas aeruginosa* as well as HPC in sediment samples at different depths (from 0-20 cm) in stations (1 to 8) varied between 25×10^3 and 51.67×10^3, 5.63×10^3 and 12.46×10^3, 17.33 and 65, 36×10^3 and 147.5×10^3 respectively. The indicators bacteria counts at S5 and S6 sites, (hereinafter termed as the polluted sites) were considerably greater than those at the other six sites. It is also possible that the types of nutrients present in two stations are different from other stations and are more easily utilized by indicators bacteria. The ecological occurrence of indicators organisms in coastal

area sediment has been documented [1, 43, 47]. In our study like many former studies [14, 23, 40], the numbers of indicator organisms in all examined sampling sites were higher for sediment samples compare to seawater samples. This may be due to the fact that sediments probably serve as a suitable environment for bacterial survival [48]. In present work, indicator bacteria were 10 to 100 times higher in sediments than in seawater samples. Crabill et al. reported that the average counts of fecal coliforms in sediment samples were 2200 times higher than the water counts [49]. In another research by Davies and Bavor, they confirmed that sediments may

Table 3 The mean, SD, minimum and maximum values of bacterial indicators levels in various stations at different depths of sediment (maximum values are expressed as bold italics; minimum values as bold underlined)

	Depth (cm)	Station 1	Station 2	Station 3	Station 4	Station 5	Station 6	Station 7	Station 8
Total coliform (MPN/100 ml)	0	*51.67 × 10³*	*75.83 × 10³*	*52.17 × 10³*	*71.67 × 10³*	*138.33 × 10³*	*131.67 × 10³*	*52.17 × 10³*	*75.83 × 10³*
	4	46 × 10³	65.66 × 10³	45.33 × 10³	60.5 × 10³	125.67 × 10³	118.33 × 10³	44.5 × 10³	62.83 × 10³
	7	38.83 × 10³	56.17 × 10³	41.33 × 10³	52.16 × 10³	107.17 × 10³	102 × 10³	40.83 × 10³	56.83 × 10³
	10	34.17 × 10³	46.33 × 10³	36.83 × 10³	43.17 × 10³	93.5 × 10³	86.5 × 10³	36.33 × 10³	48.5 × 10³
	15	29 × 10³	34 × 10³	29.33 × 10³	30.33 × 10³	89.66 × 10³	76 × 10³	29 × 10³	33 × 10³
	20	**25 × 10³**	**25.33 × 10³**	**25.67 × 10³**	**22 × 10³**	**81 × 10³**	**67.66 × 10³**	**25 × 10³**	**26 × 10³**
	Average	37.44 × 10³	50.55 × 10³	38.44 × 10³	46.63 × 10³	105.88 × 10³	97.02 × 10³	37.97 × 10³	50.5 × 10³
Fecal coliform (MPN/100 ml)	0	*12.46 × 10³*	*18.66 × 10³*	*13.73 × 10³*	*20.16 × 10³*	*31 × 10³*	*27.17 × 10³*	*11.23 × 10³*	*25.66 × 10³*
	4	11.33 × 10³	17.16 × 10³	11.9 × 10³	17.66 × 10³	29.33 × 10³	24.66 × 10³	9.8 × 10³	22.33 × 10³
	7	10.21 × 10³	14.66 × 10³	10.8 × 10³	15.73 × 10³	26.83 × 10³	23 × 10³	8.2 × 10³	19.67 × 10³
	10	9.26 × 10³	11.33 × 10³	9.56 × 10³	12.86 × 10³	24 × 10³	21.67 × 10³	7.2 × 10³	17.7 × 10³
	15	6.76	10.5 × 10³	6.83 × 10³	6.83 × 10³	21.67 × 10³	20 × 10³	5.26 × 10³	10.8 × 10³
	20	**5.63 × 10³**	**8.66 × 10³**	**5.26 × 10³**	**5.26 × 10³**	**18.33 × 10³**	**17.33 × 10³**	**3.76 × 10³**	**9 × 10³**
	Average	9.28 × 10³	13.5 × 10³	9.68 × 10³	13.09 × 10³	25.19 × 10³	22.3 × 10³	7.52 × 10³	17.42 × 10³
Pseudomonas (MPN/100 ml)	0	*65*	*108.83*	*62.6*	*73.5*	*160*	*156.6*	*39.5*	*87.6*
	4	58.83	97	52.3	62.3	135	125	37	72.6
	7	57.33	82.6	47.6	51.5	122.3	108	33.16	66.3
	10	46.1	70.8	40.5	43.3	104.5	91.6	27.6	57.16
	15	22	42.6	23.66	24	101	84	22	27
	20	**17.33**	**29.3**	**18.6**	**17.33**	**86.33**	**73**	**16**	**20. 6**
	Average	44.43	71.85	40.88	45.32	118.19	106.36	29.21	62.13
HPC (CFU/ml)	0	*147.5 × 10³*	*159.42 × 10³*	*131.67 × 10³*	*148.35 × 10³*	*224.83 × 10³*	*230.66 × 10³*	*110.5 × 10³*	*176.83 × 10³*
	4	132.5 × 10³	148 × 10³	117.33 × 10³	138 × 10³	210.17 × 10³	204.67 × 10³	102.8 × 10³	167.33 × 10³
	7	119.42 × 10³	135.5 × 10³	104.17 × 10³	126.17 × 10³	199.83 × 10³	194.5 × 10³	95.62 × 10³	161.17 × 10³
	10	109.63 × 10³	119.92 × 10³	88.83 × 10³	117.83 × 10³	186 × 10³	180.67 × 10³	89.04 × 10³	147.17 × 10³
	15	50.33 × 10³	96.33 × 10³	53 × 10³	78.66 × 10³	175 × 10³	161.67 × 10³	50.33 × 10³	98.33 × 10³
	20	**36 × 10³**	**80.67 × 10³**	**40 × 10³**	**67.66 × 10³**	**159.67 × 10³**	**149 × 10³**	**36 × 10³**	**90.33 × 10³**
	Average	99.23 × 10³	123.3 × 10³	89.17 × 10³	112.78 × 10³	192.58 × 10³	186.86 × 10³	80.71 × 10³	140.19 × 10³

have 10 to 10,000 times higher amounts of fecal indicator organisms compare with the overlying seawater [22]. This may be due to the adsorption and sedimentation tend to remove organisms from suspension and concentrate them in bottom sediments [39]. But many studies reviewed by Pachepsky and Shelton [25], Brinkmeyer et al. [26] reported no correlations between indicators bacteria in the water column and underlying sediments.

The general decay of indicators number in sediment

Our results showed that the concentration levels of bacterial indicators decreased with depth (Table 4). This may be due to the death and inactivation of bacteria with depth. Because sediments is a natural filter that ensnares environmental particulates, organic substance and

microorganisms [50, 51]. Our study are in accordance with former studies reported by Brinkmeyer et al. [26], Alm et al. [1], Haller et al. [21], and Pachepsky and Shelton [25].

As shown in Fig. 8, general decay pattern for examined bacterial indicators in sediments of Bushehr coastal areas are presented. These pattern can be useful to anticipate indicator bacteria numbers in marine environment sediments considering sediment texture and grain size. The highest percent decline of total coliform was found in depth between 0 and 4 cm, but the highest percent decline of fecal coliform, *Pseudomonas aeruginosa* and HPC bacteria were found in depth between 10 and 15 cm (Fig. 9). Haller et al. found that the concentration levels of bacterial indicators decreased with depth. They

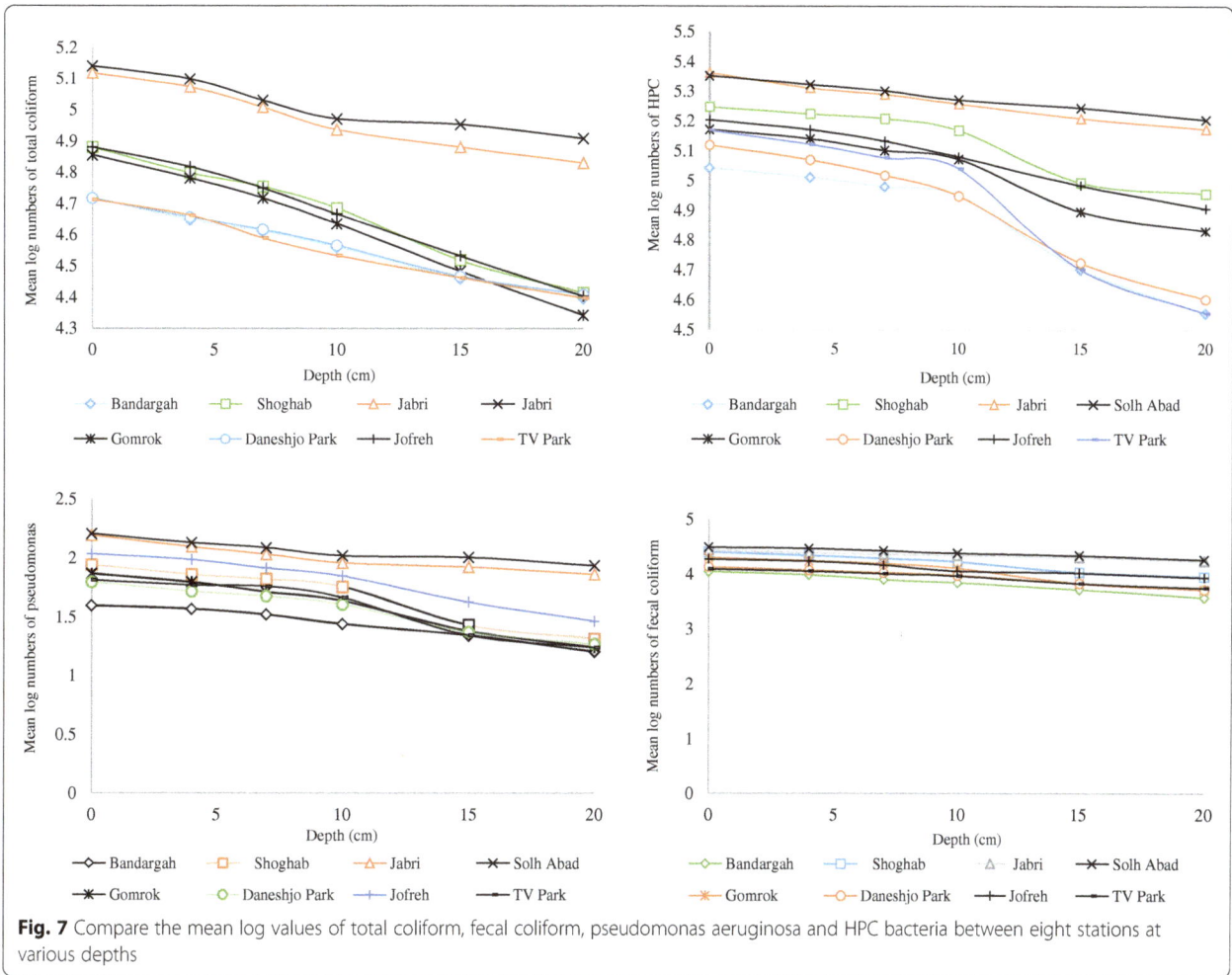

Fig. 7 Compare the mean log values of total coliform, fecal coliform, pseudomonas aeruginosa and HPC bacteria between eight stations at various depths

reported this can be due to decrease in organic matter content, they found out that organic matter content in the sediments samples reduced with depth, from 25% at 0–2 cm to 15% at 10 cm depth [21]. Brinkmeyer et al. detected a considerable reduction in the contents of indicator bacteria from the top 1 cm (10^4 to 10^5) to the deeper 15, 30, and 60 cm horizons (10^2 to 10^3) [26].

Grain size analysis
Sediment grain size analysis in Bushehr coastal areas along the Persian Gulf are shown in Table 5. As seen, sediments in five stations (Skele-Jofreh (S2), Gomrok (S4), Skele-Solh Abad (S5), Skele-Jabri (S6), and Shoghab (S8)) have a texture of silt – clay (mud) and a diameter less than 125 μm, but sediments in three stations (TV Park (S1), Daneshjo Park (S3), and Bandargah (S7)), have a texture of fine sand and a diameter between 125-250 μm. Our results revealed that the concentrations of indicators organisms were greater in muddy sediments compare with sandy ones. It is a fact

that fine-grained sediment due to the higher surface area to volume ratio has more potential to tend higher concentration levels of bacterial indicators [52–54]. In a study by Lang and Smith, it was concluded that the clay amount of soils is particularly important regarding to abundance of indicator bacteria, as clay particles protected bacteria against predators and higher availability of substrates and moisture than sand particles [55]. A higher level of *E. coli* in sites with higher percent of clay and silt and lower sand has been reported in former studies [56, 57]. Burton et al. found that *E. coli* survival was higher in sediments containing at least 25% clay (particles less than 2 mm in diameter), presumably due to enhanced attachment to the finer sediment particles [58]. In contrast to our study, Cinotto et al. reported that *E. coli* survival was higher in sediments with mainly big particles (ranging in size from 125 to 500 mm), possibly due to bigger sediment particles facilitate increased permeability and accessibility of nutrients [59].

Table 4 Mean concentration levels and mean log numbers of indicators bacteria (the average of eight stations) in sediment samples at various depths

Depth (cm)		N	Mean	Log number	Std. Deviation	Minimum	Maximum
Total coliform (MPN/100 ml)	0	48	81.17×10^3	4.909	34.65×10^3	43×10^3	180×10^3
	4	48	71.34×10^3	4.853	32.58×10^3	34×10^3	170×10^3
	7	48	61.92×10^3	4.792	27.54×10^3	33×10^3	140×10^3
	10	48	53.17×10^3	4.725	23.99×10^3	31×10^3	130×10^3
	15	48	43.78×10^3	4.641	25.32×10^3	26×10^3	120×10^3
	20	48	37.21×10^3	4.571	23.76×10^3	21×10^3	110×10^3
Fecal coliform (MPN/100 ml)	0	48	20×10^3	4.301	8.17×10^3	9×10^3	35×10^3
	4	48	18×10^3	4.255	7.68×10^3	7.9×10^3	33×10^3
	7	48	16.1×10^3	4.207	7.189×10^3	6.3×10^3	31×10^3
	10	48	14.1×10^3	4.149	6.77×10^3	4.9×10^3	27×10^3
	15	48	11.08×10^3	4.044	6.18×10^3	4.6×10^3	22×10^3
	20	48	9.16×10^3	3.961	5.59×10^3	3.1×10^3	21×10^3
Pseudomonas (MPN/100 ml)	0	48	94.23	1.974	49.656	26	240
	4	48	80.02	1.903	38.793	23	180
	7	48	71.13	1.852	34.742	22	170
	10	48	60.23	1.779	29.826	21	140
	15	48	43.29	1.636	31.403	17	130
	20	48	34.83	1.541	27.864	14	110
HPC (CFU/ml)	0	48	166.2×10^3	5.220	51.39×10^3	83×10^3	35×10^3
	4	48	152.6×10^3	5.183	45.32×10^3	76×10^3	241×10^3
	7	48	142×10^3	5.152	45.48×10^3	65×10^3	232×10^3
	10	48	129.8×10^3	5.113	44.38×10^3	56×10^3	199×10^3
	15	48	95.4×10^3	4.979	47.34×10^3	43×10^3	182×10^3
	20	48	82.4×10^3	4.916	47.62×10^3	26×10^3	172×10^3

Conclusion

The present study was the first attempt to survey indicators bacteria profiles in seawater and sediment of coastal area along the Persian Gulf. Our study revealed that there was a reverse correlation between the levels of indicator bacteria and salinity of seawater samples. The levels of indicator organisms were 10 to 100 times higher in sediments than in seawater samples. The concentration levels of indicators bacteria were higher in muddy sediments compare with sandy

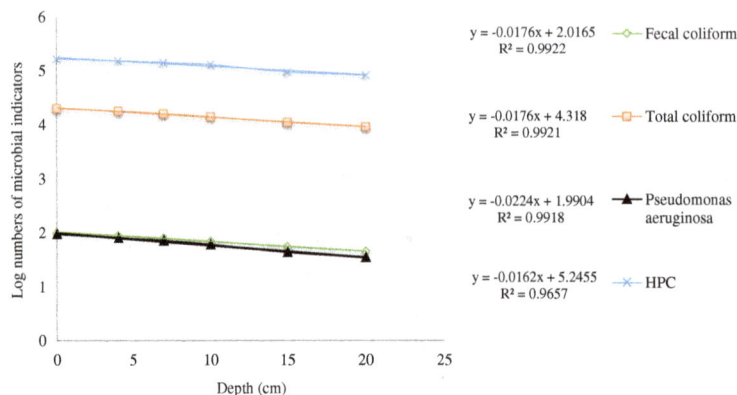

Fig. 8 General decay pattern for the microbial indicators in marine sediment

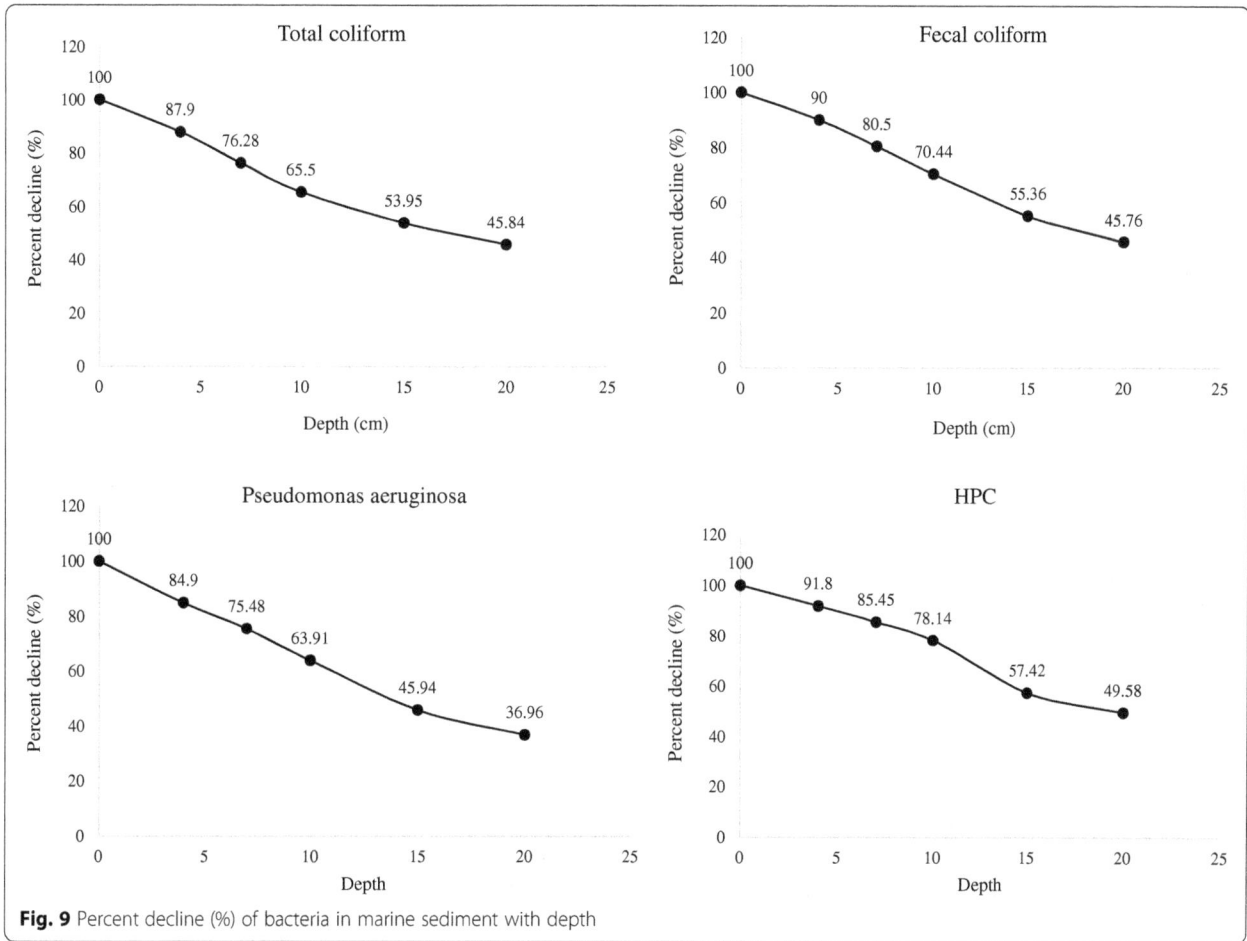

Fig. 9 Percent decline (%) of bacteria in marine sediment with depth

ones. Our results revealed that the concentration levels of bacterial indicators decreased with depth. Our presented models in this study can be useful models to anticipate indicator bacteria numbers in marine environment sediments considering sediment texture and grain size. The orders of indicator bacteria numbers in seawater and sediment samples were HPC > total coliform > fecal coliform > *pseudomonas aeruginosa*. The concentration levels of indicator bacteria in sampling stations showed that coastal areas along the Persian Gulf are facing a wide variety of anthropogenic sources. Finally monitoring and mitigation measures of marine environment particularly in places with entertainments uses are greatly suggested.

Table 5 Sediment grain size analysis in Bushehr port coasts at various stations

Station	Mesh (mm)								Texture
	4	2	1	0.5	0.25	0.125	0.063	<0.063	
1	0.934	3.728	5.756	8.938	35.327	38.158	6.794	0.437	Fine sand
2	0.283	1.412	1.226	3.796	8.443	29.172	53.848	1.820	Silt – clay
3	1.110	4.084	4.809	10.844	16.576	42.916	14.540	5.120	Fine sand
4	6.892	3.325	2.697	2.196	2.777	18.742	58.756	4.613	Silt – clay
5	1.418	1.141	0.710	0.411	0.729	31.861	64.316	0.740	Silt – clay
6	0.568	0.561	1.02	1.694	4.372	21.813	67.44	2.525	Silt – clay
7	0.221	0.132	0.676	10.328	17.910	37.431	30.254	3.046	Fine sand
8	0	0.242	0.418	0.951	4.716	39.661	51.831	2.18	Silt – clay

Abbreviations

GIS: Geographical Information System; HPC: Heterotrophic Plate Count

Acknowledgements

The authors are grateful to the Bushehr University of Medical Sciences for their financial support. This project was partly supported by Iran National Science Foundation (Research Chair Award No. 95/INSF/44913).

Funding

This study was performed as a master thesis in Environmental Health Engineering founded by Bushehr University of Medical Science.

Authors' contributions

VNK was the main investigator, collected the samples and data, and wrote the first draft of the manuscript. SD was the supervisor of study in all steps, edited and polished the final version of manuscript. IN has guided this manuscript. AO performed the statistical analysis. AV and HA were advisors of the study. RM, MK, FFG and FK conducted the experiments. All authors read and approved the final version of manuscript.

Competing interests

The authors declare that they have no competing interests.

Author details

[1]Department of Environmental Health Engineering, Faculty of Health, Bushehr University of Medical Sciences, Bushehr, Iran. [2]The Persian Gulf Marine Biotechnology Research Center, The Persian Gulf Biomedical Sciences Research Institute, Bushehr University of Medical Sciences, Boostan 19 Alley, Imam Khomeini Street, Bushehr, Iran. [3]Systems Environmental Health, Oil, Gas and Energy Research Center, The Persian Gulf Biomedical Sciences Research Institute, Bushehr University of Medical Sciences, Bushehr, Iran. [4]The Persian Gulf Tropical Medicine Research Center, The Persian Gulf Biomedical Sciences Research Institute, Bushehr University of Medical Sciences, Bushehr, Iran. [5]Environmental Health Department, School of Public Health, Iran University of Medical Sciences, Tehran, Iran. [6]The Persian Gulf Studies and Researches Center Marine Biotechnology Department, Persian Gulf University, Bushehr, Iran.

References

1. Alm EW, Burke J, Spain A. Fecal indicator bacteria are abundant in wet sand at freshwater beaches. Water Res. 2003;37(16):3978–82.
2. Gerba CP. Assessment of enteric pathogen shedding by bathers during recreational activity and its impact on water quality. Quant Microbiol. 2000; 2(1):55–68.
3. Lee SH, Levy DA, Craun GF, Beach MJ, Calderon RL. Surveillance for waterborne-disease outbreaks–United States, 1999-2000. MMWR Surveill Summ (Washington, DC: 2002). 2002; 51(8): 1-47
4. Wade TJ, Pai N, Eisenberg JN, Colford Jr JM. Do US Environmental Protection Agency water quality guidelines for recreational waters prevent gastrointestinal illness? A systematic review and meta-analysis. Environ Health Perspect. 2003;111(8):1102.
5. Feng P, Weagant SD, Grant MA, Burkhardt W, Shellfish M, Water B. BAM: Enumeration of Escherichia coli and the Coliform Bacteria. Bacteriological analytical manual. 2002;13–19.
6. United States Environmental Protection Agency. The National Seawater Quality Inventory, EPA-841-R-02-001.USEPA Office of Seawater, Washington, DC. USA: Environmental Protection Agency; 2002.
7. Carter JT, Rice EW, Buchberger SG, Lee Y. Relationships between levels of heterotrophic bacteria and water quality parameters in a drinking water distribution system. Water Res. 2000;34(5):1495–502.
8. Donovan E, Unice K, Roberts JD, Harris M, Finley B. Risk of gastrointestinal disease associated with exposure to pathogens in the water of the Lower Passaic River. Appl Environ Microbiol. 2008;74(4):994–1003.
9. Shuval H. Estimating the global burden of thalassogenic diseases: human infectious diseases caused by wastewater pollution of the marine environment. J Water Health. 2003;1(2):53–64.
10. Colford Jr JM, Wade TJ, Schiff KC, Wright CC, Griffith JF, Sandhu SK, Burns S, Sobsey M, Lovelace G, Weisberg SB. Water quality indicators and the risk of illness at beaches with nonpoint sources of fecal contamination. Epidemiology. 2007;18(1):27–35.
11. Fleisher JM, Fleming LE, Solo-Gabriele HM, Kish JK, Sinigalliano CD, Plano L, Elmir SM, Wang JD, Withum K, Shibata T, Gidley ML. The BEACHES Study: health effects and exposures from non-point source microbial contaminants in subtropical recreational marine waters. Int J Epidemiol. 2010;39(5):1291–8.
12. Benham BL, Baffaut C, Zeckoski RW, Mankin KR, Pachepsky YA, Sadeghi AM, Brannan KM, Soupir ML, Habersack MJ. Modeling bacteria fate and transport in watersheds to support TMDLs. Biol Eng Trans. 2006;49(4):987–1002.
13. Droppo IG, Krishnappan BG, Liss SN, Marvin C, Biberhofer J. Modelling sediment-microbial dynamics in the South Nation River, Ontario, Canada: Towards the prediction of aquatic and human health risk. Water Res. 2011; 45(12):3797–809.
14. Rehmann CR, Soupir ML. Importance of interactions between the water column and the sediment for microbial concentrations in streams. Water Res. 2009;43(18):4579–89.
15. Cho KH, Pachepsky YA, Kim JH, Guber AK, Shelton DR, Rowland R. Release of Escherichia coli from the bottom sediment in a first-order creek: Experiment and reach-specific modeling. J Hydrol (Amst). 2010;391(3):322–32.
16. Jamieson RC, Joy DM, Lee H, Kostaschuk R, Gordon RJ. Resuspension of Sediment-Associated in a Natural Stream. J Environ Qual. 2005;34(12):581–9.
17. Koirala SR, Gentry RW, Perfect E, Schwartz JS, Sayler GS. Temporal variation and persistence of bacteria in streams. J Environ Qual. 2008;37(4):1559–66.
18. Decamp O, Warren A. Investigation of Escherichia coli removal in various designs of subsurface flow wetlands used for wastewater treatment. Ecol Eng. 2000;14(3):293–9.
19. Droppo IG. Structural controls on floc strength and transport. Can J Civ Eng. 2004;31(4):569–78.
20. Droppo IG, King K, Tirado SM, Sousa A, Wolfaardt G, Liss SN, Warren LA. Assessing riverine sediment—pathogen dynamics: implications for the management of aquatic and human health risk. IAHS-AISH publication. 2010; 245-250
21. Haller L, Poté J, Loizeau JL, Wildi W. Distribution and survival of faecal indicator bacteria in the sediments of the Bay of Vidy, Lake Geneva, Switzerland. Ecol Indic. 2009;9(3):540–7.
22. Davies CM, Bavor HJ. The fate of stormwater-associated bacteria in constructed wetland and water pollution control pond systems. J Appl Microbiol. 2000;89(2):349–60.
23. Smith J, Edwards J, Hilger H, Steck TR. Sediment can be a reservoir for coliform bacteria released into streams. J Gen Appl Microbiol. 2008;54:173–9.
24. Byappanahalli MN, Nevers MB, Korajkic A, Staley ZR, Harwood VJ. Enterococci in the environment. Microbiol Rev. 2012;76(4):685–706.
25. Pachepsky YA, Shelton DR. Escherichia coli and fecal coliforms in freshwater and estuarine sediments. Crit Rev Environ Sci Technol. 2011;41(12):1067–110.
26. Brinkmeyer R, Amon RM, Schwarz JR, Saxton T, Roberts D, Harrison S, Ellis N, Fox J, DiGuardi K, Hochman M, Duan S. Distribution and persistence of Escherichia coli and Enterococci in stream bed and bank sediments from two urban streams in Houston, TX. Sci Total Environ. 2015;502:650–8.
27. Obiri-Danso K, Jones K. Intertidal sediments as reservoirs for hippurate negative campylobacters, salmonellae and faecal indicators in three EU recognised bathing waters in North West England. Water Res. 2000;34(2):519–27.
28. Arfaeinia H, Nabipour I, Ostovar A, Asadgol Z, Abuee E, Keshtkar M, Dobaradaran S. Assessment of sediment quality based on acid-volatile sulfide and simultaneously extracted metals in heavily industrialized area of Asaluyeh, Persian Gulf: concentrations, spatial distributions, and sediment bioavailability/toxicity. Environ Sci Pollut Res Int. 2016;9:1–20.
29. Dobaradaran S, Nabipour I, Saeedi R, Ostovar A, Khorsand M, Khajeahmadi N, Hayati R, Keshtkar M. Association of metals (Cd, Fe, As, Ni, Cu, Zn and Mn) with cigarette butts in northern part of the Persian Gulf. Tob Control. 2016.
30. Nabipour I, Dobaradaran S. Fluoride and chloride levels in the Bushehr coastal seawater of the Persian Gulf. Fluoride. 2013;46(4):204–7.
31. Dobaradaran S, Abadi DR, Mahvi AH, Javid A. Fluoride in skin and muscle of two commercial species of fish harvested off the Bushehr shores of the Persian Gulf. Fluoride. 2011;44(3):143.
32. Dobaradaran S, Naddafi K, Nazmara S, Ghaedi H. Heavy metals (Cd, Cu, Ni and Pb) content in two fish species of Persian Gulf in Bushehr Port, Iran. Afr J Biotechnol. 2013;9(37):6191–3.
33. Abadi DR, Dobaradaran S, Nabipour I, Lamani X, Ravanipour M,

Tahmasebi R, Nazmara S. Comparative investigation of heavy metal, trace, and macro element contents in commercially valuable fish species harvested off from the Persian Gulf. Environ Sci Pollut Res Int. 2015;22(9):6670–8.

34. ASTM. Standard Guide for Collection, Storage, Characterization, and Manipulation of Sediments for Toxicological Testing and for Selection of Samplers Used to Collect Benthic Invertebrates, Publ. E1391-03. Philadelphia: American Society for Testing and Materials; 2014.

35. Gy P. Sampling of particulate materials theory and practice (Vol. 6). Elsevier. USA: CDC press; 2012.

36. Buchanan JB. Sediment analysis. In: Holme NA, McIntyre AD, editors. Methods for the Study of Marine Benthos. Oxford and Edinburgh.: Blackwell Scientific Publications; 1984. p. 41–65.

37. Hamilton MJ, Hadi AZ, Griffith JF, Ishii S, Sadowsky MJ. Large scale analysis of virulence genes in Escherichia coli strains isolated from Avalon Bay, CA. Water Res. 2010;44(18):5463–73.

38. Gauthier MJ, Munro PM. Mohajer S (1987) Influence of salts and sodium chloride on the recovery of Escherichia coli from seawater. Curr Microbiol. 1987;15(1):5–10.

39. Rozen Y, Belkin S. Survival of enteric bacteria in seawater. FEMS Microbiol Rev. 2001;25(5):513–29.

40. Anderson KL, Whitlock JE, Harwood VJ. Persistence and differential survival of fecal indicator bacteria in subtropical waters and sediments. Appl Environ Microbiol. 2005;71(6):3041–8.

41. Šolić M, Krstulović N. Separate and combined effects of solar radiation, temperature, salinity and pH on the survival of faecal coliforms in seawater. Mar Pollut Bull. 1992;24(8):411–6.

42. Goyal SM, Gerba CP, Melnick JL. Occurrence and distribution of bacterial indicators and pathogens in canal communities along the Texas coast. Appl Environ Microbiol. 1977;34(2):139–49.

43. Shibata T, Solo-Gabriele HM, Fleming LE, Elmir S. Monitoring marine recreational water quality using multiple microbial indicators in an urban tropical environment. Water Res. 2004;38(13):3119–31.

44. Blaustein RA, Pachepsky Y, Hill RL, Shelton DR, Whelan G. Escherichia coli survival in waters: temperature dependence. Water Res. 2013;47(2):569–78.

45. Sampson RW, Swiatnicki SA, Osinga VL, Supita JL, McDermott CM, Kleinheinz G. Effects of temperature and sand on E. coli survival in a northern lake water microcosm. J Water Health. 2006;4(3):389–93.

46. Placha I, Venglovský J, Sasakova N, Svoboda IF. The effect of summer and winter seasons on the survival of Salmonella typhimurium and indicator micro-organisms during the storage of solid fraction of pig slurry. J Appl Microbiol. 2001;91(6):1036–43.

47. Whitman RL, Shively DA, Pawlik H, Nevers MB, Byappanahalli MN. Occurrence of Escherichia coli and enterococci in Cladophora (Chlorophyta) in nearshore water and beach sand of Lake Michigan. Appl Environ Microbiol. 2003;69:4714–9.

48. Jamieson RC, Joy DM, Lee H, Kostaschuk R, Gordon R. Transport and deposition of sediment-associated Escherichia coli in natural streams. Water Res. 2005;39(2):2665–75.

49. Crabill C, Donald R, Snelling J, Foust R, Southam G. The impact of sediment fecal coliform reservoirs on seasonal water quality in Oak Creek, Arizona. Water Res. 1999;33(9):2163–71.

50. Hijnen WAM, Schijven JF, Bonne P, Visser A, Medema GJ. Elimination of viruses, bacteria and protozoan oocysts by slow sand filtration. Water Sci Technol. 2004;50(1):147–54.

51. Hua J, An P, Winter J, Gallert C. Elimination of COD, microorganisms and pharmaceuticals from sewage by trickling through sandy soil below leaking sewers. Water Res. 2003;37(18):4395–404.

52. Piorkowski GS, Jamieson RC, Hansen LT, Bezanson GS, Yost CK. Characterizing spatial structure of sediment E. coli populations to inform sampling design. Environ Monit Assess. 2014;186(1):277–91.

53. Brennan FP, O'Flaherty V, Kramers G, Grant J, Richards KG. Long-term persistence and leaching of Escherichia coli in temperate maritime soils. Appl Environ Microbiol. 2010;76(5):1449–55.

54. Bradford SA, Simunek J, Walker SL. Transport and straining of E. coli O157: H7 in saturated porous media. Water Resour Res. 2006;42(12):1–12.

55. Lang NL, Smith SR. Influence of soil type, moisture content and biosolids application on the fate of Escherichia coli in agricultural soil under controlled laboratory conditions. J Appl Microbiol. 2007;103(6):2122–31.

56. Atwill ER, Lewis DJ, Bond RF, Pereira MD, Huerta M, Ogata SB. Protocol consideration for monitoring fecal coliform and in Northern California Estuaries. University of California, Davis School of Veterinary Medicine and University of California Cooperative Extension, Sonoma County, Santa Rosa, California. 2007

57. Craig DL, Fallowfield HJ, Cromar NJ. Use of microcosms to determine persistence of Escherichia coli in recreational coastal water and sediment and validation with in situ measurements. J Appl Microbiol. 2004;96(5):922–30.

58. Burton GA, Gunnison D, Lanza GR. Survival of pathogenic bacteria in various freshwater sediments. Appl Environ Microbiol. 1987;53(3):633–8.

59. Cinotto PJ. Occurrence of fecal-indicator bacteria and protocols for identification of fecal-contamination sources in selected reaches of the West Branch Brandywine Creek, Chester County, Pennsylvania. US Department of the Interior, US Geological Survey. 2005

18

Heavy metals removal from aqueous environments by electrocoagulation process

Edris Bazrafshan[1], Leili Mohammadi[1*], Alireza Ansari-Moghaddam[1] and Amir Hossein Mahvi[2,3,4]

Abstract

Heavy metals pollution has become a more serious environmental problem in the last several decades as a result releasing toxic materials into the environment. Various techniques such as physical, chemical, biological, advanced oxidation and electrochemical processes were used for the treatment of domestic, industrial and agricultural effluents. The commonly used conventional biological treatments processes are not only time consuming but also need large operational area. Accordingly, it seems that these methods are not cost-effective for effluent containing toxic elements. Advanced oxidation techniques result in high treatment cost and are generally used to obtain high purity grade water. The chemical coagulation technique is slow and generates large amount of sludge. Electrocoagulation is an electrochemical technique with many applications. This process has recently attracted attention as a potential technique for treating industrial wastewater due to its versatility and environmental compatibility. This process has been applied for the treatment of many kinds of wastewater such as landfill leachate, restaurant, carwash, slaughterhouse, textile, laundry, tannery, petroleum refinery wastewater and for removal of bacteria, arsenic, fluoride, pesticides and heavy metals from aqueous environments. The objective of the present manuscript is to review the potential of electrocoagulation process for the treatment of domestic, industrial and agricultural effluents, especially removal of heavy metals from aqueous environments. About 100 published studies (1977–2016) are reviewed in this paper. It is evident from the literature survey articles that electrocoagulation are the most frequently studied for the treatment of heavy metal wastewater.

Keywords: Electrocoagulation, Wastewater treatment, Heavy metals removal

Introduction

Environmental issues, mainly concerning chemical and biological water pollution, represent a key priority for civil society, public authorities and, especially, for the industrial sector. In fact, the use of water, both in urban and industrial contexts, implies its subsequent pollution: any activity, whether domestic, agricultural or industrial, produces effluents containing undesirable, and possibly toxic, pollutants. Thus, a constant effort to protect water resources is being made by the various governments, through the introduction of increasingly strict legislation covering pollutant release. In particular for liquid industrial effluents, recent restrictions impose appropriate treatments of wastewater before its release into the environment [1].

This high pollutant load poses complex and extremely varied problems, related to each particular situation. In addition, the release of organic and inorganic pollutants is not uniform (either in quality or in quantity), but always leads to the same result: toxicity for aquatic ecosystems which creates worries for the population [2].

Industrial wastewaters like electroplating or acid mine wastewaters contain various kinds of toxic substances such as cyanides, alkaline cleaning agents, degreasing solvents, oil, fat and metals [3]. Most of the metals such as copper, nickel, chromium, silver and zinc are harmful when they are discharged without treatment [3]. Heavy metals are elements having atomic weights between 63.5 and 200.6 and a specific gravity greater than 5 [4].

* Correspondence: lailimohamadi@gmail.com
[1]Health Promotion Research Center, Zahedan University of Medical Sciences, Zahedan, Iran
Full list of author information is available at the end of the article

With the rapid development of industries such as metal plating facilities, mining operations, fertilizer industries, tanneries, batteries, paper industries and pesticides, etc., heavy metals wastewaters are directly or indirectly discharged into the environment increasingly, especially in developing countries. Unlike organic contaminants, heavy metals are not biodegradable and tend to accumulate in living organisms and many heavy metal ions are known to be toxic or carcinogenic. Toxic heavy metals of particular concern in treatment of industrial wastewaters include zinc, copper, nickel, mercury, cadmium, lead and chromium. Zinc is a trace element that is essential for human health. It is important for the physiological functions of living tissue and regulates many biochemical processes. However, too much zinc can cause eminent health problems, such as stomach cramps, skin irritations, vomiting, nausea and anemia [5]. Copper does essential work in animal metabolism. But the excessive ingestion of copper brings about serious toxicological concerns, such as vomiting, cramps, convulsions, or even death [6]. Nickel exceeding its critical level might bring about serious lung and kidney problems aside from gastrointestinal distress, pulmonary fibrosis and skin dermatitis [7]. And it is known that nickel is human carcinogen. Mercury is a neurotoxin that can cause damage to the central nervous system. High concentrations of mercury cause impairment of pulmonary and kidney function, chest pain and dyspnea [8]. The classic example of mercury poisoning is Minamata Bay. Cadmium has been classified by U.S. Environmental Protection Agency as a probable human carcinogen. Cadmium exposes human health to severe risks. Chronic exposure of cadmium results in kidney dysfunction and high levels of exposure will result in death. Lead can cause central nervous system damage. Lead can also damage the kidney, liver and reproductive system, basic cellular processes and brain functions. The toxic symptoms are anemia, insomnia, headache, dizziness, and irritability, weakness of muscles, hallucination and renal damages [9]. Chromium exits in the aquatic environment mainly in two states: Cr^{3+} and Cr^{6+}. In general, Cr^{6+} is more toxic than Cr^{3+}. Cr^{6+} affects human physiology, accumulates in the food chain and causes severe health problems ranging from simple skin irritation to lung carcinoma [10]. Various regulatory bodies have set the maximum prescribed limits for the discharge of toxic heavy metals in the aquatic systems. However the metal ions are being added to the water stream at a much higher concentration than the prescribed limits by industrial activities, thus leading to the health hazards and environmental degradation (some of permissible limits and health effects of various toxic heavy metals are presented Table 1).

Heavy metals can be easily absorbed by fishes and vegetables due to their high solubility in the aquatic environments and may accumulate in the human body by means of the food chain. So these toxic heavy metals should be removed from the wastewater to protect the people and the environment. In recent years, a variety of techniques are used for heavy metals removal from water and wastewater which include ion-exchange, adsorption, chemical precipitation, membrane filtration, flocculation, coagulation, flotation and electrochemical methods [3].

Electro-coagulation is an electrochemical approach, which uses an electrical current to remove metals from solution. Electro-coagulation system is also effective in removing suspended solids, dissolved metals, tannins and dyes. The contaminants presents in wastewater are maintained in solution by electrical charges. When these ions and other charged particles are neutralized with ions of opposite electrical charges provided by electrocoagulation system, they become destabilized and precipitate in a stable form. Electrochemical methods are simple, fast, inexpensive, easily operable and eco-friendly in nature. Besides, purified water is potable, clear, colorless and odorless with low sludge production. There is no chance of secondary contamination of water in these techniques.

Electrocoagulation process (EC) has been successfully applied to remove soluble ionic species from solutions and heavy metals by various investigators [11, 12]. The EC process is based on the continuous in situ production of a coagulant in the contaminated water. It had been shown that EC is able to eliminate a variety of pollutants from wastewaters, as for example metals and arsenic [3] strontium and cesium [13], phosphate [14], sulfide, sulfate and sulfite [15], boron [16], fluoride [17], nitrate [18], chromium [19–22], cadmium [23], zinc [24], nickel [25, 26], mercury [27], cobalt [28], clay minerals [29], as well as oil [30], chemical oxygen demand [31], color [32] and organic substances [33].

The most widely used method for the treatment of metal polluted wastewater is precipitation with NaOH and coagulation with $FeSO_4$ or $Al_2(SO_4)_3$ with subsequent time-consuming sedimentation [34]. Other methods include adsorption, ion exchange and reverse osmosis [34]. Although precipitation is shown to be quite efficient in treating industrial effluents, the chemical coagulation may induce secondary pollution caused by added chemical substances [34]. These disadvantages encouraged many studies on the use of electrocoagulation for the treatment of several industrial effluents [34]. This technique does not require supplementary addition of chemicals, reduces the volume of produced sludge [33] and first economic studies indicate also a financial advantage compared to the conventional methods [35].

Table 1 Permissible limits and health effects of various toxic heavy metals

Metal contaminant	Permissible limits for industrial effluent discharge (in mg/l)						Permissible limits by international bodies (µg/l)		Health hazards
	Into inland surface waters Indian Standards: 2490 (1974)[a]	Into inland surface waters, Iranian Standards (2009)[b]	Into public sewers, Indian Standards: 3306 (1974)[a]	Into public sewers, Iranian Standards (2009)[b]	On land for irrigation, Indian Standards: 3307 (1974)[a]	On land for irrigation, Iranian Standards (2009)[b]	WHO[a]	USEPA[a]	
Arsenic	0.20	0.10	0.20	0.10	0.20	0.10	10	50	Carcinogenic, producing liver tumors, skin and gastrointestinal effects
Mercury	0.01	–	0.01	–	–	–	01	02	Corrosive to skin, eyes and muscle membrane, dermatitis, anorexia, kidney damage and severe muscle pain
Cadmium	2.00	0.10	1.00	0.10	–	0.05	03	05	Carcinogenic, cause lung fibrosis, dyspnea and weight loss
Lead	0.10	1.00	1.00	1.00	–	1.00	10	05	Suspected carcinogen, loss of appetite, anemia, muscle and joint pains, diminishing IQ, cause sterility, kidney problem and high blood pressure
Chromium	0.10	Cr^{6+} = 0.50, Cr^{3+} = 2.0	2.00	Cr^{6+} = –, Cr^{3+} = 2.0	–	Cr^{6+} = 1.00, Cr^{3+} = 2.0	50	100	Suspected human Carcinogen, producing lung tumors, allergic dermatitis
Nickel	3.0	2.0	3.0	2.0	–	2.0	–	–	Causes chronic bronchitis, reduced lung function, cancer of lungs and nasal sinus
Zinc	5.00	2.0	15.00	2.0	–	2.0	–	–	Causes short-term illness called "metal fume fever" and restlessness
Copper	3.00	1.0	3.00	1.0	–	0.2	–	1300	Long term exposure causes irritation of nose, mouth, eyes, headache, stomachache, dizziness, diarrhea

In the above Table [a]referred to Reference No. [61] and [b]referred to Reference No. [62]

EC process has the potential to extensively eliminate the disadvantages of the classical treatment techniques to achieve a sustainable and economic treatment of polluted wastewater [33, 36]. Since the turn of the 19th century, EC has been applied for wastewater treatment [37] and many studies attended to optimize the process for specific problems. Typically, empirical studies were done [34, 38]. These studies show the successful treatment of the wastewaters, however, they provide little insight into fundamental chemical and physical mechanisms [39]. Therefore, the mechanisms involved are yet not clearly understood [39]. But exactly these physicochemical mechanisms have to be understood to optimize and control the process, to allow modeling of the method and to improve the design of the system. The main objectives of the present work were to gain insight into some fundamental mechanisms and possible interactions influencing the removal process of heavy metals by electrocoagulation.

Table 2 shows the removal efficiency of heavy metals by various treatment technologies. In addition, removal of some of metals and other pollutants by EC process are presented in Table 3.

Description of electrocoagulation process

Electrocoagulation (EC) is a simple and efficient method and has been used for the treatment of many types of wastewaters such as electroplating wastewater [34], laundry wastewater [40], restaurant wastewater [38] and poultry slaughterhouse wastewater [41]. EC has been successfully used for the removal of pollutants from different industrial wastewaters (Table 4). Many studies have been reported in the literature [20, 21, 24, 42].

EC in combination with other treatment processes is a safe and effective way for the removal of pollutants. EC is an efficient technique because adsorption of hydroxide on mineral surfaces are 100 times greater on in 'situ' rather than on pre-precipitated hydroxides when metal hydroxides are used as coagulant [43]. Since the flocs formed by EC are relatively large which contain less bound water and are more stable, therefore, they can be easily removed by

Table 2 Comparison of various treatment technologies for removal of heavy metals from aqueous environments

Treatment method	Metal	pH of solution	Initial concentration (mg/l)	Efficiency (%)	References
Reverse osmosis	Ni^{2+}	3	26	98	[63]
		7	26	99	[63]
	Cu^{2+}	3	17	98	[63]
		7	17	99	[63]
	Cr	3	167	95	[63]
		7	167	99	[63]
Ultrafiltration	Ni^{2+}	7	50	99	[64]
		7	100	99	[64]
	Cu^{2+}	7	50	98	[64]
		7	100	97	[64]
	Cr	7	50	93	[64]
		7	100	76	[64]
	Ni^{2+}	-	25	100	[65]
Nanofiltration	Cu^{2+}	-	200	96	[66]
Electrocoagulation	Ni^{2+}	3	394	98	[67]
		7	394	99	[67]
	Cu^{2+}	3	45	100	[67]
		7	45	100	[67]
	Cr	3	44.5	100	[67]
		7	44.5	100	[67]
	Ni^{2+}, Zn^{2+}	6	248, 270, 282; 217, 232, 236	100	[68]
Chemical precipitation	Cu^{2+}, Zn^{2+}, Cr^{3+}, Pb^{2+}	7- 11	100 mg/L	99.3-99.6	[69]
	Cu^{2+}, Zn^{2+}, Pb^{2+}	3	0.01, 1.34, 2.3 mM	100, >94, >92	[70]
Adsorption	Pb^{2+}	4	2072	-	[71]
	Pb^{2+}	4	1036	55	[72]
	Cd^{2+}, Cr^{6+}	6	2	Cd^{2+} = 55, Cr^{6+} = 60	[22]

Table 3 Removal of heavy metals and other pollutants by EC process

References	Metals or other compounds	Concentration (mg/L)	Anode–cathode	Removal efficiency (%)
[55]	Cr^{3+}, Cr^{6+}	887.2, 1495.2	Fe-Fe	100
[67]	Cu^{2+}, Cr, Ni^{2+}	45, 44.5, 394	Al-Fe	100
[49]	Cd^{2+}	20	Al-Al	AC: 97.5, DC: 96.2
[18]	NO_3^-	150	Fe-Fe, Al-Al	90, 89.7
[23]	Pb^{2+}, Zn^{2+}, Cd^{2+}	170, 50, 1.5	Al-SS	95, 68, 66
[50]	As	150	Al-Al, Fe-Fe	93.5, 94
[26]	TOC, Ni^{2+}, Zn^{2+}	173, 248, 232	SS $_{304}$-SS $_{304}$	66, 90, 100
[73]	Humic acid	20	Fe-Fe	92.69

Nomenclature: *Cr* chromium, *Ni* nickel, *Cu* copper, *As* arsenic, *Zn* zinc, *pb* lead, *Cd* Cadmium, *Co* cobalt, *Fe* iron, *Al* aluminum, *St* steel, *SS* stainless steel

filtration. It is cost effective and easily Performance. EC needs simple equipment's and can be designed for any capacity of effluent treatment plant. Since no chemical addition is required in this process, it reduces the possibility of generation of secondary pollutants. It needs low current and therefore, can be operated by green processes, such as, solar, windmills and fuel cells [44]. It is an environment friendly technique since the electron is the main reagent and does not require addition of the reagents/chemicals. This will minimize the sludge generation to a great extent and eventually eliminate some of the harmful chemicals used as coagulants in the conventional effluent treatment methods. EC process can effectively destabilize small colloidal particles and generates lower quantity of sludge compared to other processes. The advantages of EC as compared to chemical coagulation are as follows:

1. EC requires simple equipment and is easy to operate with sufficient operational latitude to handle most problems encountered on running. Wastewater treated by EC gives pleasant/edible palatable, clear, colorless and odorless water.
2. Sludge formed by EC tends to be readily settable and easy to de-water, because its main elements/components are metallic oxides/hydroxides. Above all, it is a low sludge producing technique.

3. Flocs formed by EC are similar to chemical flocs, except that EC flocs tends to be much larger, contains less bound water, is acid-resistant and more stable and therefore, can be separated faster by filtration.
4. EC produces effluent with less total dissolved solids (TDS) content as compared with chemical treatments. If this water is reused, the low TDS level contributes to a lower water recovery cost.
5. The EC process has the advantage of removing the smallest colloidal particles, because the applied electric field sets them in faster motion, thereby facilitating the coagulation. The EC process avoids uses of chemicals and so there is no problem of neutralizing excess chemicals and no possibility of secondary pollution caused by chemical substances added at high concentration as when chemical coagulation of wastewater is used.
6. The gas bubbles produced during electrolysis can carry the pollutant to the top of the solution where it can be more easily concentrated, collected and removed. The electrolytic processes in the EC cell are controlled electrically with no moving parts, thus requiring less maintenance.

The EC technique can be conveniently used in rural areas where electricity is not available, since a solar panel

Table 4 Application of electrocoagulation process for treatment of different types of wastewater

References	Type of wastewater	Current density or current	Time (min)	pH	Anode–cathode	COD removal (%)
[47]	Olive oil mill wastewater	39.06, 78.1 and 117.18 A/m^2	60	5.2	Ti-Fe	96.14
[74]	Real dairy wastewater	5A	60	7.24	Al-Al	98.84
[75]	Slaughterhouse wastewater	5A	15	7	Al-Al	99
[76]	Carwash wastewater	5 A	15	7.65 ± 0.02	Al-Al	COD = 96.8, BOD$_5$ = 94, TSS = 98.4, MBAS = 98.6
[77]	Textile wastewater	5 A	60	7	Al-Al	98.28
[44]	Textile wastewater	-	3	10.6	Fe-Fe	84
[34]	Olive mill effluents	75 mA/cm^2	25	4-6	Al-Al	76
[37]	Industrial effluents	0.01 A/m^2	30	10.8	SS-SS	95

Nomenclature: *MS* mild steel, *SS* Stainless steel, *St* steel, *Ti* titanium, *Fe* iron, *Pt* platinum, *Cu* copper attached to the unit may be sufficient to carry out the process. Potentially recoverable metals and reuse of treated effluent are other advantages of EC. EC is an alternative to chemical precipitation for the removal of dissolved and suspended metals in aqueous solutions (see Chemical Precipitation Technology Overview). The quantity of sludge produced is lower. The floc generated is larger and heavier and settles out better than in conventional chemical precipitation processes. Since a large thickener is not required, capital costs can also be lower. The effluent generated by EC contains no added chemicals and is often of better quality, containing TDS and less colloidal particulates. Reduction of TDS has been reported at 27 %-60 %, and reduction of total suspended solids can be as great as 95 %-99 % [45].

Although EC requires energy input, it requires only low currents and can be operated using green technologies such as solar or wind power. Some of the limitations of the electrochemical coagulation are as follows [43, 46]:

1. The sacrificial anodes need to be replaced periodically.
2. EC requires minimum solution conductivity depending on reactor design, limiting its use with effluent containing low dissolved solids.
3. In case of the removal of organic compounds, from effluent containing chlorides there is a possibility of formation of toxic chlorinated organic compounds.
4. An impermeable oxide film may be formed on the cathode which may provide resistance to the flow of electric current. However, change of polarity and periodical cleaning of the electrodes may reduce this interference.
5. The high cost of electricity can result in an increase in operational cost of EC process [43].

Electrocoagulation process involves the generation of co-agulants in situ by dissolving electrically either aluminum or iron ions from aluminum or iron electrodes, respectively. In this process, the metal ions generation takes place at the anode and hydrogen gas is released from the cathode. The hydrogen gas bubbles carry the pollutant to the top of the solution where it can be more easily concentrated, collected and removed. Various reactions take place in the electrocoagulation process, where aluminum is used as the electrode:

At the anode:

$$Al \rightarrow Al^{3+}_{(aq)} + 3e \quad (1)$$

At the cathode:

$$3H_2O + 3e \rightarrow 3/2H_2 + 3OH^- \quad (2)$$

The cathode may also be chemically attacked by OH^- ions generated during H_2 evolution at high pH:

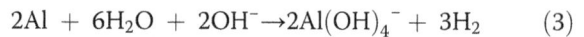

$$2Al + 6H_2O + 2OH^- \rightarrow 2Al(OH)_4^- + 3H_2 \quad (3)$$

$Al^{3+}_{(aq)}$ and OH^- ions generated by electrode reactions (1) and (2) react to form various monomeric species such as $Al(OH)^{2+}$, $Al(OH)_2^+$, $Al_2(OH)_2^{4+}$, $Al(OH)_4^-$, and polymeric species such as $Al_6(OH)_{15}^{3+}$, $Al_7(OH)_{17}^{4+}$, $Al_8(OH)_{20}^{4+}$, $Al_{13}O_4(OH)_{24}^{7+}$, $Al_{13}(OH)_{34}^{5+}$, which transform finally into $Al(OH)_3$ according to complex precipitation kinetics [43].

Freshly formed amorphous $Al(OH)_3$ "sweep flocs" have large surface areas which are beneficial for a rapid adsorption of soluble organic compounds and trapping of colloidal particles. These flocs polymerize as:

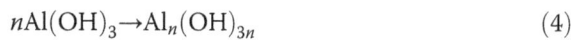

$$nAl(OH)_3 \rightarrow Al_n(OH)_{3n} \quad (4)$$

and they are easily removed from aqueous environment by sedimentation and by H_2 flotation. Secondary anodic reactions occur also during electrocoagulation process for example, in neutral and acidic chloride solutions, native and free chlorine and hypochlorite are formed which are strong oxidants. On the other hand, the aluminum hydroxide flocs normally act as adsorbents and/or traps for pollutants. Therefore, they would eliminate them from the solution [43].

In addition the main reactions occurring at the iron electrodes are:

$$Fe (s) \leftrightarrow Fe^{+3}_{aq} + 3e^- (anode) \quad (5)$$

$$3H_2O + 3e^- \leftrightarrow 3/2 H_{2g} + 3OH^-_{aq}(cathode) \quad (6)$$

In addition, Fe^{3+} and OH^- ions generated at electrode surfaces react in the bulk wastewater to form ferric hydroxide:

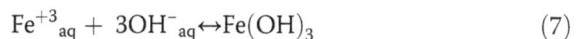

$$Fe^{+3}_{aq} + 3OH^-_{aq} \leftrightarrow Fe(OH)_3 \quad (7)$$

The suspended aluminum or iron hydroxides can remove pollutants from the solution by sorption, co-precipitation or electrostatic attraction, followed by coagulation [43].

For a particular electrical current flow in an electrolytic cell, the mass of aluminum or iron theoretically dissolved from the sacrificial anode is quantified by Faraday's law [43]:

$$w = \left[\frac{ItM}{ZF}\right] \quad (8)$$

where "w" is the amount of anode material dissolved (g), I the current (A), the electrolysis time (t), M the specific molecular weight of electrode (g/mol), Z the number of electrons involved in the reaction and F is the Faraday's constant (96485.34 C/mol). The mass of evolved hydrogen and formed hydroxyl ions can be calculated correspondingly. The amount of coagulant dosed into the solution can be increased by increasing the current and the reaction time. But increasing the current density leads to a decreased current efficiency. Some influencing factors of the EC process are current density (or applied voltage), conductivity and pH of solution, mode of operation, electrolysis time, electrode material and distance between the electrodes [43].

Batch and continuous mode of operation

It can be noticed from the literature that EC has been studied for the removal of a wide range of pollutants using batch and continuous mode of operation. (Diagram of batch and continues flow electrochemical reactor is shown in Figs. 1 and 2). A continuous system operates under steady state conditions, specially a fixed pollutant concentration and effluent flow rate. Comparably, a batch reactor's dynamic nature enables to study the range of operating conditions and is more suited for research work [45]. Continuous systems are better suited to industrial processes for large effluent volumes whereas the batch reactors are suited to laboratory and pilot plant scale applications. The continuous mode of operation is preferred due to its better control than the batch mode of operation.

Batch mode of EC reactors exhibits time dependent behavior as the coagulant is continuously generated in the reactor with the dissolution of anode. The anode material is hydrolyzed, and is capable of aggregating the pollutants. As a result, the concentration of the pollutant, coagulant, and pH keeps on changing with respect to time. A batch reactor has neither inflow nor outflow of effluent during the electrolysis time [45].

Effect of various operating parameters on pollutants removal

The efficiency of the EC process depends on many operational parameters such as conductivity of the solution, arrangement of electrode, electrode shape, type of power supply, pH of the solution, current density, distance between the electrodes, agitation speed, electrolysis time,

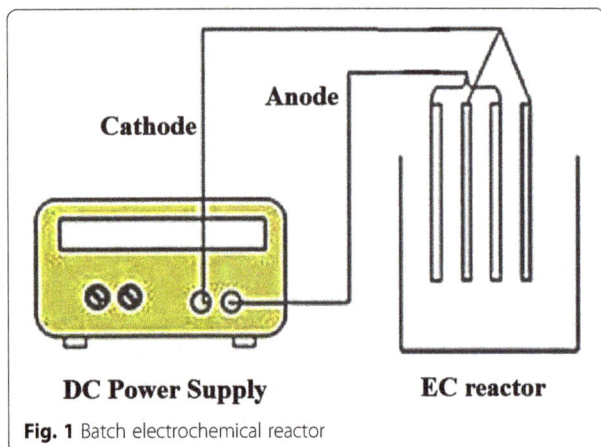
Fig. 1 Batch electrochemical reactor

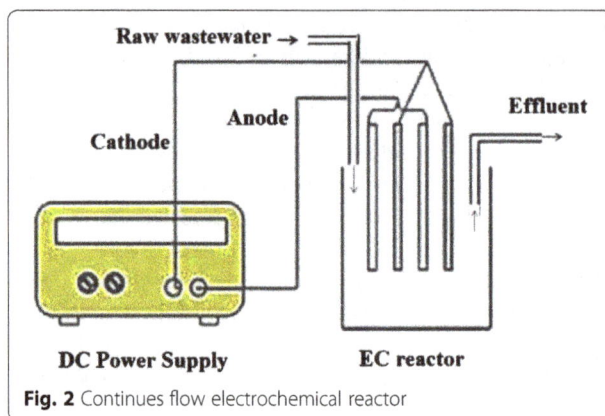
Fig. 2 Continues flow electrochemical reactor

initial pollutant concentration, retention time and passivation of the electrode.

Solution conductivity and type of power supply

Conductivity of the solution is very important parameter in electrolysis process as the removal efficiency of the pollutant and operating cost are directly related to the solution conductivity [45]. The conductivity of an electrolyte solution is a key property. In an electrochemical process, the conductivity determines the cell resistance while the properties of solvent and electrolyte determine their interaction with the electrically active species and thereby influence the electrode reactions [47].

The solution must have some minimum conductivity for the flow of the electric current. The conductivity of the low-conductivity wastewater is adjusted by adding sufficient amount of salts such as sodium chloride or sodium sulphate. There is an increase in the current density with an increase in the conductivity of the solution at constant cell voltage or reduction in the cell voltage at constant current density [48]. The energy consumption is decreased with high performance/approach solution. The energy consumption is decreased with high conductivity solution. In the EC process, there is an in situ generation of metal hydroxide ions by electrolytic oxidation of the sacrificial anode. These metal hydroxide ions act as coagulant and remove the pollutants from the solution by sedimentation. Majority of the studies reported in the literature have used direct current (DC) in the EC process. The use of DC leads to the corrosion formation on the anode due to oxidation. An oxidation layer also form on the cathode reducing the flow of current between the cathode and the anode and thereby lowering the pollutant removal efficiency [49]. These limitations of the DC electrocoagulation process have been decreased to some extent by the addition of parallel plate sacrificial electrodes in the cell configuration. Nevertheless, many have preferred the use of alternating current electrocoagulation (ACE) technology. It is believed that the ac cyclic energization retards the normal mechanisms of electrode attack that are experienced in DC electrocoagulation system, and consequently, ensure reasonable electrode life. In addition to that, since the AC electric fields in an ACE separator do not cause electrophoretic transport of the charged particles due to the frequent change of polarity, it can induce dipole–dipole interactions in a system containing non spherical charged species. Consequently, the AC electric fields may also disrupt the stability of balanced dipolar structures existing in such a system. This is, however, not possible in a DC electrocoagulation separator using DC electric fields [46].

Arrangement of electrodes

The electrode material and the connection mode of the electrodes play a significant role in the cost analysis of the EC process. Kobya et al. [50] studied the treatment of textile wastewater and compared the performances of various electrode connection modes as a function of wastewater pH, current density and operating time. They studied three different modes of electrode connection and areas follow: Monopolar electrodes in parallel connections (MP-P): The anodes and cathodes are connected in parallel due to which the current is divided between all the electrodes to the resistance of individual cells. The parallel connection needs a lower potential difference compared with serial connections [50]. Monopolar electrodes in serial connections (MP-S): In the monopolar configuration, each pair of sacrificial electrodes is internally connected with each other. The addition of the cell voltages leads to a higher potential difference for a given current. Bipolar electrode in serial connections (BP-S): In this connection mode, the outer electrodes are connected to the power supply and there is the no electrical connection between the inner electrodes [50]. Schematic diagram of EC reactor with monopolar and bipolar electrode connections is shown in Figs. 3 and 4.

The shape of the electrodes affects the pollutant removal efficiency in the EC process. It is expected that the

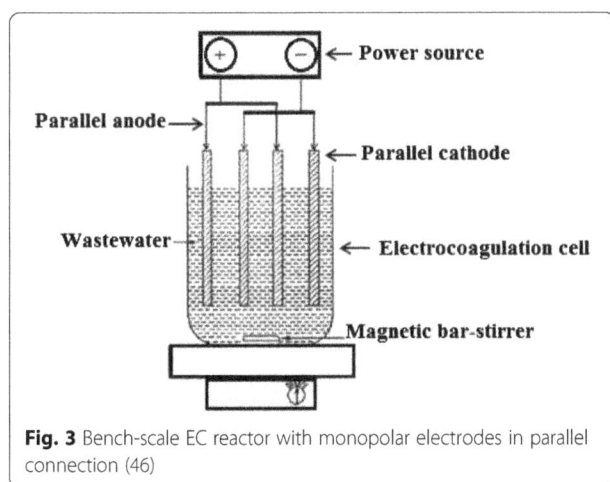

Fig. 4 Bench-scale EC reactor with bipolar electrodes in parallel connection (46)

punched holes type electrodes will result in higher removal efficiency compared to the plane electrodes. Very few studies have been reported in the literature [51] describing the effect of electrode shape on the performance of the electrostatic precipitator. Kuroda et al. [51] performed experiments using metallic electrodes with/without punched holes as a barrier discharge electrode to study the effect of electrode shape of reactor on the collector efficiency in electrostatic precipitator. They have reported higher discharge current for the electrode with punched holes than for plane electrode resulting in higher collection efficiency with punched electrode compared with plane electrode. The electric field intensity at the edge of punched holes type electrodes is higher (1.2 times) than at plane type electrode resulting in an increase in the discharge current at punched type electrode. More studies are needed to establish the effect of the electrode shape (punched hole diameter and pitch of the holes) on the EC process [51].

Current density

Current density is very important parameter in EC as it determines the coagulant dosage rate, bubble production rate, size and growth of the flocs, which can affect the efficiency of the EC. With an increase in the current density, the anode dissolution rate increases. This leads to an increase in the number of metal hydroxide flocs resulting in the increase in pollutant removal efficiency. An increase in current density above the optimum current density does not result in an increase in the pollutant removal efficiency as sufficient numbers of metal hydroxide flocs are available for the sedimentation of the pollutant [52, 53]. Effect of current density or current on removal efficiency of EC process is shown in Table 5.

Distance between the electrodes

Inter-electrode spacing is a vital parameter in the reactor design for the removal of pollutant from effluent. The inter electrode-spacing and effective surface area of electrodes

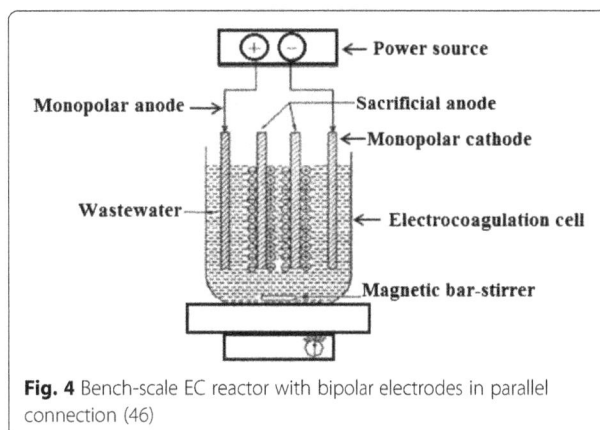

Fig. 3 Bench-scale EC reactor with monopolar electrodes in parallel connection (46)

Table 5 Effect of cell voltage (V), electrode material, electrode connection mode, current or current density, flow rate and pH on removal efficiency of heavy metals in EC process

References	Heavy metals	Current density	Cell voltage (V)	Flow rates	Optimum pH	Electrode materials	Removal efficiency (%)
[78]	Cr^{6+}	8.33 A/m²	2	1.2 m³/h	7-8.5	Fe-Fe A_M	70-85
[34]	Cu^{2+}, Zn^{2+}, Cr^{6+}	4.8A/dm²	-	10 ml/min	4	Al-Al	99, 99, 83
[19]	Cr^{6+}	30 A/m²	-	50 ml/min	5-8	Fe-Fe	80-97
[21]	Cr^{6+}	5A	40	-	10	Al-Al	99
[79]	Cr^{3+}	A_M = 10.84, Bi = 32.52 mA/cm²	-	40 ml/min	A_M = 5.5, Bi = 6	MS-MS	A_M = 90.6, Bi = 71.4
[20]	Cr^{6+}	5A	20-40	-	3	Al-Fe	99.9
[80]	Cr^{6+}	0.05, 1 A	30	-	5	Al-Al	100
[81]	Cr^{6+}	35.7 mA/ cm²	10-24	22.5 ml/min	5	Al-Al	90.4
[82]	Cr^{6+}	55.5 mA/cm²	60	12 ml/min	7	Fe-Fe, Pt Ti (platinized titanium)/Fe, Al/Al and Pt Ti/Al	65.3
[12]	Cr^{6+}	5A	40	-	3	Fe-Fe	98
[83]	Cr^{6+}	153 A/m²	15-25	-	5	Fe-Al	99
[84]	Cr^{6+}	2-25 mA/cm²	80	-	5.68	Fe-Fe	99
[55]	Cr^{3+}, Cr^{6+}	50 mA/cm²	-	-	4	Fe-Fe	100
[85]	Fe, Ni^{2+}, Cu^{2+}, Zn^{2+}, Pb^{2+}, Cd^{2+}	11.55 mA/cm²	0-30	-	7.6	Al-Al	SS = 86.5, turbidity = 81.56, BOD_5 = 83, COD = 68, color > 92.5
[86]	Cd^{2+}, Cu^{2+}	5 A	30	20 L/h	0.64	Ss-ti	Cd^{2+} = 73.8,Cu^{2+} = 98.8
[87]	Cd^{2+}	2.2, 3.5 mA/cm²	6	-	11	Al-Fe	>99.5
[88]	Cd^{2+}	0.2 A/dm²	AC:270, DC: 25	-	7	Zn-Zn	AC: 97.8, DC: 96.9
[49]	Cd^{2+}	0.2 A/dm²	DC = 25,AC = 270	-	7	Al-Al	AC: 97.5, DC: 96.2
[89]	Cd^{2+}	0.04 A/m²	70	5 ml/min	8.9	Al-Al	98.2
[90]	Zn phosphate	60.0 A/m²	30	400 mL/min	Al-Al = 5, Fe-Fe = 3	Al-Fe	max 97.8
[24]	Zn^{2+}	15 mA cm²	60	-	6	Al-Fe	>99
[42]	Zn^{2+}, Cu^{2+}	5A	40	-	7	Fe-Fe	99.99
[58]	COD, Zn^{2+}	COD = 0.90, Zn^{2+} = 0.45-1.8 A/dm²	-	-	COD = 3, ZN^{2+} = 10	Fe-graphite	COD = 88, 99.3 Zn^{2+} = 99
[91]	Ni^{2+}, Cu^{2+}	0.3 A	29	-	5	RO-Ti-Ss	Ni^{2+} = 82, Cu^{2+} = 99
[12]	Zn^{2+}, Cu^{2+}	5 A	40	-	7	Al-Al	Zn^{2+} = 99.6, Cu^{2+} = 99.9
[67]	Cu^{2+}, Cr, Ni^{2+}	10 mA/cm²	30	-	3	Fe-Al	100
[17]	Ni^{2+}	5 A	20	-	10	Fe-Fe	99.99
[50]	As			-	Fe = 6.5, Al =7	Al-Al, Fe-Fe	Fe = 93.5, Al = 95.7

Table 5 Effect of cell voltage (V), electrode material, electrode connection mode, current or current density, flow rate and pH on removal efficiency of heavy metals in EC process (Continued)

Ref.	Heavy metal	Current/current density	Al = 2.5 A/m², Fe =7.5 A/m² / Al electrode = 0.8-1.6, Fe electrode =1.5-2.3	Fe electrode = 60 ml/min, Al electrode =50 ml/min	pH	Electrode	Removal efficiency
[92]	As, Nitrite	-	NO_3=25, As^{5+}=20	2 L/h	9.5	MS-MS	NO_3=84, As^{5+}=75
[93]	As	8.86 mA/cm²	17	7 L/h	5 ± 0.2	AL-AL	89
[11]	Oil, grease, heavy metals	0.6 A/cm²	40	1 L/min	2-4	Al-Cs	Zn^{2+} = 99, Cu^{2+}, Ni^{2+} = 70, Oil and grease = 99.9, Turbidity = 99.7
[28]	Co	6.25 mA/cm²	30	-	8	Al-Al	99
[94]	Heavy metals	4 mA/cm²	30	600 mL/min	9.56	Cs-Cs	Cr^{3+} = Cu^{2+} = 100, Ni^{2+} = 99
[16]	Boron	12.5 mA/cm²	30	-	6.3	Al-Al	99.7
[95]	Ba , Zn^{2+}, Pb^{2+}	350 A/m²	30	-	10	Ss-Ss	97
[96]	Cd^{2+}	3.68 mA/ cm²	-	-	Bipolar configuration = 10.90, monopolar configuration = 9.03	AL- AL	100
[97]	Ni^{2+}	7.5 A/m²	-	6 ml/min	6	AL- AL, Fe- Fe	100
[98]	Cr^{6+}	0.55 A	20	-	1	Fe-Fe	100
[99]	Zn^{2+}, Cu^{2+}, Ni^{2+}, Ag^+, $Cr_2O_7^{2-}$	33 A/m²	30	-	9	Al-Al	>50
[100]	Cr^{6+}	50-200 A/m²	-	2.5 cm³/ s	7.5	Fe-Fe, Al-Al	40

Nomenclature: *MS* mild steel, *SS* stainless steel, *St* steel, *Ti* titanium, *Fe* iron, *Pt* platinum, *Cu* copper, *CS* carbon steel electrodes, *RuO* ruthenium oxide, *A_M* Monopolar, *B,* Bipolar

are important variable when an operational costs optimization of a reactor is needed [52]. To decrease the energy consumption (at constant current density) in the treatment of effluent with a relatively high conductivity, larger spacing should be used between electrodes. For effluent with low conductivity, energy consumption can be minimized by decreasing the spacing between the electrodes [53].

The inter electrode distance plays a significant role in the EC as the electrostatic field depends on the distance between the anode and the cathode. The maximum pollutant removal efficiency is obtained by maintaining an optimum distance between the electrodes. At the minimum inter electrode distance; the pollutant removal efficiency is low. This is due to the fact that the generated metal hydroxides which act as the flocs and remove the pollutant by sedimentation get degraded by collision with each other due to high electrostatic attraction [54]. The pollutant removal efficiency increases with an increase in the inter electrode distance from the minimum till the optimum distance between the electrodes. This is due to the fact that by further increasing the distance between the electrodes, there is a decrease in the electrostatic effects resulting in a slower movement of the generated ions. It provides more time for the generated metal hydroxide to agglomerate to form the flocs resulting in an increase in the removal efficiency of the pollutant in the solution. On further increasing the electrode distance more than the optimum electrode distance, there is a reduction in the pollutant removal efficiency. This is due to the fact that the travel time of the ions increases with an increase in the distance between the electrodes. This leads to a decrease in the electrostatic attraction resulting in the less formation of flocs needed to coagulate the pollutant [54]. The pollutant removal efficiency is low at the minimum inter electrode distance. Effect of distance between the electrodes and also type of reactor (batch or continuous) on removal efficiency of EC process are presented in Table 6.

Electrolysis time

The pollutant removal efficiency is also a function of the electrolysis time. The pollutant removal efficiency increases with an increase in the electrolysis time. But beyond the optimum electrolysis time, the pollutant removal efficiency becomes constant and does not increase with an increase in the electrolysis time. The metal hydroxides are formed by the dissolution of the anode. For a fixed current density, the number of generated metal hydroxide increases with an increase in the electrolysis time. For a longer electrolysis time, there is an increase in the generation of flocs resulting in an increase in the pollutant removal efficiency. For an electrolysis time beyond the optimum electrolysis time, the

pollutant removal efficiency does not increase as sufficient numbers of flocs are available for the removal of the pollutant [45]. Bazrafshan et al. [20] determined that Cr^{6+} reduction from synthetic chromium solution could be under legal limits as long as treatment was between 20 and 60 minutes. Effect of different electrolysis time on removal efficiency of EC process is shown in Table 6.

Effect of initial pH on the efficiency of Heavy metal removal

It has been established that pH is an important parameter influencing the performance of the electrochemical process [38]. The maximum pollutant removal efficiency is obtained at an optimum solution pH for a particular pollutant. The precipitation of a pollutant begins at a particular pH. The pollutant removal efficiency decreases by either increasing or decreasing the pH of the solution from the optimum pH. Verma et al. [55] studied the removal of hexavalent chromium from synthetic solution using EC and found that the pH of the solution has a significant effect on the Cr^{6+} removal efficiency. They performed the experiments at different pH of the synthetic solution and obtained the maximum chromium removal efficiency at the pH 4. They further reported that the pH of the synthetic solution after the EC process increased with an increase in the electrolysis time due to the generation of OH in the EC process [55].

The pH changed during batch EF, Its evolution depended on the initial pH. EF process exhibits some buffering capacity because of the balance between the production and the consumption of OH [56]. The pH has a significant influence on the coagulant species formed during coagulation processes. It also has influence on the superficial charge of the aluminum hydroxide precipitates (caused by the adsorption of ionic species) [57]. During the time-course of coagulation and EC processes, the pH changes in an opposite way and this affects significantly to the coagulant species formed, and hence to the efficiencies obtained in the removal of pollutants [57].

It cannot be said that any process is better than the other for all wastes. Under the same fluid dynamic conditions, doses of aluminum, pH, the efficiencies obtained by coagulation and EC are very similar. The pH of the waste can be a key parameter in the choice of the coagulation technology [57]. Effect of different initial pH on removal efficiency in EC process is shown in Table 5.

Cost analysis

Cost analysis plays an important role in industrial wastewater treatment procedure/method as the wastewater treatment technique should be cost attractive. The costs involved in EC include, the cost of energy consumption,

Table 6 Effect of inter electrodes distance, conductivity of solutions, energy consumption and electrolysis time on heavy metals removal efficiency in EC process

References	Heavy metal	Reactor	Electrolysis time	Inter electrode distance	Conductivity (mS/cm)	Energy consumption	Efficiency (%)
[78]	Cr^{6+}	Continuous	10-12 min	-	-	-	70-85
[34]	Cu^{2+}, Zn^{2+}, Cr^{6+}	Continuous	20 min	5 mm	-	-	99, 99, 83
[19]	Cr^{6+}	Continuous	72 min	4 mm	1.5	1 kWh/m^3	80-97
[21]	Cr^{6+}	Batch	20 min	1.5 cm	1.6	1.92-2.29 kwh/m^3	99
[79]	Cr^{3+}	Continuous	20-25 min	22 mm	5.73 , 7.36	0.1kWh/m^3	A_M = 90.6, Bi = 71.4
[20]	Cr^{6+}	Batch	20, 60 min	1.5 cm	1.6	2.11 kWh/m^3	99.9
[80]	Cr^{6+}	Batch	45 min	5 mm	20	9.0 kWh/m^3	100
[81]	Cr^{6+}	Continuous	24 min	15 mm	2	137.2 KWh/m^3	90.4
[82]	Cr^{6+}	Continuous	75 min	4 cm	2.41, 1.70	-	65.3
[12]	Cr^{6+}	Batch	60 min	1.5 cm	1.6	35.06 kwh/g	98
[83]	Cr^{6+}	Batch	25 min	1.5 cm	0.59- 3.4	16.3 kWh/m^3	99
[84]	Cr^{6+}	Batch	5-10 min	0.3 cm	365	38 kWh/m^3	99
[55]	Cr^{3+}, Cr^{6+}	Batch	15 min	0.5 cm	-	-	100
[85]	Fe, Ni^{2+}, Cu^{2+}, Zn^{2+}, Pb^{2+}, Cd^{2+}	Batch	10 min	1 cm	2.1	-	SS = 86.5, Turbidity = 81.56, BOD_5 = 83, COD = 68, Color > 92.5
[19]	Cd^{2+}	Batch	20 min	1.5 cm	-	9.37 kwh/kg	>99
[86]	Cd^{2+}, Cu^{2+}	Continuous	120 min	1.5 cm	-	10.99 kWh/kg	Cd^{2+} = 73.8,Cu^{2+} = 98.8
[87]	Cd^{2+}	Batch	10 min	-	1.05- 5.22	-	>99.5
[88]	Cd^{2+}	Batch	30 min	5 mm	-	AC:0.6, DC: 1.2 kWh/m^3	AC: 97.8, DC: 96.9
[49]	Cd^{2+}	Batch	AC: 30, DC: 45 min	5 mm	-	AC:0.4, DC:1 kWh/ kg	AC: 97.5, DC: 96.2
[89]	Cd^{2+}	Continuous	200 min	1 cm	1.06	-	98.2
[90]	Zn phosphate	Batch and continuous	15 min = Fe electrode, 25 min = Al electrode	Batch = 11, continuous = 20 mm	Batch = 5.1-5.3, Continuous = 4.8- 4.9	Al electrode =0.18–11.29, Fe electrode= 0.24-8.47 kWh/m^3	Max 97.8
[24]	Zn^{2+}	Batch	10 min	11 mm	3000 µS/cm	3.3 kWh/kg	>99
[42]	Zn^{2+}, Cu^{2+}	Batch	60 min	1.5 cm	1.6	Zn^{2+} = 22.31, Cu^{2+} = 35.63KWh/g	99.99
[58]	COD, Zn^{2+}	Batch	50 min	16 mm	0.49	1.7 kWh/kg	COD = 88, 99.3, Zn^{2+} = 99
[91]	Ni^{2+}, Cu^{2+}	Batch	60 min	1 cm	634 µS/cm	-	Ni^{2+} = 82, Cu^{2+} = 99
[12]	Zn^{2+}, Cu^{2+}	Batch	60 min	1.5 cm	1.6	Zn^{2+} = 19.98, Cu^{2+} = 35.06 kWh	Zn^{2+} = Cu^{2+} = 99.9
[67]	Cu^{2+}, Cr, Ni^{2+}	Batch	20 min	10 mm	2	10.07 kWh/m^3	100
[17]	Ni^{2+}	Batch	20, 40 min	1 cm	1.6	9.37 kWh/kg	99.9
[50]	As	Continuous	Fe electrode =12.5, Al electrode =15 min	13 mm	1.55	Fe electrode =0.015, Al electrode =0.032 kWh/m^3	Fe = 93.5, Al = 95.7
[92]	As^{5+},NO_2^-	Continuous	120 min	7 cm	-	-	Nitrite =84, As^{5+} = 75

Table 6 Effect of inter electrodes distance, conductivity of solutions, energy consumption and electrolysis time on heavy metals removal efficiency in EC process (*Continued*)

Ref	Pollutant	Mode	Time	Distance	Conductivity	Energy consumption	Removal efficiency
[93]	As	Continuous	30 min	1.2 cm	1700 ± 37 µS/cm	3.03 kWh/m³	89
[11]	Oil, grease, heavy metals	Continuous	105 s	-	-	Cu^{2+}, Ni^{2+} = 0.166, Zn^{2+} = 0.117, Oil and grease = 0.116, Turbidity = 0.116, Oil and grease = 0.117 kwh/m³	Zn^{2+} = 99, Cu^{2+}, Ni^{2+} = 70, Oil and grease = 99.9, Turbidity = 99.7
[28]	Co	Batch	15 min	-	6.5	3.3 kwh/m³	99
[94]	Heavy metal	Continuous	45 min	15 mm	8.9 ± 0.2	6.25 kWh/m3	Cr^{3+} = Cu^{2+} = 100, Ni^{2+} = 99
[16]	Boron	Batch	89 min	0.5 cm	30,000 mS/cm	2.4 kWh/m3	99.7
[95]	Ba, Zn^{2+}, Pb^{2+}	Continuous	20 min	10 mm	-	14 kWh/m3	97
[96]	Cd^{2+}	Batch	5 min	0.5 cm	1.176 mS/cm	1.6 kW h m 3	100
[97]	Ni^{2+}	Continuous	20 min	10 mm	1 mS/cm	-	100
[98]	Cr^{6+}	Batch	14 min	0.87 cm	adjusted	0.007 kWh/g	100
[99]	Zn^{2+}, Cu^{2+}, Ni^{2+}, Ag^+, $Cr_2O_7^{2-}$	Batch	30 min	5 mm	20 mS/cm	-	>50
[100]	Cr^{6+}	Continuous	60 min	20 mm	2.4 mS/cm	-	40

cost of the dissolved electrode (electrode consumption) and the cost of addition of any external chemical (for increasing the solution conductivity or varying the pH of the solution).

Electrode consumption can calculate by equation 8 which presented earlier. In addition, electrical energy consumption is a very important economical parameter in the electrocoagulation process and can calculated using the following equation [33]:

$$E = \left[\frac{UIt}{1000\ V} \right] \tag{9}$$

where E is the energy consumption (kWh/m^3), U is the applied voltage (V), I is the current intensity (A), t is the electrocoagulation time (h), and V is the volume of the treated wastewater (m^3).

The detailed calculation of operating cost for the treatment of fluoride containing drinking water using EC has been reported by Ghosh et al. [58]. Espinoza-Quinones et al., (2009) studied the removal of organic and inorganic pollutants from a wastewater of lather finishing industrial process using EC. They found the EC to be cheaper compared to the conventional method. The operational cost for the EC was found to be US $ 1.7 per cubic meter of the treated tannery effluent as compared to the cost of US $ 3.5 per cubic meter of the treated effluent for conventional methods [59]. Similarly Bayramoglu et al. [60] have been reported that the operating cost of chemical coagulation is 3.2 times as high as that of EC for the treatment of textile wastewater.

Conclusions

The rapid urbanization and industrialization in the developing countries are creating high levels of water pollution due to harmful industrial effects and sewage discharges. The characteristics of industrial effluents in terms of nature of contaminates, their concentrations, treatment technique and required disposal method vary significantly depending on the type of industry. Further, the choice of an effluent treatment technique is governed by various parameters such as contaminants, their concentration, volume to be treated and toxicity to microbes. Electrocoagulation is a treatment process that is capable of being an effective treatment process as conventional methods such as chemical coagulation. Having observed trends over the last years, it has been noted that electrocoagulation is capable of having high removal efficiencies of color, chemical oxygen demand (COD), biochemical oxygen demand (BOD_5) and achieving a more efficient treatment processes quicker than traditional coagulation and inexpensive than other methods of treatment such as

ultraviolet (UV) and ozone. Unlike biological treatment which requires specific conditions, therefore limiting the ability to treat many wastewaters with high toxicity, xenobiotic compounds, and pH, electrocoagulation can be used to treat multifaceted wastewaters, including industrial, agricultural, and domestic. Continual research using this technology will not only improve new modeling techniques can be used to predict many factors and develop equations that will predict the effectiveness of treatment.

Electrocoagulation is an attractive method for the treatment of various kinds of wastewater, by virtue of various benefits including environmental capability, versatility, energy efficiency, safety, selectivity and cost effectiveness. The process is characterized by simple equipment, easy operation, less operating time and decreased amount of sludge which sediments rapidly and retain less water. However, further studies needs to be performed to study the effect of shape and geometry of the electrodes (punched hole and pitch of the holes) to possibly improve the pollutant removal efficiency. Efforts should be made to study the phenomena of electrode passivation to reduce the operating cost of the EC process. Most of the studies reported in the literature have been carried out at the laboratory scale using synthetic solutions. Efforts should be made to perform EC experiments at pilot plant scale using real industrial effluent to explore the possibility of using EC for treatment of real industrial effluents.

Competing interests
The authors declare that they have no competing interests.

Authors' contribution
EB and LM were performed design, analysis and writing the study. AA and AM contributed in writing of the manuscript. All authors read and approved the final manuscript.

Author details
[1]Health Promotion Research Center, Zahedan University of Medical Sciences, Zahedan, Iran. [2]Department of Environmental Health Engineering, School of Public Health, Tehran University of Medical Sciences, Tehran, Iran. [3]Center for Solid Waste Research, Institute for Environmental Research, Tehran University of Medical Sciences, Tehran, Iran. [4]National Institute of Health Research, Tehran University of Medical Sciences,, Tehran, Iran.

References
1. Bertrand S, Giuseppe T, Jeremie C, Jean-François M, Sophie G, Pierre-Marie B, et al. Heavy metal removal from industrial effluents by sorption on cross-linked starch: Chemical study and impact on water toxicity. J Environ Manage. 2011;92:765–72.
2. Blais JF, Dufresne S, Mercier G. State of the art of technologies for metal removal from industrial effluents. J Wat Sci. 1999;12/4:687–711.
3. Hunsom M, Pruksathorn K, Damronglered S, Vergnes H, Duverneuil P. Electrochemical treatment of heavy metals (Cu2+, Cr6+, Ni2+) from industrial effluent and modeling of copper reduction. Water Res. 2005;39:610–16.

4.   Srivastava NK, Majumder CB. Novel biofiltration methods for the treatment of heavy metals from industrial wastewater. J Hazard Mater. 2008;28, 151(1):1–8.

5.   Oyaro N, Juddy O, Murago ENM, Gitonga E. The contents of Pb, Cu, Zn and Cd in meat in Nairobi. Kenya Int J Food Agric Environ. 2007;5:119–21.

6.   Paulino AT, Minasse FAS, Guilherme MR, Reis AV, Muniz EC, Nozaki J. Novel adsorbent based on silkworm chrysalides for removal of heavy metals from wastewaters. J Colloid Interf Sci. 2006;301:479–87.

7.   Borba CE, Guirardello R, Silva EA, Veit MT, Tavares CRG. Removal of nickel^{2+} ions from aqueous solution by biosorption in a fixed bed column: experimental and theoretical breakthrough curves. Bio Chem Eng J. 2006;30:184–91.

8.   Namasivayam C, Kadirvelu K. Uptake of mercury^{2+} from wastewater by activated carbon from unwanted agricultural solid by- product: coirpith. Carbon. 1999;37:79–84.

9.   Naseem R, Tahir SS. Removal of Pb^{2+} from aqueous solution by using bentonite as an adsorbent. Water Res. 2001;35:3982–6.

10. Khezami L, Capart R. Removal of chromium (VI) from aqueous solution by activated carbons: Kinetic and equilibrium studies. J Hazard Mater. 2005;123:223–31.

11. Rincon GJ, Motta EJL. Simultaneous removal of oil and grease, and heavy metals from artificial bilge water using electrocoagulation/flotation. J Environ Manage. 2014;144:42–50.

12. Nouri J, Mahvi AH, Bazrafshan E. Application of electrocoagulation process in removal of zinc and copper from aqueous solutions by aluminum electrodes. Int J Environ Res. 2010;4:201–8.

13. Kamaraj R, Vasudevan S. Evaluation of electrocoagulation process for the removal of strontium and cesium from aqueous solution. Che Eng Res Des. 2015;93:522–30.

14. Mahvi AH, Ebrahimi SJA, Mesdaghinia A, Gharibi H, Sowlat MH. Performance evaluation of a continuous bipolar electrocoagulation/electrooxidation-electroflotation (ECEO–EF) reactor designed for simultaneous removal of ammonia and phosphate from wastewater effluent. J Hazard Mater. 2011;192(3):1267–74.

15. Apaydin KU, Gonullu MT. An investigation on the treatment of tannery wastewater by electrocoagulation. Global Nest J. 2009;11:546–55.

16. Mohamed-Hasnain I, Ezerie HE, Zubair A, Saleh FM, Shamsul R, Mohamed K. Boron removal by electrocoagulation and recovery. Water Res. 2014;51:113–23.

17. Bazrafshan E, Ownagh K, Mahvi AH. Application of electrocoagulation process using iron and aluminum electrodes for fluoride removal from aqueous environment. E- J Chem. 2012;9(4):2297–308.

18. Malakootian M, Yousefi N, Fatehizadeh A. Survey efficiency of electrocoagulation on nitrate removal from aqueous solution. Int J Environ Sci Technol. 2011;8(1):107–14.

19. Bazrafshan E, Mahvi AH, Nasseri S, Mesdaghinia AR, Vaezi F, Nazmara S. Removal of cadmium from industrial effluents by electrocoagulation process using iron electrodes. Iran J Environ Health Sci Eng. 2006;3(4):261–6.

20. Bazrafshan E, Mahvi AH, Naseri S, Mesdaghinia AR. Performance evaluation of electrocoagulation process for removal of chromium^{6+} from synthetic chromium solutions using iron and aluminum electrodes. Turk J Eng Environ Sci. 2008;32:59–66.

21. Mahvi AH, Bazrafshan E. Removal of cadmium from industrial effluents by electrocoagulation process using aluminum electrodes. World Appl Sci J. 2007;2(1):34–9.

22. Gholami Borujeni F, Mahvi AH, Nejatzadeh-Barandoozi F. Removal of heavy metal ions from aqueous solution by application of low cost materials. Fresen Environ Bull. 2013;22(3):655–8.

23. Pociecha M, Lestan D. Using electrocoagulation for metal and chelant separation from washing solution after EDTA leaching of Pb, Zn and Cd contaminated soil. J Hazard Mater. 2010;174(1–3):670–8.

24. Kobya M, Demirbas E, Sozbir M. Depolarization of aqueous reactive dye Remazol Red 3B by electrocoagulation. Color Technol. 2010;126(5):282–8.

25. Mansoorian HJ, Rajabizadeh A, Bazrafshan E, Mahvi AH. Practical assessment of electrocoagulation process in removing nickel metal from aqueous solutions using iron-rod electrodes. Desalin Water Treat. 2012;44:29–35.

26. Kabdaşlı I, Vardar B, Alaton AI, Tünay O. Effect of dye auxiliaries on color and COD removal from simulated reactive dye bath effluent by electrocoagulation. Chem Eng J. 2009;148:89–96.

27. Chaturvedi SI. Mercury removal using Al-Al electrodes by electrocoagulation. IJMER. 2013;3(1):109–15.

28. Shafaei A, Pajootan E, Nikazar M, Arami M. Removal of Co^{2+} from aqueous solution by electrocoagulation process using aluminum electrodes. Desalination. 2011;279(1–3):121–6.

29. Holt PK, Barton GW, Mitchell CA. Deciphering the science behind electrocoagulation to remove clay particles from water. Water Sci Technol. 2004;50(12):177–84.

30. Pouet MF, Grasmick A. Urban wastewater treatment by electrocoagulation and flotation. Water Sci Technol. 1995;31(3–4):275–83.

31. Jung KW, Hwang MJ, Park DS, Ahn KH. Combining fluidized metal-impregnated granular activated carbon in three-dimensional electrocoagulation system: Feasibility and optimization test of color and COD removal from real cotton textile wastewater. Sep Purif Technol. 2015;146:154–67.

32. Adjeroud N, Dahmoune F, Merzouk B, Leclerc JP, Madani K. Improvement of electrocoagulation–electroflotation treatment of effluent by addition of Opuntia ficus indica pad juice. Sep Purif Technol. 2015;144:168–76.

33. Bazrafshan E, Biglari H, Mahvi AH. Phenol removal by electrocoagulation process from aqueous solutions. Fresen Environ Bull. 2012;21(2):364–71.

34. Adhoum N, Monser L, Bellakhal N, Belgaied JE. Treatment of electroplating wastewater containing Cu^{2+}, Zn^{2+}and Cr^{6+} by electrocoagulation. J Hazard Mater. 2004;B 112(3):207–13.

35. Meunier N, Drogui P, Montane C, Hausler R, Mercier G, Blais JF. Comparison between electrocoagulation and chemical precipitation for metals removal from acidic soil leachate. J Hazard Mater. 2006;137:581–90.

36. Yu MJ, Koo JS, Myung GN, Cho YK, Cho YM. Evaluation of bipolar electrocoagulation applied to biofiltration for phosphorus removal. Water Sci Technol. 2005;51(10):231–9.

37. Matteson MJ, Dobson RL, Glenn J, Robert W, Kukunoor NS, Waits I, et al. Electrocoagulation and separation of aqueous suspensions of ultrafine particles. Colloid Surface A. 1995;104(1):101–9.

38. Chen X, Chen G, Yue PL. Separation of pollutants from restaurant wastewater by electrocoagulation. Sep Purif Technol. 2000;19(1–2):65–76.

39. Chen G. Electrochemical technologies in wastewater treatment. Sep Purif Technol. 2004;38:11–41.

40. Janpoor F, Torabian A, Khatibikamal V. Treatment of laundry wastewater by electrocoagulation. J Chem Technol Biotechnol. 2011;86(8):1113–20.

41. Kobya M, Senturk E, Bayramoglu M. Treatment of poultry slaughterhouse wastewaters by electrocoagulation. J Hazard Mater. 2006;133(1–3):172–6.

42. Bazrafshan E, Mahvi AH, Zazoli MA. Removal of zinc and copper from aqueous solutions by electrocoagulation technology using iron electrodes. Asian J Chem. 2011;23(12):5506–10.

43. Mollah MYA, Morkovsky P, Gomes JAG, Kesmez M, Parga J, Cocke DL. Fundamentals, present and future perspectives of electrocoagulation. J Hazard Mater. 2004;B114(1–3):199–210.

44. Zaroual Z, Azzi M, Saib N, Chainet E. Contribution to the study of electrocoagulation mechanism in basic textile effluent. J Hazard Mater. 2006;B131:73–8.

45. Khandegar V, Saroha AK. Electrocoagulation for the treatment of textile industry effluent- A review. J Environ Manage. 2013;128:949–63.

46. Mollah M, Schennach R, Parga JR, Cocke DL. Electrocoagulation (EC) science and applications. J Hazard Mater. 2001;84:29–41.

47. Yazdanbakhsh AR, Massoudinejad MR, Arman K, Aghayani E. Investigating the Potential of Electrocoagulation- Flotation (ECF) Process for Pollutants Removal from Olive Oil Mill Wastewater. J Appl Environ Biol Sci. 2013;3(3):22–8.

48. Merzouk B, Madani K, Sekki A. Using electrocoagulation- electroflotation technology to treat synthetic solution and textile wastewater.two case studies. Desalination. 2010;250(5):573–7.

49. Vasudevan S, Lakshmi J, Sozhan G. Effects of alternating and direct current in electrocoagulation process on the removal of cadmium from water. J Hazard Mater. 2011;192(1):26–34.

50. Kobya M, Ulu F, Gebologlu U, Demirbas E, Oncel MS. Treatment of potable water containing low concentration of arsenic with electrocoagulation: different connection modes and Fe-Al electrodes. Sep Purif Technol. 2011;77(3):283–93.

51. Kuroda Y, Kawada Y, Takahashi T, Ehara Y, Ito T, Zukeran A, et al. Effect of electrode shape on discharge current and performance with barrier discharge type electrostatic precipitator. J Electrostat. 2003;57(3):407–15.

52. Bukhari AA. Investigation of the electrocoagulation treatment process for the removal of total suspended solids and turbidity from municipal wastewater. Bioresour Technol. 2008;99(5):914–21.

53. Vik EA, Carlson DA, Eikum AS, Gjessing ET. Electrocoagulation of potable water. Water Res. 1984;18(11):1355–60.

54. Aoudj S, Khelifa A, Drouiche N, Belkada R, Miroud D. Simultaneous removal of chromium (VI) and fluoride by electrocoagulation- electroflotation: Application of a hybrid Fe-Al anode. Chem Eng J. 2015;267:153–62.

55. Verma SK, Khandegar V, Saroha AK. Removal of chromium from electroplating industry effluent using electrocoagulation. J Hazard Toxic Radio Waste. 2013;17(2):146–52.

56. Chen G. Electrochemical technologies in wastewater treatment. Sep Purif Technol. 2004;38:11–41.

57. Canizares P, Jiménez C, Martínez F, Rodrigo M, Saez C. The pH as a key parameter in the choice between coagulation and electrocoagulation for the treatment of wastewaters. J Hazard Mater. 2009;163:158–64.

58. Ghosh P, Samanta AN, Ray S. Reduction of COD and removal of Zn^{2+} from rayon industry wastewater by combined electrofenton treatment and chemical precipitation. Desalination. 2011;266:213–17.

59. Espinoza- Quinones FR, Fornari MMT, Modenes AN, Palacio SM, DaSilva FG, Szymanski N, et al. Pollutant removal from tannery effluent by electrocoagulation. Chem Eng J. 2009;151(1–3):59–65.

60. Bayramoglu M, Eyvaz M, Kobya M. Treatment of the textile wastewater by electrocoagulation economical evaluation. Chem Eng J. 2007;128:155–61.

61. Orisakwe OE, Nduka JK, Amadi CN, Dike DO, Bede O. Heavy metals health risk assessment for population via consumption of food crops and fruits in Owerri, South Eastern, Nigeria. Chem Cent J. 2012;6:77.

62. Institute of Standards and Industrial Research of Iran (ISIRI). (2009). Drinking water: Physical and chemical specifications. 5th edition. Standard No. 1053.

63. Ozaki H, Sharmab K, Saktaywirf W. Performance of an ultra- low- pressure reverse osmosis membrane (ULPROM) for separating heavy metal: effects of interference parameters. Desalination. 2002;144:287–94.

64. Barakat MA, Schmidt E. Polymer- enhanced ultrafiltration process for heavy metals removal from industrial wastewater. Desalination. 2010;256:90–3.

65. Kryvoruchko A, Yurlova L, Kornilovich B. Purification of water containing heavy metals by chelating- enhanced ultrafiltration. Desalination. 2002;144:243–8.

66. Qdaisa HA, Moussa H. Removal of heavy metals from wastewater by membrane processes: a comparative study. Desalination. 2004;164:105–10.

67. Akbal F, Camci S. Copper, chromium and nickel removal from metal plating wastewater by electrocoagulation. Desalination. 2011;269:214–22.

68. Kabdasli I, Arslan T, Olmez HT, Alaton AI, Tünay O. Complexing agent and heavy metal removals from metal plating effluent by electrocoagulation with stainless steel electrodes. J Hazard Mater. 2009;15, 165(1–3):838–45.

69. Chen QY, Luo Z, Hills C, Xue G, Tyrer M. Precipitation of heavy metals from wastewater using simulated flue gas: sequent additions of fly ash, lime and carbon dioxide. Water Res. 2009;43:2605–14.

70. Alvarez MT, Crespo C, Mattiasson B. Precipitation of Zn^{2+}, Cu^{2+} and Pb^{2+} at bench- scale using biogenic hydrogen sulfide from the utilization of volatile fatty acids. Chemosphere. 2007;66:1677–83.

71. Inglezakis VJ, Grigoropoulou HP. Modeling of ion exchange of Pb^{2+} in fixed beds of clinoptilolite. Micropor Mesopor Mat. 2003;61:273–82.

72. Inglezakis VJ, Stylianou MA, Gkantzou D, Loizidou MD. Removal of Pb^{2+} from aqueous solutions by using clinoptilolite and bentonite as adsorbents. Desalination. 2007;210:248–56.

73. Bazrafshan E, Biglari H, Mahvi AH. Humic acid removal from aqueous environments by electrocoagulation process using iron electrodes. E J Chem. 2012;9(4):2453–61.

74. Bazrafshan E, Moein H, Kord Mostafapour F, Nakhaie S. Application of electrocoagulation process for dairy wastewater treatment. J Chem. 2013;Article. ID 640139:8.

75. Bazrafshan E, Farzadkia M, Kord Mostafapour F, Ownagh KA, Mahvi AH. Slaughterhouse wastewater treatment by combined chemical coagulation and electrocoagulation process. PLOS ONE. 2012;7(6):1–8.

76. Bazrafshan E, KordMostafapour F, Soori MM, Mahvi AH. Application of combined chemical coagulation and electrocoagulation process to carwash wastewater treatment. Fresen Environ Bull. 2012;21(9a):2694–701.

77. Bazrafshan E, Mahvi AH, Zazouli MA. Textile wastewater treatment by electrocoagulation process using aluminum electrodes. Iran J Health Sci. 2014;2(1):16–29.

78. Osipenko VD, Pogorelyi PI. Electrocoagulation neutralization of chromium containing effluent. Metallurgist. 1977;21(9–10):44–5.

79. Golder AK, Samanta AN, Ray S. Removal of trivalent chromium by electrocoagulation. Sep Purif Technol. 2007;53(1):33–41.

80. Heidmann I, Calmano W. Removal of Cr^{6+} from model wastewaters by electrocoagulation with Fe electrodes. Sep Purif Technol. 2008;61:15–21.

81. Bhatti MS, Reddy AS, Thukral AK. Electrocoagulation removal of Cr^{6+} from simulated wastewater using response surface methodology. J Hazard Mater. 2009;172:839–46.

82. Mouedhen G, Feki M, De Petris WM, Ayedi HF. Electrochemical removal of Cr^{6+} from aqueous media using iron and aluminum as electrode materials: towards a better understanding of the involved phenomena. J Hazard Mater. 2009;168(2–3):983–91.

83. Keshmirizadeh E, Yousefi S, Rofouei MK. An investigation on the new operational parameter effective in Cr^{6+} removal efficiency: a study on electrocoagulation by alternating pulse current. J Hazard Mater. 2011;190(1–3):119–24.

84. Bassam AA. Electrocoagulation of chromium containing synthetic wastewater using monopolar iron electrodes. Damascus University J. 2012;28(2):79–89.

85. Merzouk B, Gourich B, Sekki A, Madani K, Chibaned M. Removal turbidity and separation of heavy metals using electrocoagulation-electroflotation technique a case study. J Hazard Mater. 2009;164:215–22.

86. BashaAhmed C, Bhadrinarayana NS, Anantharaman N. Heavy metal removal from copper smelting effluent using electrochemical cylindrical flow reactor. J Hazard Mater. 2008;152:71–8.

87. Chien H, Luke C, Chen-Lu Y. Effect of anions on electrochemical coagulation for cadmium removal. Sep Purif Technol. 2009;65:137–46.

88. Vasudevan S, Lakshmi J. Effects of alternating and direct current in electrocoagulation process on the removal of cadmium from water e a novel approach. Sep Purif Technol. 2011;80(3):643–51.

89. Saber EM, ElSayed MN, Ibrahim HH, Osama AD, Abdykalykova R, Beisebekov M. Removal of Cadmium Pollutants in Drinking Water Using Alternating Current Electrocoagulation Technique. G J E R. 2013;7(3):45–51.

90. Kobya M, Demirbas E, Dedeli A, Sensoy MT. Treatment of rinse water from zinc phosphate coating by batch and continuous electrocoagulation processes. J Hazard Mater. 2009;173(1–3):326–34.

91. Khelifa A, Moulay S, Naceur AW. Treatment of metal finishing effluents by the electroflotation technique. Desalination. 2005;181:27–33.

92. Kumar NS, Goel S. Factors influencing arsenic and nitrate removal from drinking water in a continuous flow electrocoagulation (EC) process. J Hazard Mater. 2010;173(1–3):528–33.

93. Emilijan M, Srdjan R, Jasmina A, Aleksandra T, Milena M, Mile K, et al. Removal of arsenic from groundwater rich in natural organic matter (NOM) by continuous electrocoagulation/flocculation (ECF). Sep Purif Technol. 2014;136:150–6.

94. Al-Shannag M, Al-Qodah Z, Bani-Melhem K, Rasool-Qtaishat M, Alkasrawi M. Heavy metal ions removal from metal plating wastewater using electrocoagulation: Kinetic study and process performance. Chem Eng J. 2015;260:749–56.

95. Mota IO, Castro JA, Casqueira RG, Junior AG. Study of electroflotation method for treatment of wastewater from washing soil contaminated by heavy metals. J Mater Res Technol. 2015;4(2):109–13.

96. Brahmi Kh, Bouguerra W, Hamrouni B, Elaloui E, Loungou M, Tlili Z. Investigation of electrocoagulation reactor design parameters effect on the removal of cadmium from synthetic and phosphate industrial wastewater. Arab J of Chem. 2015; http://dx.doi.org/10.1016/j.arabjc.2014.12.012.

97. Jun L, Yan L, Mengxuan Y, Xiaoyun M, Shengling L. Removing heavy metal ions with continuous aluminum electrocoagulation: A study on back mixing and utilization rate of electro-generated Al ions. Chem Eng J. 2015;267:86–92.

98. El-Taweel Y, Nassef E, Elkheriany I, Sayed D. Removal of Cr(VI) ions from waste water by electrocoagulation using iron electrode. Egypt J Petrol. 2015; http://dx.doi.org/10.1016/j.ejpe.2015.05.011.

99. Heidmann I, Calmano W. Removal of Zn(II), Cu(II), Ni(II), Ag(I) and Cr(VI) present in aqueous solutions by aluminium electrocoagulation. J Hazard Mater. 2008;152:934–41.

100. Zongo I, Leclerc JP, Maiga HA, Wéthé J, Lapicque F. Removal of hexavalent chromium from industrial wastewater by electrocoagulation: A comprehensive comparison of aluminium and iron electrodes. Sep Purif Technol. 2009;66:159–66.

PERMISSIONS

LIST OF CONTRIBUTORS

Hesham M Sharaf
Department of Zoology, Faculty of Science, Zagazig University, Zagazig, Egypt

Abdalla M Shehata
Chemistry Department, Faculty of Science, Suez Canal University, Al-arish, Egypt

Seyed Kamran Foad Marashi and Hamid-Reza Kariminia
Department of Chemical and Petroleum Engineering, Sharif University of Technology, P.O. Box 11155–9465, Azadi Ave., Tehran, Iran

Shengmou Huang and Gan Lin
School of chemical engineering and food science, Hubei University of Arts and Science, Xiangyang 441053, China

Karina Bustos-Ramírez
Centro Conjunto de Investigación en Química Sustentable, UAEM-UNAM.
Km.12 de la carretera Toluca-Atlacomulco, San Cayetano 50200, Estado de México, México
División de Estudios de Posgrado e Investigación, Instituto Tecnológico de Querétaro, Av. Tecnológico s/n Esq. M. Escobedo, Col. Centro Histórico, 76000 Querétaro, México
Centro de Física Aplicada y Tecnología Avanzada, Universidad Nacional Autónoma de México, Boulevard Juriquilla 3001, 76230 Querétaro, México

Carlos Eduardo Barrera-Díaz and Reyna Natividad-Rangel
Centro Conjunto de Investigación en Química Sustentable, UAEM-UNAM.
Km.12 de la carretera Toluca-Atlacomulco, San Cayetano 50200, Estado de México, México

Miguel De Icaza-Herrera
Centro de Física Aplicada y Tecnología Avanzada, Universidad Nacional Autónoma de México, Boulevard Juriquilla 3001, 76230 Querétaro, México

Ana Laura Martínez-Hernández and Carlos Velasco-Santos
División de Estudios de Posgrado e Investigación, Instituto Tecnológico de Querétaro, Av. Tecnológico s/n Esq. M. Escobedo, Col. Centro Histórico, 76000 Querétaro, México
Centro de Física Aplicada y Tecnología Avanzada, Universidad Nacional Autónoma de México, Boulevard Juriquilla 3001, 76230 Querétaro, México

Marta Gmurek, Magdalena Olak-Kucharczyk and Stanisław Ledakowicz
Faculty of Process and Environmental Engineering, Department of Bioprocess
Engineering, Lodz University of Technology, Wolczanska 213, 90-924 Lodz, Poland

Kamyar Yaghmaeian, Reza Khosravi Mashizi , Amir Hossein Mahvi, Mahmood Alimohammadi and Shahrokh Nazmara
Department of Environmental Health Engineering, School of Public Health, Tehran University of Medical Sciences, Tehran, Iran

Simin Nasseri
Department of Environmental Health Engineering, School of Public Health, Tehran University of Medical Sciences, Tehran, Iran
Center for Water Quality Research (CWQR), Institute for Environmental Research (IER), Tehran University of Medical Sciences, Tehran, Iran

Reza Rezaee, Ramin Nabizadeh and Shahrokh Nazmara
Department of Environmental Health Engineering, School of Public Health, Tehran University of Medical Sciences, Tehran, Iran

Simin Nasseri
Department of Environmental Health Engineering, School of Public Health, Tehran University of Medical Sciences, Tehran, Iran
Center for Water Quality Research, Institute for Environmental Research, Tehran University of Medical Sciences, Tehran, Iran

Amir Hossein Mahvi
Department of Environmental Health Engineering, School of Public Health, Tehran University of Medical Sciences, Tehran, Iran
Center for Solid Waste Research, Institute for Environmental Research, Tehran University of Medical Sciences, Tehran, Iran
National Institute of Health Research, Tehran University of Medical Sciences, Tehran, Iran

Seyyed Abbas Mousavi
Department of Chemical and Petroleum Engineering, Sharif University of Technology, Tehran, Iran

Alimorad Rashidi
Nanotechnology Research Center, Research Institute of Petroleum Industry (RIPI), Tehran, Iran

Ali Jafari
Department of Environmental Health Engineering, School of Public Health, Lorestan
University of Medical Sciences, Khorramabad, Iran

Gholamreza Bonyadinejad
Environment Research Center and Department of Environmental Health Engineering, School of Health, Isfahan University of Medical Sciences (IUMS), Isfahan 81676-36954, Iran
Student Research Center, School of Health, IUMS,Isfahan, Iran

Mohsen Khosravi
Nanotechnology Department, University of Isfahan, Isfahan 81744-73441, Iran.

Afshin Ebrahimi and Roya Nateghi
Environment Research Center and Department of Environmental Health Engineering, School of Health, Isfahan University of Medical Sciences (IUMS),Isfahan 81676-36954, Iran

Seyed Mahmood Taghavi-Shahri
Research Center for Environmental Pollutants, Qom University of Medical Sciences, Qom 37136-49373, Iran

Hamed Mohammadi
Environment Research Center and Department of Environmental Health Engineering, School of Health, Isfahan University of Medical Sciences (IUMS),Isfahan 81676-36954, Iran
Student Research Center, School of Health, IUMS, Isfahan, Iran

Emel Kocak
Department of Environmental Engineering, Faculty of Civil Engineering, Yildiz
Technical University, Istanbul, Turkey

Adel A.S. Banana
Environment Engineering Department, Subrata College, University of Zawia, Zawia, Libya

R. M. S. Radin Mohamed
Faculty of Civil & Environment Engineering, UTHM, Parit Raja,Malaysia

A. A. S. Al-Gheethi
Faculty of Civil & Environment Engineering, UTHM, Parit Raja, Malaysia
High Institute of health sciences, Sana'a, Yemen

Mohammad Ali Baghapour and Amir Fadaei Nobandegani
Department of Environmental Health Engineering, School of Health, Shiraz University of Medical Sciences, Shiraz, IR, Iran

Nasser Talebbeydokhti
Department of Civil Engineering, College of Engineering, Shiraz University, Shiraz, IR, Iran

Somayeh Bagherzadeh
Department of Hydrogeology, Ab Ati Pazhooh Consulting Engineers, Shiraz, IR, Iran

Ata Allah Nadiri and Maryam Gharekhani
Department of Earth Science, Faculty of Science, University of Tabriz,Tabriz, East Azarbaijan, IR, Iran

Nima Chitsazan
Research Engineer at EnTech Engineering,PC11 broadway 21st floor, New York, NY 10004, USA

S. Ahmad Mokhtari
Research Center for Environmental Health Technology, School of Public Health, Iran University of Medical Sciences, Tehran, Iran
Department of Environmental Health Engineering, School of Public Health, Ardabil University of Medical Sciences,Ardabil, Iran

Mehdi Farzadkia, Ali Esrafili, Roshanak Rezaei Kalantari, Ahmad Jonidi Jafari,Majid Kermani and Mitra Gholami
Research Center for Environmental Health Technology, School of Public Health, Iran University of Medical Sciences, Tehran, Iran

Department of Environmental Health Engineering, School of Public Health, Iran University of Medical Sciences, Tehran, Iran

Allahbakhsh Javid and Hamed Gharibi
School of Public Health, Shahroud University of Medical Sciences, Shahroud, Iran

Alireza Mesdaghinia and Mahmood Alimohammadi
Department of Environmental Health Engineering, School of public Health, Tehran University of Medical Sciences, Tehran, Iran

Simin Nasseri
Department of Environmental Health Engineering, School of public Health, Tehran University of Medical Sciences, Tehran, Iran
Center for Water Quality Research, Institute for Environmental Research, Tehran University of Medical Sciences, Tehran, Iran

Amir Hossein Mahvi
Department of Environmental Health Engineering, School of public Health, Tehran University of Medical Sciences, Tehran, Iran
Center for Solid Waste Research, Institute for Environmental Research, Tehran University of Medical Sciences, Tehran, Iran

Hesham Dahshan, Ayman Mohamed Megahed, Amr Mohamed Mohamed Abd-Elall, Mahdy Abdel-Goad Abd-El-Kader, Ehab Nabawy and Mariam Hassan Elbana
Department of Veterinary Public Health, Faculty of Veterinary Medicine, Zagazig University, Zagazig, Sharkia governorate, Egypt

Vahid Noroozi Karbasdehi, Masoumeh Ravanipour, Mozhgan Keshtkar and Roghayeh Mirahmadi
Department of Environmental Health Engineering, Faculty of Health, Bushehr University of Medical Sciences, Bushehr, Iran

Sina Dobaradaran
Department of Environmental Health Engineering, Faculty of Health, Bushehr University of Medical Sciences, Bushehr, Iran
The Persian Gulf Marine Biotechnology Research Center, The Persian Gulf Biomedical Sciences Research Institute, Bushehr University of Medical Sciences, Boostan 19 Alley,Imam Khomeini Street, Bushehr, Iran

Systems Environmental Health, Oil, Gas and Energy Research Center, The Persian Gulf Biomedical Sciences Research Institute, Bushehr University of Medical Sciences, Bushehr, Iran

Iraj Nabipour and Afshin Ostovar
The Persian Gulf Tropical Medicine Research Center, The Persian Gulf Biomedical Sciences Research Institute, Bushehr University of Medical Sciences, Bushehr, Iran.

Amir Vazirizadeh
The Persian Gulf Studies and Researches Center Marine Biotechnology Department, Persian Gulf University, Bushehr, Iran

Shahrokh Nazmara
School of Public Health,Tehran University of Medical Sciences, Tehran, Iran

Mohsen Noorinezhad
Ecology Department, Iranian Shrimp Research Institute, Bushehr, Iran

Elham Kazemi, Hamid Karyab and Mohammad-Mehdi Emamjome
Department of Environmental Health Engineering, School of Health, Qazvin University of Medical Sciences, Qazvin, Iran
Bahonar Blvd, College of Medical Sciences, Qazvin, Iran

Vahid Noroozi Karbasdehi, Roghayeh Mirahmadi, Mozhgan Keshtkar, Fatemeh Faraji Ghasemi and Farzaneh Khalifei
Department of Environmental Health Engineering, Faculty of Health, Bushehr University of Medical Sciences, Bushehr, Iran

Sina Dobaradaran
Department of Environmental Health Engineering, Faculty of Health, Bushehr University of Medical Sciences, Bushehr, Iran
The Persian Gulf Marine Biotechnology Research Center, The Persian Gulf Biomedical Sciences Research Institute, Bushehr University of Medical Sciences, Boostan 19 Alley,Imam Khomeini Street, Bushehr, Iran
Systems Environmental Health, Oil, Gas and Energy Research Center, The Persian Gulf Biomedical Sciences Research Institute, Bushehr University of Medical Sciences, Bushehr, Iran

Iraj Nabipour and Afshin Ostovar
The Persian Gulf Tropical Medicine Research Center, The Persian Gulf Biomedical Sciences Research Institute, Bushehr University of Medical Sciences, Bushehr, Iran

Hossein Arfaeinia
Environmental Health Department, School of Public Health, Iran University of Medical Sciences, Tehran, Iran

Amir Vazirizadeh
The Persian Gulf Studies and Researches Center Marine Biotechnology Department, Persian Gulf University, Bushehr,Iran

Edris Bazrafshan, Leili Mohammadi and Alireza Ansari-Moghaddam
Health Promotion Research Center, Zahedan University of Medical Sciences, Zahedan, Iran

Amir Hossein Mahvi
Department of Environmental Health Engineering, School of Public Health, Tehran University of Medical Sciences, Tehran, Iran
Center for Solid Waste Research, Institute for Environmental Research, Tehran University of Medical Sciences, Tehran, Iran
National Institute of Health Research,Tehran University of Medical Sciences,, Tehran, Iran

Index

www.ingramcontent.com/pod-product-compliance
Lightning Source LLC
Chambersburg PA
CBHW082011190326
41458CB00010B/3157